New Directions in Civil Engineering

Series Editor
W. F. CHEN
Purdue University

Zdeněk P. Bažant and Jaime Planas
Fracture and Size Effect in Concrete and Other Quasibrittle Materials

W.F. Chen and Seung-Eock Kim
LRFD Steel Design Using Advanced Analysis

W.F. Chen and E.M. Lui
Stability Design of Steel Frames

W.F. Chen and K.H. Mossallam
Concrete Buildings: Analysis for Safe Construction

W.F. Chen and S. Toma
Advanced Analysis of Steel Frames: Theory, Software, and Applications

W.F. Chen and Shouji Toma
Analysis and Software of Cylindrical Members

Y.K. Cheung and L.G. Tham
Finite Strip Method

Hsai-Yang Fang
Introduction to Environmental Geotechnology

Yuhshi Fukumoto and George C. Lee
Stability and Ductility of Steel Structures under Cyclic Loading

Ajaya Kumar Gupta
Response Spectrum Method in Seismic Analysis and Design of Structures

C.S. Krishnamoorthy and S. Rajeev
Artificial Intelligence and Expert Systems for Engineers

Boris A. Krylov
Cold Weather Concreting

Pavel Marek, Milan Guštar and Thalia Anagnos
Simulation-Based Reliability Assessment for Structural Engineers

N.S. Trahair
Flexural-Torsional Buckling of Structures

Senol Utku
Theory of Adaptive Structures: Incorporating Intelligence into Engineered Products

Jan G.M. van Mier
Fracture Processes of Concrete

S. Vigneswaran and C. Visvanathan
Water Treatment Processes: Simple Options

Senol Utku, Sc.D.
Fellow-ASCE Professor of
Civil Engineering and of Computer Science

Theory of ADAPTIVE STRUCTURES

Incorporating Intelligence into Engineered Products

CRC Press
Taylor & Francis Group
Boca Raton London New York

CRC Press is an imprint of the
Taylor & Francis Group, an informa business

CRC Press
6000 Broken Sound Parkway, NW
Suite 300, Boca Raton, FL 33487

270 Madison Avenue
New York, NY 10016

2 Park Square, Milton Park
Abingdon, Oxon OX14 4RN, UK

Library of Congress Cataloging-in-Publication Data

Utku, Senol
Theory of adaptive structures : incorporating intelligence into engineered products / Senol Utku.
p. cm. -- (New directions in civil engineering)
Includes bibliographical references.
ISBN 0-8493-7431-6
1. Structural control (Engineering) 2. Smart structures.
I. Title. II. Series.
TA654.9.U855 1998
624.1'7—dc21 97-47326
CIP

International Standard Book Number 0-8493-7431-6
Library of Congress Card Number 97-47326

The author wishes to acknowledge his gratitude to the following people and organizations:

Felicia Shapiro, Associate Editor, Engineering and Chemistry, CRC Press

Mimi Williams and other professionals at CRC Press who were involved with the production of the book

Jet Propulsion Laboratory for the cover photos of the Goldstone RF antenna, and the Pathfinder lander on Mars

John Hancock Mutual Life Insurance for the cover photo of the John Hancock Tower in Boston

Jennifer Lewis Bass of Duke University Class '92 (civil engineering) for the cover photo of her as a Duke University soccer player

Thomas Hill of Duke University Class '93 (history) for the cover photo of him as a Duke University basketball player

Ayda Utku (Duke University Class '87) of Text Noumena, Durham, NC, for the conceptual design of the book cover.

To the memories of

Professor Mustafa Inan, Istanbul Technical University

Professor Charles H. Norris, University of Washington

my most inspiring mentors.

Preface

Since the Eighties there has been an ever increasing interest in actively controlling the behavior of engineering systems. This is partly due to the impact of advances in materials technology on sensors and actuator technologies, and partly due to the availability of very powerful and reliable microprocessors at costs that were unimaginably low in the preceding decade. Engineering systems where the *mechanics of deformable bodies* is either the dominant discipline or the supporting one are also orienting themselves to benefit from these developments. Pressures stemming from such reorientations have forced the engineers and engineering scientists, in the area of deformable bodies, to revive and broaden the activities that did not find applications in the era of large simulations for predicting the behavior under well studied, nevertheless imagined, circumstances.

Specifically, structural engineering is gearing itself to meet the new challenges created by the advances in sensor, actuator, and microprocessor technologies. In the Fifties and the Sixties, the introduction of digital computers invited structural engineers to examine their analysis methods, and develop software for those methods most suited to digital computation. It was an all out effort that culminated in the *displacement finite element method* which benefited not only the structural engineering industry, the aerospace industry, and the sciences that required computer simulations, but also the computer industry itself. This created a need for larger and faster computers, and established trends in which number crunching methodologies and aids were developed. It was a rewarding symbiosis that benefited both the engineering industry and sciences and the large scale computer industry.

However, over the decades, the impact of computers on engineering methods became so dominant that even in academia approaches that did not fit quite well with large scale numerical simulation, such as the force method, had to be abandoned or atrophied due to lack of interest and funding. By the end of the Sixties, huge engineering simulation softwares dominated not only the large computer centers but also the thinking of engineers both in academia and in industry.

The adaptive structures technology created an environment where the analysis, not the computation, of structural response due to actuator inserted deformations became important. The problems related to the placement, the operation in real time, and the energy consumption of the actuators require the review and the broadening of theories long dormant due to the emphasis placed in the numerical simulations of structural behavior by the *displacement finite element method.*

This book is intended to furnish the basic theory needed by the modern structural engineer in the design and the control of discrete parameter adaptive structures. It is motivated by the efforts of the author and his students at Duke University, those of his colleagues at the Jet Propulsion Laboratory, the works of many investigators who have been contributing actively in the last six International Conferences on Adaptive Structures (the first in Maui in 1990, the second in Nagoya in 1991, the third in San Diego in 1992, the fourth in Cologne in 1993, the fifth in Sendai in 1994, and the sixth in Key West in 1995), and the works presented in the First World Conference on Structural Control in Pasadena in 1994.

The book is not intended to give an historical account of the works of countless number of workers in the field. There are many interesting state-of-the-art review papers[1] which may establish a starting point for the history minded reader in order to trace the origins and the trends in the field of adaptive structures.

---

[1] See for example, "Adaptive Structures Research at ISAS, 1984-1990," Koryo Miura, *Journal of Intelligent Material Systems and Structures*, Vol. 3, pp. 54-74, January 1992.

"Adaptive Structures in Japan," Senol Utku and Ben K. Wada, *Journal of Intelligent Material Systems and Structures*, Vol.4, pp.437-451, October 1991.

"Adaptive Structures in Europe," E. J. Breitbach, pp.32-48, *Second Joint Japan/US Conference on Adaptive Structures, Nov. 12-14, '91, Nagoya, Japan*, Matsuzaki, Wada (eds.), Technomic Publishing Co., Lancaster, PA, 1992.

"Control-Structure Interaction Research at NASA Langley Research Center," Willard W. Anderson, Jerry R. Newsom, pp.43-55, *Fourth International Conference on Adaptive Structures, Nov. 2-4, '91, Cologne, Federal Republic of Germany*, Breitbach, Wada, Natori (eds.), Technomic Publishing Co., Inc., Lancaster, PA, 1994.

"Second Generation of Active Structural Control in Civil Engineering," G. W. Housner, T. T. Soong, and S. F. Masri, panel 3-18, Vol. 1, *Proceedings of First World Conference on Structural Control*, Los Angeles, CA, 3-5 August, 1994.

"Future Direction on Research and Development of Seismic-Response-Controlled Structure," Takuji Kobori, panel 19-31, Vol. 1, *Proceedings of First World Conference on Structural Control*, Los Angeles, CA, 3-5 August, 1994.

The book consists of 12 chapters. Chapter 1 is an introduction to adaptive structures. Chapters 2-5 are on the design and the control of discrete parameter adaptive structures subjected to static disturbances and control loads. Chapters 6-11 are on the design and the control of discrete parameter adaptive structures subjected to dynamic disturbances and control loads. Chapter 12 is an introduction to distributed parameter adaptive structures.

In Chapter 2, the incremental excitation-response relations of discrete parameter structures are derived, assuming linear elastic and time-invariant behavior. Chapter 3 deals with the active control of some of the response quantities in the static case. Statically determinate discrete parameter adaptive structures are discussed in Chapter 4 because of their important role in maintaining the pre-control values of many of the unobserved response components, during the control of the observed response components. This chapter also discusses trusses as slow moving mechanical manipulators and methods for real time computation of Jacobian matrix of controlled nodal position vectors with respect to actuator induced element deformations. Due to diversified design considerations, most engineering structures are statically indeterminate. The energy efficient control of such structures is studied in Chapter 5. The actuator placement problem in the design, and the actuator selection problem in the operation of adaptive structures, are discussed extensively in Chapters 4 and 5, from the standpoint of control robustness and control energy minimization.

Chapter 6 discusses for the dynamic case the excitation-response relations of discrete parameter adaptive structures, assuming linear, elastic, and time-invariant behavior. Chapter 7 is about inverse relations for the dynamic case. Chapter 8 discusses active vibration control in discrete parameter adaptive structures for the autonomous case, i.e., for the case where no appreciable disturbances exist during control. Chapter 9 is on the active vibration control in discrete parameter adaptive structures for the nonautonomous case, i.e., for the case where disturbances are active during the control. Chapter 10 is on vibration control in buildings subjected to wind excitations, and Chapter 11 is on active vibration control in buildings subjected to seismic excitations. Chapter 12 is an introduction to distributed parameter adaptive structures.

The text is written for practicing engineers, engineering seniors, and first year graduate students. Many of its subjects have been covered in piecemeal fashion in the courses taught by the author since the late Eighties, and some are introduced in Part II of *Elementary Structural Analysis*, 4th ed., Utku et al., McGraw-Hill Publishing Co., New York, 1991. The material is taught by the author in *The Theory of Adaptive Structures* course, which is a new course instituted in 1995 at the author's university for the benefit of all first year engineering graduate students and seniors who may be interested in the subject. It is intended to support the current trend of incorporating intelligence into engineered products.

Preface

The author is thankful for the countless number of good and intelligent people who helped him form his professional identity and output. For this book, he wishes to acknowledge the diligent and creative work of his students at Duke University, and the support of the Jet Propulsion Laboratory through his long time colleague and friend Ben K. Wada. He is forever grateful to his wife and colleague Bisulay, his daughter and associate Ayda, and his son and counselor Sinan, for their tangible and intangible assistance and loyal support in nursing the development of this book. In many ways the book may be considered a result of the inspirations the author received from his mentors at his two alma maters: Istanbul Technical University and Massachusetts Institute of Technology.

Senol Utku
15 December 1997

Contents

List of Figures

List of Tables

1
Introduction

1.1 History

To understand the events taking place in its environment, humankind created the sciences, and to mold its environment to its benefit, it has engineering. An engineer tries to control things under the scrutiny of society. Specifically, a structural engineer tries to control the response of a structure to the loads that it will be subjected to during its design life.

The function of a structure usually determines the general features of its geometry; the environment and the current technology determine its material. The past experience with the environment and the structure's function define the loads. Considering the loads with worst case scenarios, a structural engineer designs the structure by determining the free geometrical parameters of the structure's material volume such that the structural response never exceeds the limits set to ensure the structure's functionality and integrity. The structure is constructed according to its design and left unattended to complete its design life. In modern parlance, this is called *passive design.* Historically this has been the only design paradigm available to the structural engineers.

The experience with space structures since the early Sixties has shown that a passively designed space structure can be monitored continuously, and its response can be controlled actively if the means to do so are provided during the construction. With the advances in sensor, actuator, and microprocessor technologies, it became economically feasible to actively control some of the responses of a passively designed space structure in real time. Initially, the telemetry data obtained from spacecraft was evaluated by human operators on the earth's surface, and corrective actions were relayed to the spacecraft through radio signals. Later, using the reliable excitation-response relations of the spacecraft and various feedback control algorithms, on-board microprocessors took over the job of the human operator. In order to keep the structure at the close proximity of its nominal state, computers continuously evaluate the outputs of sensors which monitor deviations from the nominal state, and issue commands to the actuators when required to eliminate the deviations. This modern approach may be called design by incorporating intelligence into the system. It is very likely that *design by incorporating intelligence into the engineering products* will be the new design paradigm for all engineering branches in the coming millennium.

The new design paradigm of incorporating intelligence into engineering products is basically the result of the availability of extremely capable microprocessors at a minute fraction of the total cost of many engineering products, and the result of advances in sensor and actuator technologies. This situation encourages the engineers to use active means in controlling the response of their products in addition to the passive means. Modern passive design techniques rely heavily on digital simulations of products' behavior under speculated loading conditions. By incorporating the microprocessors into engineering products for intelligent and user-friendly behavior, a new area opens for microprocessors beyond their current use in numerical simulations.

An adaptive structure may be considered as an intelligent variant of its passively designed counterpart. It is passively designed considering most of the loading scenarios; however, for many other loading scenarios, active means are considered and incorporated into the design in order to control their effects. For example, a passively designed building may be equipped with active means to compensate for those loading cases that building codes may have missed or not treated in depth. The term "*active means*" refers to "*actuators and sensors and an on-board microprocessor loaded with appropriate software*".

The deformable mirror used in adaptive optics technology is a good example of an adaptive structure. The deformable mirror in the telescope changes its geometry in real time to compensate the effects of low frequency atmospheric distortions on the wave front of the incoming light beam. There are many other examples of adaptive structures that are being implemented or considered for space and earth-bound structures. The applications in space structures include precision shape control of antenna surfaces for collection of electromagnetic waves, the precision control of distance between the signal reception points of space interferometers, and vibration control in space structures where the attenuation of transient vibrations are too slow for the efficient operation of the spacecraft systems. The applications in earthbound structures include the elimination of unwanted vibrations from the dynamic response of buildings subjected to wind and seismic excitations, deformation control in precision structures, stress control in composites, and many others.

1.2 Definitions

An *adaptive structure* is an engineering structure whose response to excitations can be controlled in real time by the insertion of internal deformations through appropriately placed actuators that are part of the structure. This is similar to what was adopted in the biological world billions of years ago, where change in geometric configurations, motility, and motion take place by deformation insertion in the cells. Nature is using adaptive structures concepts in all life forms[1] all the time without interruption after the failure in the truly rotating flagella of the pro-caryotic e-coli bacterium many billions of years ago.[2]

A structure is identified by its physical properties at its *at-rest state*,[3] that is, the state when no loads are acting. These properties are categorized as *structural parameters.* For example, in a truss structure with uniform bars, the coordinates of the joints in a global coordinate system, the cross-sectional area, the unit mass, and the Young's modulus of each bar, the damping constants, and the locations and the directions of prescribed nodal displacements constitute the list of its structural parameters. In adaptive structures, the locations and attributes of actuators and sensors are also part of its structural parameters. A structure is called *time-invariant*, if the structural parameters do not change as a function of time.

The *excitations* are usual loads that a structure may be subjected to during its design life, such as dead and live loads due to the earth's gravitational acceleration, wind loads, acoustic loads, impactive or impulsive type surface traction loads due to fluids and solids that are in contact with the structure, thermal loads, loads due to fabrication errors, loads due to support settlements, seismic loads, loads due to ocean waves, loads due to attached machinery, loads due to changes in material properties and geometry, etc. In adaptive structures, the actuator induced effects are also considered excitations.

The state of the structure under dead loads (such as weight in earth-bound structures) and permanent live loads (such as centrifugal forces in a spinning structure with a constant velocity) is its *nominal state.* As datum

---

[1]See, for example, "Exoskeletal Sensors for Walking," Sasha N. Zill and Ernst-August Seyfarth, *Scientific American*, July 1996.

[2]For a good account of this, the reader may refer to "Why the Wheels Won't Go" by Michael La Barbera, *The American Naturalist*, v.121, n.3, pp.395-408, 1983.

[3]The term "*state*" is used in this chapter to mean the "*complete state*"which includes other system attributes beyond "*position*" and "*momentum.*"

for response measurements, depending on the application, either the nominal state or the at-rest state is taken as the *reference state*. In adaptive structures, the reference state is usually taken as the nominal state. Unusual circumstances aside, the reference state is a *stable equilibrium state*, namely, the structure returns to its reference state if an external agent pulls it to another state in the vicinity of the reference state and then releases it.

When the structure at its reference state is subjected to a load, it assumes another state. The difference between these two states is called the *response* of the structure to that load. The quantities measuring this difference are called response quantities corresponding to the excitation.

The equations relating the measures of response $\boldsymbol{\psi}$ to the excitation measures $\mathbf{f}$ are called *excitation-response relations*. Formally, these relations may be shown by

$$\mathcal{L}[\boldsymbol{\psi}] = \mathbf{f} \tag{1.1}$$

where $\mathcal{L}$ is a matrix operator that can be algebraic, differential, integral, or a combination of these.

In the theory of structures, the excitation response relations are a consequence of the following laws and the rules governing the mechanics of deformable bodies:

- Newton's law (or d'Alembert Principle) for force equilibrium,
- rules of Euclidean geometry,
- constitutive laws of the material.

The measures of state $\boldsymbol{\psi}$ of an engineering structure consist of the following three group of entities:

- displacements $\mathbf{d}$ (or nodal deflections $\boldsymbol{\xi}$),
- stresses $\boldsymbol{\sigma}$ (or internal forces in structural elements $\mathbf{s}$),
- strains $\boldsymbol{\epsilon}$ (or internal deformations of structural elements $\mathbf{v}$).

In adaptive structures, one or more response components are monitored, i.e., measured, continuously. The devices that make the measurements are the *sensors*. When the measures of monitored response quantities exceed the bounds that are defined *a priori* by the structure's designer, they are eliminated by subjecting the structure to an additional set of excitations which may minimally alter the existing stress (or internal force) state and the overall momentum of the structure. The devices that create and apply these additional excitations onto the structure are *actuators*. They may be incorporated in the structure during its construction, or may be added on later.

In adaptive structures, actuators are internal deformation inducing devices, such as turnbuckles in truss bars, shape-memory alloy or piezoelectric fibers or laminate in composites, heating elements, etc. When activated, these devices change the lengths of the bars or the strain state of the composite, i.e., they insert *internal deformations* into the structure. These types of actuators, as discussed later in this chapter, are quite different from the external-force inducing actuators, such as pyrotechnic devices, electro-magnetic force inducers, and pressure or force exerting mechanisms.

1.3 Types of Structures

In order to analyze the problems of a structural system quantitatively one has to idealize it as a simpler model by preserving its basic features, and ignoring its less important attributes. The idealization depends on the objectives of the analysis, as well as the geometry, the material, the supports, and the loads of the structure. This book is confined to engineering structures where the structural elements closely represent their idealized counterparts, and they are assembled together to meet the simplifications of the idealized model. For example, a truss structure consists of uniform bar elements that can be represented by their centroidal axes, and the bar elements are connected such that the connections can be idealized as frictionless joints at the intersection points of the centroidal axes. Engineering structures are constructed such that they can be represented by their idealized models.

Depending upon the type of functions used in describing their states, structures are classified as *discrete parameter structures* and *distributed parameter structures*.

In a *distributed parameter structure*, response ψ can be represented by $\hat{n}$-tuples that are functions of one or more spatial coordinates $\mathbf{x}$ and they may or may not be functions of time t. For example, by denoting the stress state by $\boldsymbol{\sigma} = \boldsymbol{\sigma}(\mathbf{x}, t)$, the strain state by $\boldsymbol{\varepsilon} = \boldsymbol{\varepsilon}(\mathbf{x}, t)$, and the displacement state by $\mathbf{d} = \mathbf{d}(\mathbf{x}, t)$, the complete state may be expressed as

$$\psi = \psi(x, t) = \left\{ \begin{array}{c} \mathbf{d}(\mathbf{x}, t) \\ \boldsymbol{\sigma}(\mathbf{x}, t) \\ \boldsymbol{\varepsilon}(\mathbf{x}, t) \end{array} \right\} \tag{1.2}$$

where $\mathbf{x}$ takes all values within the material volume. The distributed parameter structures include beams, arches, cables, plates, shells, and solids which are sketched in Fig.1.1.

In a *discrete parameter structure*, response ψ can be represented by a real $\hat{n}$-tuple where the entries are constants or variables that are functions of time only. For example, when ψ is a function of time t, by denoting the list of internal forces by $\mathbf{s} = \mathbf{s}(t)$, the list of element deformations by $\mathbf{v} = \mathbf{v}(t)$, and the list of joint deflections by $\boldsymbol{\xi} = \boldsymbol{\xi}(t)$, the complete state

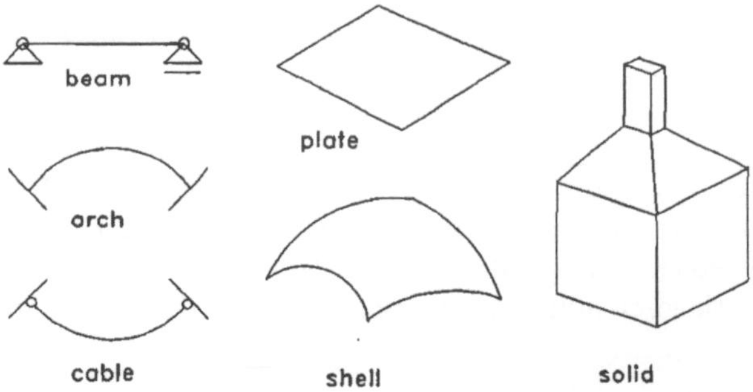

FIGURE 1.1. Examples of distributed parameter structures

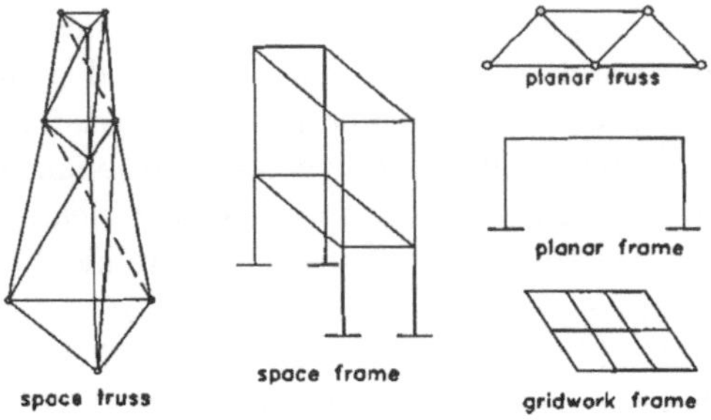

FIGURE 1.2. Examples of discrete parameter structures

may be expressed as

$$\psi = \psi(t) = \left\{ \begin{array}{c} \xi(t) \\ \mathbf{s}(t) \\ \mathbf{v}(t) \end{array} \right\} \tag{1.3}$$

The discrete parameter structures include planar and space trusses, planar and space frames, and gridwork frames which are depicted in Fig.1.2.

By suitable mathematical tools, such as finite differences, finite elements, weighted residual methods, etc., a distributed parameter structure can be approximated by a discrete parameter structural model. These models are called *discretized distributed parameter structures.* In this book, the distributed parameter structures and their discretization methods[4] are not discussed. However, a short introduction to distributed parameter structures is given in Chapter 12.

[4]For a good treatment of discretization methods, see *Engineering Analysis - A Survey of Numerical Procedures*, Stephen H. Crandall, McGraw-Hill Book Company, Inc., 1956.

1.4 Types of Problems

When the response ψ is not a function of time t the problem is of *static* type. In static problems, the operator $\mathcal{L}$ and loading $\mathbf{f}$ in Eq.(1.1) are not functions of time t.

When the response ψ is a function of the time t, the problem is of *dynamic* type. In dynamic problems, the operator $\mathcal{L}$ and loading $\mathbf{f}$ in Eq.(1.1) are functions of the time variable t.

The operator $\mathcal{L}$ in the excitation response relations given by Eq.(1.1) is basically a non-linear operator. However, due to difficulties in handling the equations involving non-linear operators and, perhaps more importantly, in order to ensure the predictability of structural response throughout the design life, when one has the opportunity, one linearizes them. Writing the equilibrium equations for the geometric configuration of the reference state, using the linearized strain-displacement relations, and keeping the stresses within the proportional limits of the constitutive laws of material ensure that the excitation-response relations are linear, i.e., $\mathcal{L}$ operator is linear. The response predictions by the linearized relations should be acceptably close to the actual response. The designer can ensure this by adjusting the structural parameters. A structural analysis problem is called *linear* when operator $\mathcal{L}$ in Eq.(1.1) is linear.

For the linear and time invariant problems of discrete parameter or discretized distributed parameter structures Eq.(1.1) may be rewritten as

$$\grave{\mathbf{A}}\psi = \mathbf{f} \tag{1.4}$$

where $\grave{\mathbf{A}}$ is a constant real $\grave{n} \times \grave{n}$ matrix operator with some time differentiation entries in dynamic case, and ψ is a real $\grave{n}$-tuple ($\grave{\mathbf{A}} \in \mathrm{R}^{\grave{n}\times\grave{n}}$ and $\psi \in \mathrm{R}^{\grave{n}\times 1}$). In engineering structures $\grave{\mathbf{A}}$ is not singular, i.e., mapping from ψ to $\mathbf{f}$ or $\mathbf{f}$ to ψ is a one-to-one onto map. This means that there is a unique response ψ for each excitation $\mathbf{f}$.

Most of the discussions in this book are for the time-invariant linear problems of adaptive engineering structures which are either of discrete parameter type or of discretized distributed parameter type.

1.5 Qualitative Analysis in State Space

The totality of responses, defined by Eq.(1.4) as a function of loading $\mathbf{f}$, defines a real $\grave{n}$-dimensional vector space when $\psi \in \mathrm{R}^{\grave{n}\times 1}$. This vector space is called the *state space* of the structure. In the state space, each point represents one response ψ. If we use the same rules in measuring distances and angles as in the three-dimensional Euclidean space, the vector space, hence the state space, is called Euclidean. The state space representation enables one to discuss the structural behavior from a geometrical point of view which is adopted throughout this book.

The square of the distance of a point in the state space from the origin, i.e., from the nominal state, is closely related with the internal energy of the structure at that state. The larger the distance, the larger the structure's internal energy. In the static case, the internal energy of the structure is its total strain energy, and in the dynamic case it is the sum of the total strain energy and the total kinetic energy of the structure at that state.

The successive states of a structure, initially at its nominal state and then subjected to a prescribed time-dependent loading, trace a continuous curve in the state space, starting from the point which represents the nominal state. If we define the point representing the nominal state as the zero point of the vector space, then the curve is called the *trajectory* of responses corresponding to the loading. We may select a set of basis vectors, preferably a set of orthonormal basis vectors, in the vector space to describe the trajectory. We may consider the orthonormal basis vectors as the generalization of the unit vectors of a Cartesian coordinate system in the three-dimensional Euclidean space. The origin in the vector space is the zero point representing the nominal state. Since the number of response components $\hat{n}$ is also the dimension of the space, it is very hard to plot the trajectory when we have more than three response components. However, if the basis vectors are orthonormal, we can use planes defined by any two of the basis vectors to plot the normal projection of the trajectory. Suppose we choose the plane defined by the first two basis vectors as the projection plane. Then the projected curve on this plane represents the time variation of the response components along the first two basis vectors.

1.5.1 Example of Static Transient Loading

Consider a small constant load that is applied very slowly on a supported structure. The term "*supported*" means the structure cannot undergo rigid body motions under the loading, and the term "*very slowly*" refers to a situation where the material particles never attain large enough accelerations due to the loading to cause inertial forces. This is called *static loading.* The trajectory for the static loading is a line segment between origin O and the final state P. The projection of the trajectory on the plane defined by the ith and the jth basis vectors, i.e., (i, j)-plane, is the line segment OP' (prime indicates projection) as shown in Fig.1.3. Points on line segment OP correspond to the successive states of the structure as the load increases from its zero value at point O to its final value at point P. Suppose the load is applied in ten minutes at a constant rate. Then the middle point M of line segment OP is the state at the fifth minute of the loading. Point P corresponds to the state at the tenth minute of the loading and thereafter. If this small load stays on the structure another ten minutes and then dissipates slowly to zero at the thirtieth minute following the start of the loading event, after staying at P ten minutes, the trajectory will follow the loading trajectory in the reverse direction to end up at point O.

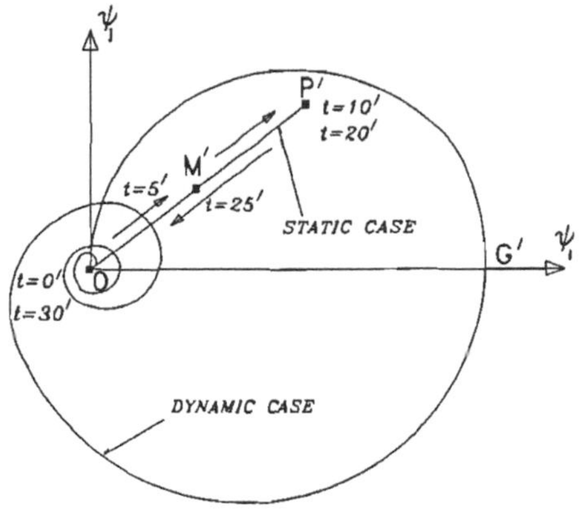

FIGURE 1.3. State space representation of response in a plane

Imagine that the structure is an antenna structure, and the small static loading is a thermal load. It is very likely that during the thirty minutes of a slow loading/unloading episode, the antenna surface would be distorted considerably, stopping the functions of the antenna system. If the antenna structure were an adaptive structure, by slowly introducing compensating internal deformations, the shape of the surface could be preserved at its nominal state and the function of the antenna system would not be interrupted at all. Suppose the first ten components of the response define distortions of the antenna's surface. Then the projection of the trajectory on the planes defined by any two of the ten corresponding basis vectors would never leave the origin O.

1.5.2 *Example of Dynamic Transient Loading*

Suppose the same constant load mentioned in the previous subsection is applied and then removed from the structure in one second, instead of thirty minutes. This is a case of *dynamical loading* where the suddenness of the loading causes the material particles to acquire accelerations, thus giving rise to inertial forces, and time dependent response. Since the structure is supported, it cannot undergo rigid body motions; it can only vibrate before it returns back to its equilibrium state at point O. The trajectory in this case is an outward curve from point O to point G, followed by some inwardly spiraling curve from point G to point O. The outward curve is a result of a sudden increase of the total internal energy which pushes the state at O to a higher energy state at G. If it were not for the structural damping, the total energy would stay the same, and the structure would assume different successive states, all with the same internal energy. If the square of the distance from O represents the internal energy, then the trajectory would be on the surface of a hypersphere centered at O, with radius equal to the square root of the internal energy. However, if damping exists, the internal energy of the structure will gradually dissipate; hence, the trajectory will inwardly spiral towards point O. How fast the inward spiraling curve reaches point O depends on the damping properties of the structure. Theoretically, to reach the equilibrium state at point O may take infinite time. The projection of the trajectory on the (i, j)-plane is sketched by $OG'O$ curve in Fig.1.3. Time t is the implicit parameter of the curve. Point G is the point of the trajectory which is farthest from point O. The time to reach point G is the *rise-time*, and the length $\overline{OG}$ corresponds to the *overshoot.*

Imagine that the structure is an antenna structure, and the dynamic load represents some small object hitting the antenna and causing it to vibrate. It is very likely that the vibrations are large enough to stop the functioning of the antenna system. If the antenna were a space antenna, due to small damping, the vibrations would continue for a long time. However, if the antenna structure were an adaptive structure, by introducing compensating internal deformations, many of the undesired vibration modes would be damped out quickly, and the functioning of the antenna system would not be interrupted. Suppose the first ten components of the response represent the participation factors of the undesired vibration mode shapes. Then the projection of the trajectory on the planes defined by any two of the ten corresponding basis vectors (i.e., the undesired vibration mode shapes) would be in the close neighborhood of origin O.

1.6 Constituents of Adaptive Structures

As mentioned earlier an adaptive structure is a conventional *structure* which is equipped with *sensors* and *actuators* that work under the control of an on-board *microprocessor.* These constituents are discussed below to the extent required by the discussions in the rest of the book.

1.6.1 Structure

The structure is a discrete parameter structure which is time-invariant unless otherwise stated. It consists of M number of structural elements brought together at N number of nodal points such that the structural elements are in contact with each other only at the nodes. The points of a structural element that can be attached to a node are called vertices. A discrete parameter structure and its global right-handed Cartesian reference frame are sketched in Fig.1.4. Note that the coordinate axes are labelled with X, Y, Z. For identification purposes, nodes and structural elements are labelled with integers, such that the labels for the nodes run from one to N, and the labels for the structural elements run from one to M. Node labels

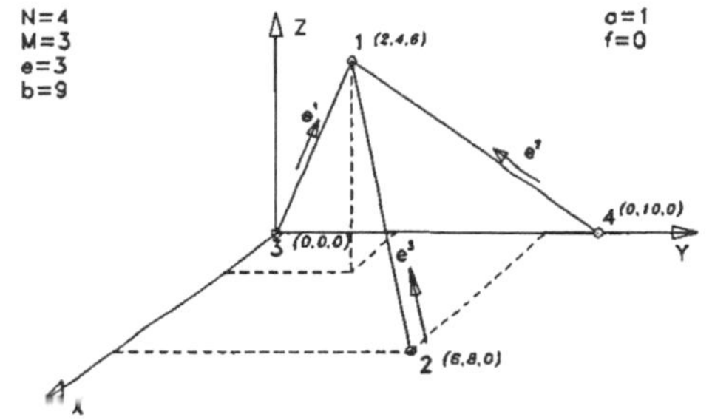

FIGURE 1.4. A discrete parameter structure and its global reference frame

TABLE 1.1. Deflections of Node i

Structure	ξ_i^T	e
Space frame	$d_{iX}, d_{iY}, d_{iZ}, r_{iX}, r_{iY}, r_{iZ}$	6
Planar frame	d_{iX}, d_{iY}, r_{iZ}	3
Gridwork frame	r_{iX}, r_{iY}, d_{iZ}	3
Space truss	d_{iX}, d_{iY}, d_{iZ}	3
Planar truss	d_{iX}, d_{iY}	2

X, Y, Z: global	*d: displacement*
i: node label	*r: rotation*

appear in the subscript of nodal quantities, and element labels appear in the superscripts of elemental quantities.

When a discrete parameter structure is loaded, its nodes move. The quantitative description of a node's movement in a global reference frame as an e-tuple is its deflection. The quantity e is called the *deflection degrees of freedom* of the node. It is different for different structures. For example e is 6 for space frame nodes. The six components of the deflection of a space frame node consist of three displacements along and three rotations about the global reference frame axes. The number, the direction, and the physical character of the deflection components for the structures studied in this book are given in Table 1.1 where X, Y, Z are the labels of global axes (see Fig.1.4).

By using special devices between a node and the supporting body, any number of deflection components at any node can be made to take the values dictated by the supporting body. The total number of such nodal deflection components is shown by symbol b which represents the *number of deflection constraints.*

In an N-node discrete parameter structure with e-number of deflection components per node, and b-number of deflection constraints, the number

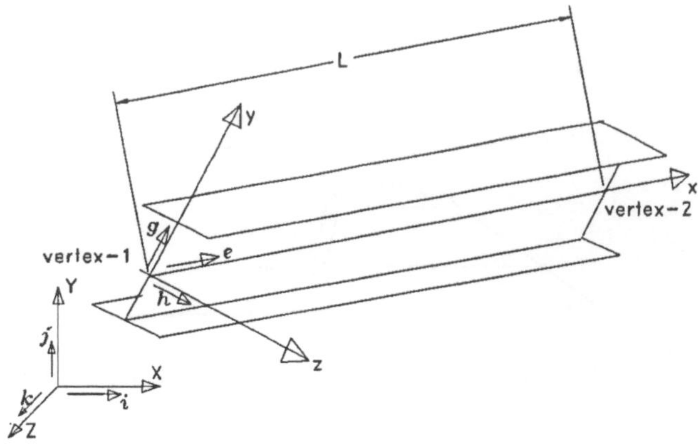

FIGURE 1.5. A one-dimensional structural element and its local reference frame

of deflection components is

$$n = Ne - b \tag{1.5}$$

The complete list of the nodal deflection components may be shown by $\boldsymbol{\xi} \in \mathrm{R}^{n\times 1}$ where the ith partition $\boldsymbol{\xi}_i$ lists the deflections of node i.

The end points of a one-dimensional structural element are its vertices. A one-dimensional structural element has two vertices. Although a discrete parameter structure may contain structural elements with two or more vertex points, in this work we assume that all structural elements are one-dimensional with a straight centroidal axis, as in trusses and frames shown in Fig.1.2. If n^i denotes the number of vertices of the ith structural element, then:

$$n^i = 2\ ,\ for\ i = 1, \cdots, M \tag{1.6}$$

is assumed. Each structural element has a right-handed Cartesian reference frame of its own, where the origin is located at the first vertex, the first coordinate axis x is coincident with the centroidal axis of the structural element and heads towards the second vertex, and the other two coordinate axes, y and z, lie in the cross-section. Fig.1.5 depicts a one-dimensional structural element with its local reference frame. The local reference frames are differentiated from each other by the labels of their structural elements: $(x^i, y^i, z^i) = (x, y, z)^i$, $i = 1, \cdots, M$.

The internal forces develop in a structural element as a result of the loads and the movements of its vertices. Referring to the local reference frame, the a number of quantities, defined as the sum of the stresses and their first moments (about the centroid) that act on the cross-section at the second vertex of a structural element, constitutes the *element forces* of the element. For example, in a space frame element, the element forces consist of the normal force, the two shear forces, the torsional moment and

TABLE 1.2. Element Forces of the k'th Structural Element

Structure	${s^k}^T$	a
Space frame	$N^k, Q^k, Q'^k, T^k, M'^k, M^k$	6
Planar frame	N^k, Q^k, M^k	3
Gridwork frame	T^k, M'^k, Q'^k	3
Space truss	N^k	1
Planar truss	N^k	1

X, Y, Z: global	$N^k = F_x;\ T^k = M_x$
x, y, z: local	$Q^k = F_y;\ M'^k = M_y$
F: force	$Q'^k = F_z;\ M^k = M_z$
M: moment	*k: element label*

the two bending moments, all acting at the second vertex. For the ith space frame element, the element forces are represented by the six-tuple s^i.

By controlling the way a vertex is attached to a node, any or all of the element forces of a structural element can be made zero. For example in truss structures the vertices of elements are attached to the nodes such that the cross-sections at the vertices are free to rotate; hence, no moments and shear forces can develop in the elements due to the movements of the vertices. Therefore, for the ith truss element, the list of element forces s^i is a single scalar, representing the axial force. In Table 1.2, the number, the direction, and the physical meaning of element force components in various types of structures considered in this book are given. In the table, x, y, z are the axis labels of the local reference frame, and X, Y, Z are the labels of the axes of the global reference frame (see Figs.1.5 and 1.4).

When the structure falls into one of the types listed in the table, then the number of element force components per structural element remains the same for all elements. However, in frame structures, for some elements, the vertex-to-node connections may be altered in order to make certain element force components zero. These may be considered as element force constraints. The total number of element force components that are made zero in this way is the number of element force constraints. It is represented by symbol f.

The internal force response may be represented by s where the ith partition is s^iwhich lists the element forces of the ith structural element. Vector s possesses

$$m = Ma - f \tag{1.7}$$

number of real components, i.e., $s \in R^{m\times 1}$.

The internal deformation response may be represented by v where the ith partition is v^i which lists the element deformations of the ith structural element. Vector v also possesses m real components, i.e., $v \in R^{m\times 1}$. When

TABLE 1.3. Element Deformations of the k'th Structural Element

Structure	$\mathbf{v}^{k^T}$	a
Space frame	$\Delta d_x^k, \Delta d_y^k, \Delta d_z^k, \Delta r_x^k, \Delta r_y^k, \Delta r_z^k$	6
Planar frame	$\Delta d_x^k, \Delta d_y^k, \Delta r_z^k$	3
Gridwork frame	$\Delta r_x^k, \Delta r_y^k, \Delta d_z^k$	3
Space truss	Δd_x^k	1
Planar truss	Δd_x^k	1

X, Y, Z: *global*	x, y, z: *local*; k : *element label*
d: *displacement*	$\Delta d = d_{vertex-2} - d_{vertex-1}$
r: *rotation*	$\Delta r = r_{vertex-2} - r_{vertex-1}$

the structure is loaded, the cross-section at the second vertex of a structural element moves relative to the one at the first vertex. The relative movement is described in the local reference frame of the element, and is shown by the a-tuple $\mathbf{v}^i$ which represents the *element deformations* of the ith structural element. In Table 1.3, the number and directions of the element deformation components are given for the structures considered in this book. In the table x, y, z are the labels of the axes of local reference frame, whereas X, Y, Z are the labels of axes of the global reference frame (see Figs. 1.4 and 1.5). For any structural element, the number and the directions of the element force components coincide with those of the element deformation components (compare Tables 1.2 and 1.3), such that, for the ith element, the inner product of $< \mathbf{s}^i, \mathbf{v}^i >$ may be considered as the work done by forces listed in $\mathbf{s}^i$ under the deflections listed in $\mathbf{v}^i$, when $\mathbf{s}^i$and $\mathbf{v}^i$ are caused by independent agents.

For this structure the total response ψ would be as displayed in Eq.(1.3), consisting of $n + 2m$ real components, i.e., $\psi \in \mathrm{R}^{(n+2m)\times 1}$.

1.6.2 Sensors

Any number of components of response vector ψ displayed in Eq.(1.3) may be measured in real time by devices called sensors. Since the components of ψ are of different physical character, so are the sensors. There are already many sensors in the market place, such as displacement sensors, deformation sensors, strain sensors, stress sensors, force sensors, velocity sensors, and acceleration sensors, etc. Since modern sensors are usually electrical devices, their sampling rates and precisions are high although, for some, frequent calibration may be required. This book is not about sensors. In the treatment of this book, it is assumed that, when a response quantity needs to be monitored at a given rate, there is one or more physical sensors to do the job economically, without delay and error.

1.6.3 Actuators

In adaptive structures, the actuators are devices that can insert deformations into structural elements. They are called deformation inserting actuators. They should not be confused with external force exerting actuators such as the attitude control jets or pyrotechnic devices used in spacecrafts which always create internal forces in the structure, whereas a deformation inserting actuator may or may not create internal forces. In fact, any structure may be designed not to resist the slow insertion of element deformations through a deformation inserting actuator. A simple example of deformation inserting actuator is a turnbuckle imbedded in a truss as a truss bar, as depicted in Fig.1.6. In case (a) of the figure, the truss is statically determinate and not loaded; the insertion of deformation through the turnbuckle will not be resisted by the structure. In case (b) of the figure, the truss is statically indeterminate and not loaded; the insertion of deformations will be resisted by the structure.

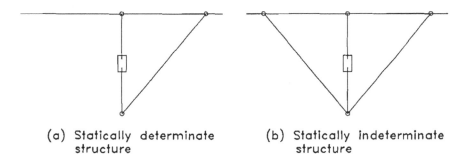

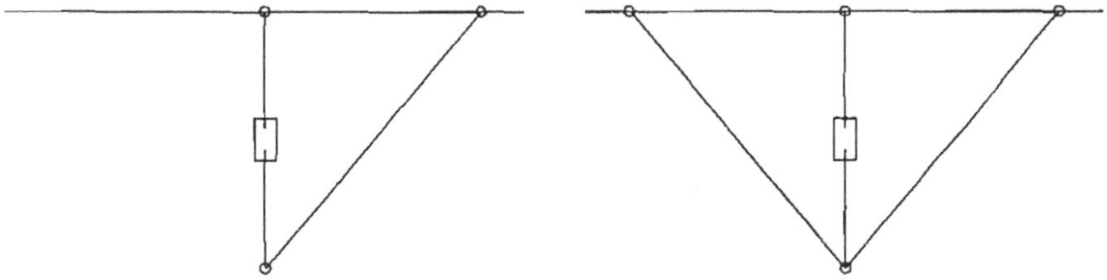

FIGURE 1.6. A simple turnbuckle as a deformation inserting actuator

For the purposes of this book, a structure is an adaptive structure only when its actuators are of deformation inserting type. In the discussions, it is assumed that the actuators conform with this definition. Since the actuators are part of the structure and the structure is assumed time-invariant in most of the discussions, then the actuators should be also time invariant, i.e., their properties should not change with time. There are many modern actuators in the market place, but only a few fulfill the expectations of the time invariant theory. This situation may pose a challenge to actuator manufacturers who are interested in providing actuators for the time-invariant adaptive structures.

The element deformations inserted by the actuators are called *prescribed element deformations.* For the ith structural element, the prescribed element deformations are represented by $\mathbf{v}_o^i$. The complete list of prescribed element deformations of an adaptive structure is represented by $\mathbf{v}_o$ which is an m-tuple where m is as defined in Eq.(1.7). The ith partition of $\mathbf{v}_o$ is

$\mathbf{v}_o^i$. Note that $\mathbf{v}_o$ and $\mathbf{v}_o^i$ are, respectively, the prescribed versions of $\mathbf{v}$ and $\mathbf{v}^i$ which are discussed in Subsection 1.6.1.

When a component of $\mathbf{v}_o^i$ is zero, it means that no deformation is inserted in that direction. If all entries of $\mathbf{v}_o$ are zero, then it means that the adaptive structure is not being controlled. If all but one entry of $\mathbf{v}_o$ are zero, then the adaptive structure either possesses one actuator, or it may possess many but only one is active.

Due to their simplicity and effectiveness, presently, most actuators are of "axial length change" inserting type. These types of actuators are especially useful in truss structures and in those frame structures where internal axial forces dominate the internal force response. However, when a need arises, the manufacture of other types of deformation inserting actuators is possible.

Throughout the book, $\mathbf{v}_o$ is used to represent the physical controls that bring the adaptive structure to a desired state. Since prescribed element deformations may be caused also by other effects, such as by fabrication errors in the structural elements, by thermal loads, or by support settlements, we will use $\mathbf{v}_{of}$, $\mathbf{v}_{ot}$, or $\mathbf{v}_{os}$, respectively, for the latter. When these are also present, the actuator-created prescribed element deformations are denoted by $\mathbf{v}_{oc}$.

1.6.4 Microprocessors and Their Software

These are digital computers which are loaded with appropriate software and attached to the structure. In order to keep the structure at close proximity to its nominal state at all times, they evaluate the sensor outputs and issue commands to the actuators at appropriate sampling rates. The sampling rate, i.e., the number of samplings per unit time, depends on the adaptability criteria prescribed for the structure. This, of course, is closely related with the function of the structure. For example, if the adaptive structure is the flexible reflector of an earthbound telescope that uses the adaptive optics technology to compensate for the atmospheric disturbances of frequency 0.2 Hz or less, then, perhaps, a sampling rate of 20 Hz may be appropriate. This corresponds to a sampling interval of 50 milliseconds. The microprocessor should be capable of doing all the necessary computations and data processing during this time interval.

The serial execution times of tasks that need to be completed in a sampling interval may vary between large margins, depending upon the number of sensors, the number of actuators, and the complexity of the algorithms that analyze the inputs from the sensors and then compute and transmit the required actuation amounts to the actuators. The computer can be a less powerful microprocessor working in uni-processing mode, or it can be an assembly of many powerful microprocessors working in a multi-processing mode. The tasks in a sampling interval can be of diverse granularity and seriality. The realizability of an adaptive structure is very much hinged to the completion of these tasks within the sampling interval and with acceptable accuracy. The fact that the tasks may have to be completed within milliseconds or less is a perpetual challenge for the development of fast machines or machine configurations and efficient serial or parallel algorithms.

This book discusses some of these tasks and their computational complexity. It is assumed that the computer attached to the structure can perform these tasks with the speed and the accuracy required by the adaptability criteria prescribed for the structure.

1.7 Objectives, Scope, and Outline

The primary objective of this book is to provide a basic theory for actively changing the state of a discrete parameter engineering structure, from a known state A to a partially or completely known neighboring state B, by using deformation inserting actuators. If state B is in the close neighborhood of state A one can drive the structure from A to B in one incremental step or otherwise many successive incremental steps may be needed to reach the state at B. Time invariance and linear behavior are assumed for the structure only during an incremental step.

The structure can be brought from state A to state B by either following a partially prescribed trajectory or some unknown optimal trajectory that needs to be determined. As the discussions will reveal, there can be infinitely many trajectories between states A and B. In actively changing the state, the most important consideration is to minimize the necessary energy. This point is a major concern in this book. It is shown that, when options are available, the determination of the trajectory, and also the important attributes of the adaptive structure, such as the number, the location, and the amount of the deformation insertion for the actuators, can be achieved by minimizing the energy.

Although it is discussed sufficiently, this book is not about the application of classical control theories on structures where all the attributes of the structure and its force applying (not deformation inserting) actuators are assumed known *a priori*. In most of such applications, the treatment boils down to finding the controls, usually by heuristics and seldom by sound mathematics, to drive the structure from a known undesired state to its nominal state, usually ignoring the effects of on-going disturbances. Such works are usually referred to as *structural control* and appear to be inspired essentially by the successful works on *attitude control of rigid bodies* that took place during and after World War II.

This book is about how to transform a conventional engineering structure into an adaptive structure so that it needs minimal energy to achieve its prescribed level of adaptability and uses *deformation inserting actuators* instead of *force applying actuators*. In the jargon of *structural control*, the structure and its sensors and actuators constitute the *plant*. The book's primary goal is the determination of the parameters of the plant so that the control objectives are attained with minimal energy. The application of classical control theories, which try to optimize the performance once the plant is defined, is only secondary to this goal.

Since the actuators of adaptive structures are of the deformation inserting type, the engineer, who is trained in the area of *structural control* and now working in the area of adaptive structures, is confronted with the classical problem of *geometric compatibility of deformations*, often without realizing it. Due to the dominance of the *displacement method* in current engineering curricula, it is very likely that the engineer is not aware of the *geometric compatibility requirement of deformations*, since there is no such requirement in the displacement method. However, the fact remains that if the inserted deformations are not geometrically compatible, then one needs additional energy in driving the actuators in order to overcome the resistance of the structure to such insertions. Few present day investigators in *structural control* are concerned with this problem. This book is intended to remedy this situation.

The treatment of the book falls into two groups. In the first group, the discussions are for driving an adaptive structure from a known state A to a partially prescribed state B, statically; that is, slowly so that no appreciable inertial forces develop during the process. In the second group, the discussions are for dynamic cases where the driving of the structure by actuators and other loads is fast enough to cause the development of appreciable inertial forces in the structure. Chapters 2-5 constitute the first group, which is for the static case. Chapters 6-11 constitute the second group, which is for the dynamic case. The problems in the static case are piecewise linear and time-invariant. The problems in the dynamic case are continuously linear and time-invariant. Chapters 2-11 are for discrete parameter adaptive structures. Chapter 12 is an introduction to distributed parameter adaptive structures.

2

Incremental Excitation-Response Relations, Static Case

The material in this chapter is basically a summary of an earlier treatment.[1] For the possibility of nonlinear analysis, the treatment is in incremental form here.

2.1 Basic Definitions

The discussions of this chapter are for obtaining mathematical relations between small excitations and corresponding response quantities in a discrete parameter structure. Application times and magnitudes of the excitations are such that during the loading episode the structure remains time invariant, and its excitation-response relations are linear.

2.1.1 Structure

As discussed in Section 1.6, the structure itself is the most important component of an adaptive structure. The structure is a conventional discrete parameter engineering structure which is a truss or a frame, either in plane

---

[1] *Elementary Structural Analysis, 4th ed.*, Senol Utku, Charles Head Norris, and John Benson Wilbur, Part II, McGraw-Hill, Inc., 1991; *Instructors Manual to Accompany Elementary Structural Analysis*, Senol Utku, McGraw-Hill Publishing Company; *LADS: Linear Analysis of Discrete Structures, A Study Manual for Elementary Structural Analysis, 4th ed, Utku-Norris-Wilbur*, Senol Utku, Dr. Utku and Associates, Durham, NC 27705-5754.

or in space. It has M structural elements ($M > 0$) which are joined together at their vertices at N nodes ($N > 1$). However, in a two bar planar truss $M = 2$, and $N = 3$, these quantities are of the order of thousands for many other types of structures, such as the fuselage of an aircraft, or the stiffening girder of a suspension bridge, or the main frame of a skyscraper, or the supporting structure of an antenna surface. The labels of the nodes and the elements are sequential positive integers. As shown in Fig. 1.4, a global right handed Cartesian reference frame, with axes X, Y, Z, is used to describe the nodal quantities such as position vectors, deflections, and nodal forces. The descriptions, in the global reference frame, of the unit vectors along axes X, Y, Z, are $\mathbf{i} = [1,0,0]^T$, $\mathbf{j} = [0,1,0]^T$, $\mathbf{k} = [0,0,1]^T$, respectively. The global reference frame is an inertially fixed reference frame.

Unless otherwise stated, each structural element is a uniform bar or beam with a straight centroidal axis. The end points of a structural element are its vertices. As shown in Fig. 1.5, each structural element possesses a local right-handed Cartesian reference frame which is placed at the first vertex, with the first local axis x overlapping the centroidal axis and heading towards the second vertex. The other two local axes, y and z, are in the cross-sectional plane and usually coincident with the principal axes of the cross-section. The descriptions, in the global reference frame, of the unit vectors along axes x, y, z are $\mathbf{e} = [e_X, e_Y, e_Z]^T$, $\mathbf{g} = [g_X, g_Y, g_Z]^T$, $\mathbf{h} = [h_X, h_Y, h_Z]^T$, respectively. For the kth structural element, the local reference frame axes and their unit vectors carry the label k in the superscript, i.e., $(x, y, z)^k$ or (x^k, y^k, z^k), and $[\mathbf{e}^k, \mathbf{g}^k, \mathbf{h}^k]$ or $[\mathbf{e}, \mathbf{g}, \mathbf{h}]^k$. Since the local reference frames are attached to the structural elements, their orientations relative to the global reference frame change as the structure moves from one deflection state to another.

For defining the geometry at a reference state, one needs the position vectors of the nodes and the node labels of the vertices of elements. Matrix $\grave{\mathbf{X}} \in \mathrm{R}^{3\times N}$ defines nodal position vectors such that $\grave{\mathbf{x}}_i$, the ith column of $\grave{\mathbf{X}}$, is the description, in the global reference frame, of the position vector of node i. The *connectivity matrix* $\mathbf{J} \in (\mathrm{Int})^{2\times M}$ defines the position of structural elements relative to the nodes such that $\mathbf{j}^k$ lists for the kth structural element the node label of the first vertex in j_1^k and the node label of the second vertex in j_2^k. The information in $\mathbf{J}$ does not change in successive states of the structure, whereas the information in $\grave{\mathbf{X}}$ needs to be updated at the end of each incremental step. When $\grave{\mathbf{X}}$ and $\mathbf{J}$ are available, then for any structural element k, the length l^k and the description, in the global reference frame, of the first local axis unit vector $\mathbf{e}^k$ can be computed from

$$\left.\begin{aligned} l^k &= \left\| \grave{\mathbf{x}}_{j_2^k} - \grave{\mathbf{x}}_{j_1^k} \right\|_E \\ \mathbf{e}^k &= (\grave{\mathbf{x}}_{j_2^k} - \grave{\mathbf{x}}_{j_1^k})/l^k \end{aligned}\right\} \quad \text{for } k = 1, \cdots, M \tag{2.1}$$

For planar trusses and frames, and gridwork frames, the third local axis is always coincident with the third global axis, i.e.,

$$\mathbf{h}^k = \mathbf{k} \ , \ \text{for } k = 1, \cdots, M \tag{2.2}$$

For space trusses and frames, one has to define $\mathbf{h}^k$ (or $\mathbf{g}^k$) individually for all elements. Note that for all truss and frame elements

$$\mathbf{g}^k = \mathbf{h}^k \times \mathbf{e}^k \ , \ \text{for } k = 1, \cdots, M \tag{2.3}$$

The geometric definition of a truss structure is completed by providing the cross-sectional areas of the bars. In planar frames in addition to the cross-sectional areas, one has to provide the second area moment about the third local axis for each frame element. In space frames, in addition to these, the second area moment about the second local axis and the torsional constant should be provided for all elements. For gridwork frames, only the torsional constant and the second area moment about the second local axis are needed for all elements.

2.1.2 Excitations

These are the agents that cause the structure to change its state. In this chapter, it is assumed that the excitations are applied statically, i.e., no material particle gains discernible accelerations due to the application of the excitations. After Chapter 5, this restriction is removed. The determination of the static response is very important in engineering design, since it defines the final equilibrium state of the structure when its vibrational kinetic energy due to the sudden loading is dissipated by the internal friction forces. Note that the term "load" is being used here for the term "excitation", since the former has the implication of statical application.

In the static analysis of a discrete parameter structure, it is convenient to think that the loading is achieved in two steps. In the first step all nodes are first locked in their reference positions and then the loads are applied. The locked-in state is called the *kinematically determinate state*, and the forces that are developed at the nodes are called the *holding forces.* The methods of elementary structural analysis enable one to compute the holding forces and also the internal forces due to force-type element loads, one element at a time. In the second step, by the slow removal of the holding forces, the joints are unlocked. As a result, the nodes move to their final positions, along a straight line trajectory in the state space, and the structure assumes its new equilibrium state.

The nodal loads include all prescribed concentrated forces and moments, and also prescribed displacements and rotations, acting on the nodes. The elemental loads are the loads acting on the structural elements, not on the nodes. They consist of concentrated or distributed loads acting on the elements, and the loads associated with fabrication errors, shrinkage, thermal effects, and deformation inserting actuators of the elements. If there are no element loads, then the reference state and the kinematically determinate state of the structure are identical.

Because of their physical character, the loads can be categorized in two groups: force-type or deflection-type. Force-type loads are concentrated or distributed prescribed forces that act either on the nodes or on the elements. Loads due to support settlements are of nodal deflection-type, and loads associated with shrinkage, fabrication errors, thermal effects, deformation inserting actuators are of elemental deformation-type.

In the analysis of adaptive structures, it is convenient to treat the concentrated or distributed force-type elemental loads separately, as in the moment distribution methods of elementary structural analysis. When the structure is in the kinematically determinate state, the internal forces and also the fixed-end reactions acting on the vertices, as a result of these force-type elemental loads, can be computed easily, one element at a time. In the second step of loading mentioned above, the force-type elemental loads and the corresponding internal forces are ignored, but the negatives of the computed fixed-end reactions acting on the vertices are considered at the nodes. The final response of the structure is the sum of the second step response plus the internal force response computed earlier for the force-type elemental loads, one element at a time. This process is shown in Fig.2.1. In part (a) of the figure, a portal frame is shown with its elemental and nodal force-type loads. In (b), the internal forces (only moments) in the kinematically determinate state are shown. In (c) the second step loading and the corresponding moment response are shown. In (d) the final response is shown as the sum of responses in (b) and (c). In the figure the internal moment responses are sketched in the form of moment diagrams.

In the discussions of this book, the force-type nodal loads are the prescribed concentrated forces and moments acting at the nodes plus the negatives of the *fixed-end reactions acting on the elements due to force-type element loads.* The description of the force-type nodal loads, in the global reference frame, of node i is denoted by $\mathbf{p}_i$ which is an e-tuple if the node is not restrained. The number, the direction, and the physical meaning of the components of $\mathbf{p}_i$ are given in Table 2.1 for various frame and truss structures.

2.1.3 Incremental Response

The response of a structure to a given incremental loading is its incremental response which consists of incremental nodal deflections $\Delta\boldsymbol{\xi}$, incremental elemental deformations $\Delta\mathbf{v}$, and incremental element forces $\Delta\mathbf{s}$.

The changes in the position of a node are described by its displacements and rotations. These quantities when described in the global reference frame X, Y, Z are the deflections of the node. For the ith node they are denoted by $\boldsymbol{\xi}_i$ which is an e-tuple if none of the deflection components are restrained by

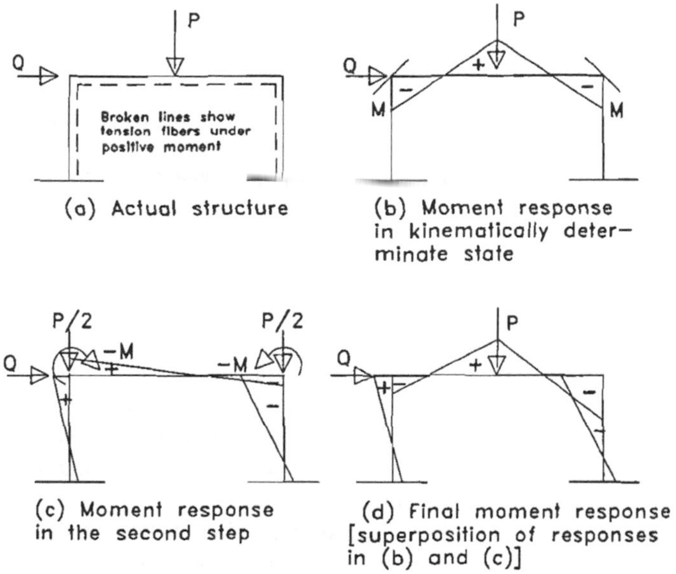

FIGURE 2.1. Handling of force-type elemental and nodal loads

TABLE 2.1. Force-type Loads of a Node

Structure	$\mathbf{p}_i^T$	e
Space frame	$P_{iX}, P_{iY}, P_{iZ}, M_{iX}, M_{iY}, M_{iZ}$	6
Planar frame	P_{iX}, P_{iY}, M_{iZ}	3
Gridwork frame	M_{iX}, M_{iY}, P_{iZ}	3
Space truss	P_{iX}, P_{iY}, P_{iZ}	3
Planar truss	P_{iX}, P_{iY}	2

X, Y, Z : *global axes*	P : *force*
i : *node label*	M : *moment*

a support. The values of e and the meanings of the entries of $\boldsymbol{\xi}_i$ are given in Table 1.1 for various types of frame and truss structures. The incremental nodal deflection of node i is $\Delta\boldsymbol{\xi}$ and the complete list of incremental nodal deflections is

$$\Delta\boldsymbol{\xi}^T = [\Delta\boldsymbol{\xi}_1^T, \Delta\boldsymbol{\xi}_2^T, \cdots, \Delta\boldsymbol{\xi}_N^T] \tag{2.4}$$

For an N-node structure with b number of deflection constraints, one may observe that $\Delta\boldsymbol{\xi} \in \mathrm{R}^{n\times 1}$, where $n = Ne - b$.

The way that structural elements are connected to nodes determine if the structure is a truss or a frame. In truss structures, the connections are with frictionless joints such that the cross-sections at the vertices are free to rotate, whereas in frame structures the connections are rigid such that the vertex cross-sections rotate together with their nodes. In frame structures, however, by using special connection devices, a deflection component of the cross-section at a vertex can be made independent of that of the node; then no force (or moment) can develop between the vertex and the node in the direction of the component. These are the element force constraints, and should be furnished for each structural element that has this type of special vertex-to-node connection.

Since nodes and vertices are connected, when nodes move, the cross-section at the second vertex of an element moves and rotates relative to the cross-section at the first vertex. The quantities describing these movements in the local reference frame x, y, z of the element are the element deformations. For an incremental loading, for the kth structural element they are shown by a-tuple $\Delta\mathbf{v}^k$, if no special release mechanism exists between the vertex and the node to nullify the corresponding internal force component (i.e., if no element force constraint exists). The value of a and the meanings of the entries of $\Delta\mathbf{v}^k$ are given in Table 1.3 for various types of frame and truss structures. The list of these quantities for the whole structure is the element deformations of the structure:

$$\Delta\mathbf{v}^T = [\Delta\mathbf{v}^{1^T}, \Delta\mathbf{v}^{2^T}, \cdots, \Delta\mathbf{v}^{M^T}] \tag{2.5}$$

For an M-element structure with f number of element force constraints, one may observe that $\Delta\mathbf{v} \in \mathrm{R}^{m\times 1}$, where $m = Ma - f$.

The internal forces develop as a result of element deformations and element loads. The quantities describing internal forces at the second vertex, in the local reference frame x, y, z of the structural element, are the element forces. For an incremental loading, for the kth structural element they are shown by a-tuple $\Delta\mathbf{s}^k$, if none of them is made zero by element force constraints. The value of a and the meanings of the entries of $\Delta\mathbf{s}^k$ are given in Table 1.2 for various types of frame and truss structures. The list of these quantities for the whole structure is the element forces of the structure:

$$\Delta\mathbf{s}^T = [\Delta\mathbf{s}^{1^T}, \Delta\mathbf{s}^{2^T}, \cdots, \Delta\mathbf{s}^{M^T}] \tag{2.6}$$

For an M-element structure with f number of element force constraints, one may observe that $\Delta\mathbf{s} \in R^{m\times 1}$, where $m = Ma - f$.

The deflection response $\Delta\boldsymbol{\xi}$, the element deformations response $\Delta\mathbf{v}$, and the element forces response $\Delta\mathbf{s}$ together constitute the complete incremental response $\Delta\boldsymbol{\psi}$ of the discrete parameter structure:

$$\Delta\boldsymbol{\psi}^T = [\Delta\boldsymbol{\xi}^T, \Delta\mathbf{s}^T, \Delta\mathbf{v}^T] \tag{2.7}$$

Since $\Delta\mathbf{s}, \Delta\mathbf{v} \in R^{m\times 1}$ and $\Delta\boldsymbol{\xi} \in R^{n\times 1}$, one observes that $\Delta\boldsymbol{\psi} \in R^{(2m+n)\times 1}$.

2.1.4 Excitation-Response Relations

When loads $\Delta\mathbf{f}$ are applied on a discrete parameter structure described above, the structure responds by changing its state. The new state is identified by the changes in internal forces $\Delta\mathbf{s}$, in internal deformations $\Delta\mathbf{v}$, and in the location and orientation of its nodal points $\Delta\boldsymbol{\xi}$, all relative to the reference state, i.e., the state of the structure before the application of the current incremental loads. The reference state is a *stable equilibrium state* which can be the at-rest state where the measures of all past excitations and responses are zero, or it can be any other equilibrium state where the measures are not all zero. Unless otherwise stated, the reference state is taken as a stable equilibrium state.

When loads $\Delta\mathbf{f}$ are applied statically to the structure at its reference state, the response $\Delta\boldsymbol{\psi}$ takes place. The relationship between excitations $\Delta\mathbf{f}$ and the response $\Delta\boldsymbol{\psi}$ is called the excitation-response relation. It is algebraic and can be linearized as

$$\grave{\mathbf{A}}\Delta\boldsymbol{\psi} = \Delta\mathbf{f} \tag{2.8}$$

From the discussions above, it is clear that $\Delta\boldsymbol{\psi} \in R^{(n+2m)\times 1}$. Since for each excitation $\Delta\mathbf{f}$, there is a unique response $\Delta\boldsymbol{\psi}$, one can conclude that $\grave{\mathbf{A}}$ should be a nonsingular square matrix, i.e., $\grave{\mathbf{A}} \in R^{(n+2m)\times(n+2m)}$ and not rank deficient. Therefore one must have $n+2m$ linearly independent scalar equations in the excitation-response relations. By using *Newton's laws*, the rules of *Euclidean geometry*, and *Hooke's law*, explicit expressions for the linearized excitation-response relations as stated formally in Eq.(2.8) can be developed. This is done in the remainder of this chapter.

2.2 Equilibrium of Forces

When the loads are applied slowly on the structure at the reference state, it moves to a new equilibrium state along a straight line trajectory in the state space. The new state is defined by the response quantities $\Delta\boldsymbol{\psi}$ as defined in Eq.(2.7). Since it is an equilibrium state, the internal force components $\Delta\mathbf{s}$

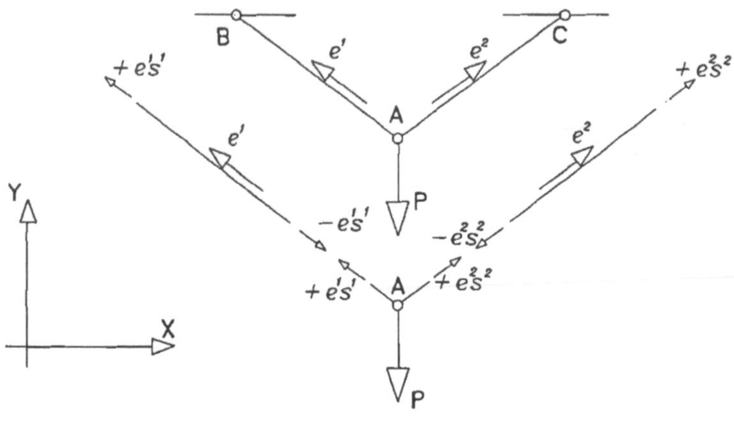

FIGURE 2.2. Free-body diagrams of nodes and elements

of $\Delta\psi$ must satisfy the force equilibrium requirements per Newton's laws; namely, every material particle must be in static equilibrium under the internal and the external forces that act on it. For the discrete parameter structures, this may be restated by saying that every structural element and every node must be in static equilibrium under the effect of their internal and external forces. In Fig. 2.2, the free-body diagrams of the nodes and the elements of a structure are shown. In the equilibrium state, the external and the internal forces acting on any of the nodes or elements must have a zero force resultant and a zero moment resultant.

2.2.1 Equilibrium of Structural Elements

From the free body diagram of a structural element, one may observe that it may be under the effect of force-type elemental loads and its vertex forces. As discussed in Subsection 2.1.2, the force-type element loads are in static equilibrium with the fixed-end reactions which are determined so that the force equilibrium requirements are also satisfied. Removing the force-type element loads and the corresponding fixed-end reactions from the free-body diagram, one is left with the vertex forces due to non-force-type element loads and nodal deflections. For the kth structural element, the forces at the second vertex, as described in the local reference frame of the element, are listed in a-tuple $\Delta\mathbf{s}^k$ which represents the element forces of the kth element. Table 1.2 shows that $\Delta\mathbf{s}^k$ lists for the second vertex of the kth structural element, axial force ΔN^k in trusses; axial force ΔN^k, shear force ΔQ^k and bending moment ΔM^k in planar frames; axial force ΔN^k, two shear forces ΔQ^k and $\Delta Q'^k$, torsional moment ΔT^k and two bending moments ΔM^k and $\Delta M'^k$ in space frames; and torsional moment ΔT^k, bending moment $\Delta M'^k$, and shear force $\Delta Q'^k$ in gridwork frames. One can ensure the static equilibrium of the structural element by taking

TABLE 2.2. Transformation Matrices for Second Vertex Forces

Structure	$\mathbf{T}_2^k$ Matrix	
Space frame	$\begin{bmatrix} \mathbf{e} & \mathbf{g} & \mathbf{h} & \cdot & \cdot & \cdot \\ \cdot & \cdot & \cdot & \mathbf{e} & \mathbf{g} & \mathbf{h} \end{bmatrix}^k$	$\in R^{6\times 6}$
Planar frame	$\begin{bmatrix} \mathbf{e} & \mathbf{g} & \mathbf{k} \end{bmatrix}^k$	$\in R^{3\times 3}$
Gridwork frame	$\begin{bmatrix} \mathbf{e} & \mathbf{g} & \mathbf{k} \end{bmatrix}^k$	$\in R^{3\times 3}$
Space truss	$[\mathbf{e}]^k$	$\in R^{3\times 1}$
Planar truss	$[\mathbf{e}]^k$	$\in R^{2\times 1}$

the first vertex forces to be in static equilibrium with those listed in $\Delta \mathbf{s}^k$. The next paragraph shows how this is done.

Let $\Delta \mathbf{q}_2^k$ denote the description of element forces $\Delta \mathbf{s}^k$ in the global reference frame. Since the descriptions, in the global reference frame, of the unit vectors of the local reference frame are $\mathbf{e}^k, \mathbf{g}^k, \mathbf{h}^k$, one may easily obtain for $\Delta \mathbf{q}_2^k$ expressions of the type

$$\Delta \mathbf{q}_2^k = \mathbf{T}_2^k \Delta \mathbf{s}^k \tag{2.9}$$

For example, for the space truss element, since $\Delta \mathbf{q}_2^k = \mathbf{e}^k \Delta s_1^k$, then $\mathbf{T}_2^k = \mathbf{e}^k$ where matrix $\mathbf{T}_2^k$ is the matrix transforming the element forces $\Delta \mathbf{s}^k$ into the global description of the second vertex forces for the kth structural element $\Delta \mathbf{q}_2^k$. Table 2.2 gives $\mathbf{T}_2^k$ matrices for various truss and frame elements.

Let $\Delta \mathbf{q}_1^k$ denote the description, in the global reference frame, of the forces at the first vertex. By writing the force and moment equilibrium equations in the free body diagram of the structural element subjected to vertex forces $\Delta \mathbf{q}_1^k$ at the first vertex and $\Delta \mathbf{q}_2^k = \mathbf{T}_2^k \Delta \mathbf{s}^k$ at the second vertex, one may obtain for $\Delta \mathbf{q}_1^k$ expressions of the type:

$$\Delta \mathbf{q}_1^k = \mathbf{T}_1^k \Delta \mathbf{s}^k \tag{2.10}$$

For example, for the space truss element, the force equilibrium in the axial direction yields $\Delta \mathbf{q}_1^k = -\mathbf{e}^k \Delta s_1^k$, hence $\mathbf{T}_1^k = -\mathbf{e}^k$. Table 2.3 gives $\mathbf{T}_1^k$ matrices for various truss and frame elements, where L^k is the length of the kth element.

Note that matrices ($\mathbf{T}_l^k$, for $l = 1, 2$ and $k = 1, 2, \cdots, M$) $\in R^{e\times a}$ so long as no element force constraints exist. If they exist, one deletes the columns in $\mathbf{T}_1^k$ and $\mathbf{T}_2^k$ matrices corresponding to the zeroed element force components. This requires that the vertex with the element force constraint be chosen as the first vertex.

In conclusion, as long as the vertex forces at the first vertex are taken as in Eq.(2.10), for $k = 1, 2, \cdots, M$, then the equilibrium of structural elements is ensured.

TABLE 2.3. Transformation Matrices for First Vertex Forces

Structure	$\mathbf{T}_1^k$ Matrix	
Space frame	$-\begin{bmatrix} \mathbf{e} & \mathbf{g} & \mathbf{h} & \cdot & \cdot & \cdot \\ \cdot & L\mathbf{h} & -L\mathbf{g} & \mathbf{e} & \mathbf{g} & \mathbf{h} \end{bmatrix}^k$	$\in R^{6\times6}$
Planar frame	$-\begin{bmatrix} \mathbf{e} & \mathbf{g}+L\mathbf{k} & \mathbf{k} \end{bmatrix}^k$	$\in R^{3\times3}$
Gridwork frame	$-\begin{bmatrix} \mathbf{e} & \mathbf{g} & \mathbf{k}-L\mathbf{g} \end{bmatrix}^k$	$\in R^{3\times3}$
Space truss	$-[\mathbf{e}]^k$	$\in R^{3\times1}$
Planar truss	$-[\mathbf{e}]^k$	$\in R^{2\times1}$

2.2.2 Equilibrium of Nodes

The internal and the external forces acting on a node must be in static equilibrium. This must hold for every node in the equilibrium state. The external forces acting on a node are the force-type nodal loads plus the negatives of the fixed-end reactions acting on the elements due to force-type element loads. The internal forces acting on a node are the negatives of the vertex forces of the elements incident to the node, as shown in Fig. 2.2. The internal forces may consist of two parts: one part due to current loading, and another part, called preexisting internal forces, due to earlier loadings before the current loading is applied. Let $\Delta\mathbf{q}$ and $\grave{\mathbf{q}}$ denote these two parts. Preexisting internal forces are not conservative, i.e., they change their orientations together with the element during the loading. Depending upon the magnitudes of the nodal deflections, the difference between the preexisting internal forces at the reference state and those at the final equilibrium state defines $\grave{\mathbf{q}}$, which should be considered in the nodal force equilibrium equations. In this subsection, the nodal equilibrium equations are obtained by assuming that the expressions for $\grave{\mathbf{q}}$ are available. The expressions for $\grave{\mathbf{q}}$ are developed in the next subsection.

For convenience, all forces are described in the global reference frame. Let $\Delta\mathbf{q}_l^k$ and $\grave{\mathbf{q}}_l^k$ denote the descriptions, in the global reference frame, of the vertex forces and the effects of preexisting vertex forces, respectively, acting on the lth vertex of the kth structural element. The kth element is between nodes j_1^k and j_2^k [i.e., $(1,k)$ and $(2,k)$ entries of element connectivity matrix $\mathbf{J}$]. The equilibrium equations of node i may be written as

$$\Delta\mathbf{p}_i - \sum_{k=1}^{M}\sum_{l=1}^{2}\delta_{i,j_l^k}(\Delta\mathbf{q}_l^k+\grave{\mathbf{q}}_l^k) = \mathbf{o} \tag{2.11}$$

where $\Delta\mathbf{p}_i$, i.e., the ith partition of $\Delta\mathbf{p}$, is the force-type external load acting at node i, and δ is the Kronecker delta which is 1 if $i = j_l^k$, otherwise 0. One may use Eq.(2.9 or 2.10) and Table 2.2 or 2.3 to express this equation as

$$\Delta \mathbf{p}_i - \sum_{k=1}^{M}\sum_{l=1}^{2} \delta_{i,j_l^k}(\mathring{\mathbf{q}}_l^k + \mathbf{T}_l^k \Delta \mathbf{s}^k) = \mathbf{o} \tag{2.12}$$

In this part of the development, let us assume that there are no deflection and element force constraints, i.e., $b = f = 0$. Then

$$\Delta \mathbf{s}^k = \mathbf{I}_k^{'T} \Delta \mathbf{s} \tag{2.13}$$

where $\mathbf{I}_k^{'T}$ is the transpose of the kth column partition of the Math order identity matrix partitioned a column at a time, and

$$\Delta \mathbf{p} = \sum_{i=1}^{N} \mathbf{I}_i \Delta \mathbf{p}_i \tag{2.14}$$

where $\mathbf{I}_i$ is the ith column partition of the Neth order identity matrix partitioned e column at a time. By first substituting $\mathbf{s}^k$ from Eq.(2.13) into Eq.(2.12) and then using $\Delta \mathbf{p}_i$ from the latter in Eq.(2.14), one may obtain after contractions

$$\Delta \mathbf{p} = \mathring{\mathbf{q}} + \sum_{k=1}^{M}\sum_{l=1}^{2} \mathbf{I}_{j_l^k} \mathbf{T}_l^k \mathbf{I}_k^{'T} \Delta \mathbf{s} \tag{2.15}$$

where

$$\mathring{\mathbf{q}} = \sum_{k=1}^{M}\sum_{l=1}^{2} \mathbf{I}_{j_l^k} \mathring{\mathbf{q}}_l^k \tag{2.16}$$

In these equations the identity of $\left[\sum_{i=1}^{N} \delta_{i,j_l^k} \mathbf{I}_i\right] = \mathbf{I}_{j_l^k}$ is used (matrix $\mathbf{I}_{j_l^k}$ is the matrix $\mathbf{I}_i$ defined above, for $i = j_l^k$). By defining the coefficient matrix of $\mathbf{s}$ in Eq.(2.15) as

$$\mathbf{B}^o = \sum_{k=1}^{M}\sum_{l=1}^{2} \mathbf{I}_{j_l^k} \mathbf{T}_l^k \mathbf{I}_k^{'T} \tag{2.17}$$

the former may be restated as

$$\Delta \mathbf{p} = \mathring{\mathbf{q}} + \mathbf{B}^o \Delta \mathbf{s} \tag{2.18}$$

These are the nodal force equilibrium equations of a structure when there are no nodal deflection constraints, i.e., $b = 0$, and no element force constraints, i.e., $f = 0$.

There are Ne scalar equations in Eq.(2.18), each for one nodal deflection direction of the unconstrained structure, involving the Ma number of element force components listed in $\Delta \mathbf{s}^T = [\Delta \mathbf{s}^{1^T}, \Delta \mathbf{s}^{2^T}, \cdots, \Delta \mathbf{s}^{M^T}]$. Partitioning the equations such that each entry of $\mathbf{B}^o$ is an $e \times a$ submatrix,

TABLE 2.4. Algorithm to Generate B-zero Matrix

No	Statement
1	$\mathbf{B}^o \leftarrow 0$
2	*for* $m = 1$ *to* M
3	*for* $k = 1$ *to* 2
4	$l \leftarrow j_k^m$
5	*generate* $\mathbf{T}_k^m$
6	$\mathbf{B}_{l,m}^o \leftarrow \mathbf{T}_k^m$

Eq.(2.18) may be displayed as

$$\begin{Bmatrix} \Delta\mathbf{p}_1 \\ \Delta\mathbf{p}_2 \\ \vdots \\ \Delta\mathbf{p}_N \end{Bmatrix} = \begin{Bmatrix} \mathring{\mathbf{q}}_1 \\ \mathring{\mathbf{q}}_2 \\ \vdots \\ \mathring{\mathbf{q}}_N \end{Bmatrix} + \begin{bmatrix} \mathbf{B}_{1,1}^o & \mathbf{B}_{1,2}^o & \cdots & \mathbf{B}_{1,M}^o \\ \mathbf{B}_{2,1}^o & \mathbf{B}_{2,2}^o & \cdots & \mathbf{B}_{2,M}^o \\ \vdots & \vdots & & \vdots \\ \mathbf{B}_{N,1}^o & \mathbf{B}_{N,2}^o & \cdots & \mathbf{B}_{N,M}^o \end{bmatrix} \begin{Bmatrix} \Delta\mathbf{s}^1 \\ \Delta\mathbf{s}^2 \\ \vdots \\ \Delta\mathbf{s}^M \end{Bmatrix} \tag{2.19}$$

where the ith row is the e number of force equilibrium equations of node i. Since, in general, much fewer than M structural elements are incident at node i, many of the submatrices in the ith row partition of $\mathbf{B}^o$ are zero. Observe that the contributions of the kth element forces $\Delta\mathbf{s}^k$ to the nodal force equilibrium equations are $\mathbf{B}_k^o\Delta\mathbf{s}^k$, where $\mathbf{B}_k^o$ is the kth column partition of $\mathbf{B}^o$. Since the kth element is connected to nodes j_1^k and j_2^k, then

$$\mathbf{B}_k^o = \begin{bmatrix} \vdots \\ \mathbf{T}_1^k \\ \vdots \\ \mathbf{T}_2^k \\ \vdots \end{bmatrix} \begin{matrix} \vdots \\ \longleftarrow \text{node } j_1^k \\ \vdots \\ \longleftarrow \text{node } j_2^k \\ \vdots \end{matrix} \quad , \text{ for } k = 1, \cdots, M \tag{2.20}$$

which shows that the kth column partition of $\mathbf{B}^o$, i.e., $\mathbf{B}_k^o$ contains all $e \times a$ zero submatrices, except for the j_1^kth and j_2^kth row partitions which contain matrices $\mathbf{T}_1^k$ and $\mathbf{T}_2^k$. The display in Eq.(2.20) assumes that $j_1^k < j_2^k$; otherwise one needs to interchange $\mathbf{T}_1^k$ and $\mathbf{T}_2^k$ in the display. Clearly, there are only two nonzero submatrices in any one column partition of $\mathbf{B}^o$. It is very sparse, and its sparsity increases with increasing M and N. The ratio of the nonzero entries to the total number entries is $2/(MN)$. The basic information to generate $\mathbf{B}^o$ matrix consists of $\mathbf{J}$ and $\mathbf{T}_l^k$ for $k = 1, \cdots, M$ and $l = 1, 2$, which needs much smaller storage than that required by $\mathbf{B}^o$ itself. The algorithm to generate $\mathbf{B}^o$ according to its definition in Eq.(2.17) is given in Table 2.4.

The nodal force equilibrium equations of a structure with deflection and force constraints, i.e., when $b \neq 0$ and $f \neq 0$, can be obtained from the nodal force equilibrium equations without the constraints. Once the equations without the constraints are obtained, the element force constraints can be imposed on them by deleting the zero element forces from the list of element forces $\Delta\mathbf{s}$, and their corresponding columns from $\mathbf{B}^o$. If there are f number of element force constraints, then f number of entries in $\Delta\mathbf{s}$ and corresponding f number of columns of $\mathbf{B}^o$ should be deleted. Similarly, the deflection constraints can be imposed one at a time, by removing the equilibrium equation corresponding to the constrained direction. The equilibrium equation of a constrained direction involves an unknown reaction which develops as a result of the constraining mechanism, i.e., the support. An equilibrium equation with a reaction brings no new information for the determination of the response. When the response is computed, b number of reactions can be computed by using the b number of removed equations. If a constrained direction is not one of the nodal deflection directions defined by the global reference frame axes, then the equilibrium equation in the global direction which is closest to the restrained direction is removed, and the equations corresponding to the retained directions are replaced by a weighted sum of the equilibrium equations of the restrained node with weights implied by the deflection constraint. If there are b number of deflection constraints, then b number of scalar equations are deleted such that there should be $Ne - b$ number of linearly independent nodal force equilibrium equations. After the imposition of element force and deflection constraints, the nodal force equilibrium equations become

$$\grave{\mathbf{q}} + \mathbf{B}\Delta\mathbf{s} = \Delta\mathbf{p} \tag{2.21}$$

where $\mathbf{B} \in \mathrm{R}^{n \times m}$, $n = Ne - b$ and $m = Ma - f$. In computer applications, instead of physical deletion, entities may be removed logically by negating their row and/or column labels.[2]

Expressions for the preexisting internal force effects $\grave{\mathbf{q}}$ are developed in the next subsection.

2.2.3 Effect of Preexisting Internal Forces

In order to see the effect of preexisting internal forces consider the planar truss structure consisting of two identical collinear bars of length L each, as shown in Fig. 2.3. If there is no initial internal force in the bars, the truss cannot carry the concentrated load P at joint B. However, if the

[2]This approach may be used when dealing with small problems. In larger problems where the storage is critical, by proper logic, one may generate the final matrix directly. For an example see *ELAS - A General-Purpose Computer Program for the Equilibrium Problems of Linear Structures* (*vol. 1, User's Manual*, Senol Utku and Fevzican A. Akyuz, February 1968; *vol.2, Documentation of the Program*, Senol Utku, October 1969), *Technical Report 32-1240*, NASA, Jet Propulsion Laboratory, Pasadena, CA. [The software was distributed under code name *ELAS* to the general public through Com-

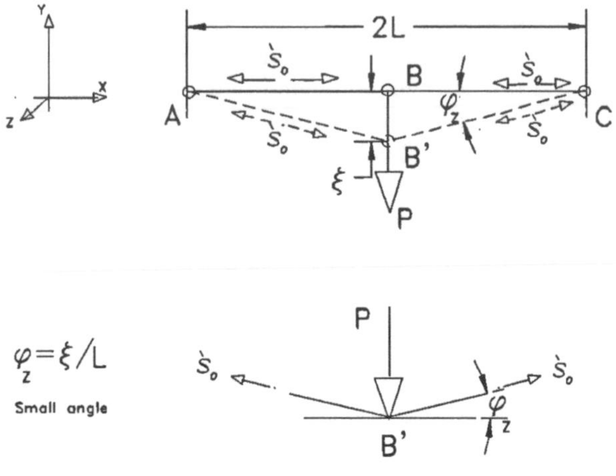

FIGURE 2.3. Effect of preexisting internal forces on equilibrium

bars are stretched initially with an axial force $\grave{s}_o$, there will be a stable equilibrium state marked with the new position B' of the loaded joint. Assuming small angles, the transverse equilibrium of the loaded node in the deformed configuration gives

$$\frac{2\grave{s}_o}{L}\Delta\xi + [0,0]\Delta \mathrm{s} = P \tag{2.22}$$

where $\Delta\xi = BB'$, and $\Delta \mathrm{s}^T = [\Delta s^1, \Delta s^2]$ is the list of bar forces due to the load P. Since there will be almost no element deformation if $\left|\frac{P}{2\grave{s}_o}\right| << 1$, $\Delta \mathrm{s} = \mathrm{o}$. The *axial stiffness* of a bar of length L, cross-sectional area A, and Young's modulus E is EA/L. As this example shows, the *transverse stiffness* of the bar with preexisting axial force $\grave{s}_o = A\sigma_w$ (where σ_w is the working axial stress) is $\sigma_w A/L$. The ratio of the transverse stiffness to the axial one is σ_w/E which is very small in magnitude. For example, for steel, the magnitude of this ratio is of the order of $1/1000$. However, as this reasoning implies that it may be ignored in many applications, the effect of preexisting internal forces on nodal force equilibrium equations will be considered in this discussion.

The comparison of Eq.(2.22) with Eq.(2.21) shows that

$$\grave{\mathrm{q}} = \grave{\mathrm{K}}\Delta\boldsymbol{\xi} \tag{2.23}$$

is the form of the preexisting internal force effects. By using this in Eq.(2.21), the general form of the linearized nodal force equilibrium equations may

puter Software Management and Information Center (*COSMIC*), University of Georgia, Athens, GA.]

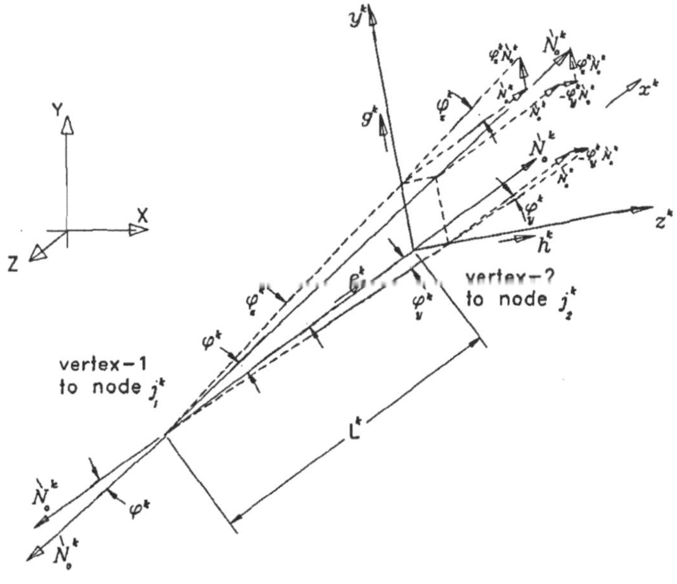

FIGURE 2.4. Unbalanced internal forces due to orientation change

be stated as

$$\grave{\mathbf{K}}\Delta\boldsymbol{\xi} + \mathbf{B}\Delta\mathbf{s} = \Delta\mathbf{p} \tag{2.24}$$

where the ($n = Ne - b$)th order coefficient matrix of $\Delta\boldsymbol{\xi}$, i.e., $\grave{\mathbf{K}}$, is a function of preexisting internal forces $\grave{\mathbf{s}}_o$. Matrix $\grave{\mathbf{K}}$ has various names: *geometric matrix* or *preexisting stress matrix*, or *stiffness matrix due to preexisting stresses.* As will be shown, it is a real symmetric matrix. The rest of this subsection develops expressions for $\grave{\mathbf{K}}$.

As in cable structures, the preexisting axial force in structural elements may resist or assist the orientation changes due to the current loading. In Fig.(2.4), the free body diagram of the kth element of a space frame is shown. It is a uniform element of length L^k under the effect of preexisting element forces $\grave{\mathbf{s}}_o^{kT} = [\grave{N}_o^k, 0, 0, 0, 0, 0]$, where $\grave{N}_o^k$ is the preexisting axial force. The space frame element is connected to nodes j_1^k and j_2^k. When the frame is subjected to the current loading, the nodes will be displaced by $\boldsymbol{\xi}_{j_1^k}$ and $\boldsymbol{\xi}_{j_2^k}$. Because of these deflections, the element will not only be stretched but also rotate about the axes of the local reference frame. Because the nodal deflections are of small magnitude, the small rotations about local y and z axes may be computed as

$$\begin{aligned} \varphi_y^k &= -[\mathbf{h}^{kT}, \mathbf{o}^T](\Delta\boldsymbol{\xi}_{j_2^k} - \Delta\boldsymbol{\xi}_{j_1^k})/L^k \\ \varphi_z^k &= [\mathbf{g}^{kT}, \mathbf{o}^T](\Delta\boldsymbol{\xi}_{j_2^k} - \Delta\boldsymbol{\xi}_{j_1^k})/L^k \end{aligned} \tag{2.25}$$

These rotations are such that the following are acceptable approximations:

$$\begin{aligned} s_y &= \sin\varphi_y^k \cong \varphi_y^k \\ s_z &= \sin\varphi_z^k \cong \varphi_z^k \\ c_y &= \cos\varphi_y^k \cong 1 \\ c_z &= \cos\varphi_z^k \cong 1 \end{aligned} \tag{2.26}$$

The negative of the quantity $\{(\grave{\mathbf{s}}_o^k$ acting on the rotated element) - ($\grave{\mathbf{s}}_o^k$ acting on element at reference state)$\}$ is the description, in the local reference frame, of the preexisting internal force effect due to the orientation change of the element that should be considered in the equilibrium equations of node j_2^k. Its description in the global reference frame is $\grave{\mathbf{q}}_2^k$. It follows from its definition above:

$$\grave{\mathbf{q}}_2^k = \begin{bmatrix} \mathbf{Q} & \cdot \\ \cdot & \mathbf{Q} \end{bmatrix} \begin{bmatrix} (\mathbf{R}_y - \mathbf{I}) + (\mathbf{R}_z - \mathbf{I}) & \cdot \\ \cdot & (\mathbf{R}_y - \mathbf{I}) + (\mathbf{R}_z - \mathbf{I}) \end{bmatrix} \grave{\mathbf{s}}_o^k \tag{2.27}$$

where $\mathbf{Q} = [\mathbf{e}^k, \mathbf{g}^k, \mathbf{h}^k]$, $\mathbf{I}$ is the third order identity matrix, and $\mathbf{R}_y$ and $\mathbf{R}_z$ are rotation matrices about the y and z axes, respectively, of the local reference frame.

$$\mathbf{R}_y = \begin{bmatrix} c_y & \cdot & s_y \\ \cdot & 1 & \cdot \\ -s_y & \cdot & c_y \end{bmatrix} \cong \begin{bmatrix} 1 & \cdot & \varphi_y^k \\ \cdot & 1 & \cdot \\ -\varphi_y^k & \cdot & 1 \end{bmatrix} \tag{2.28}$$

$$\mathbf{R}_z = \begin{bmatrix} c_z & -s_z & \cdot \\ s_z & c_z & \cdot \\ \cdot & \cdot & 1 \end{bmatrix} \cong \begin{bmatrix} 1 & -\varphi_z^k & \cdot \\ \varphi_z^k & 1 & \cdot \\ \cdot & \cdot & 1 \end{bmatrix} \tag{2.29}$$

Using these in Eq.(2.27), the latter may be rewritten as

$$\grave{\mathbf{q}}_2^k = \begin{bmatrix} \mathbf{Q}\mathbf{R}_y' & \cdot \\ \cdot & \mathbf{Q}\mathbf{R}_y' \end{bmatrix} \grave{\mathbf{s}}_o^k \varphi_y^k + \begin{bmatrix} \mathbf{Q}\mathbf{R}_z' & \cdot \\ \cdot & \mathbf{Q}\mathbf{R}_z' \end{bmatrix} \grave{\mathbf{s}}_o^k \varphi_z^k \tag{2.30}$$

where

$$\mathbf{R}_y' = \begin{bmatrix} \cdot & \cdot & 1 \\ \cdot & \cdot & \cdot \\ -1 & \cdot & \cdot \end{bmatrix} \text{ and } \mathbf{R}_z' = \begin{bmatrix} \cdot & 1 & \cdot \\ -1 & \cdot & \cdot \\ \cdot & \cdot & \cdot \end{bmatrix} \tag{2.31}$$

By using the definitions of φ_y^k and φ_z^k from Eq.(2.25), one may rewrite Eq.(2.30) as

$$\grave{\mathbf{q}}_2^k = \frac{\grave{N}_o^k}{L^k} \begin{bmatrix} \mathbf{h}^k \mathbf{h}^{kT} + \mathbf{g}^k \mathbf{g}^{kT} & \cdot \\ \cdot & \cdot \end{bmatrix} (\Delta \boldsymbol{\xi}_{j_2^k} - \Delta \boldsymbol{\xi}_{j_1^k}) \tag{2.32}$$

Observing that the preexisting forces at the first vertex are the negatives of those in the second vertex, i.e., $\grave{\mathbf{q}}_1^k = -\grave{\mathbf{q}}_2^k$, Eq.(2.32) may be generalized to cover both vertices as

$$\grave{\mathbf{q}}_l^k = \sum_{l'=1}^{2} \mathbf{G}_{ll'}^{\grave{N}_o^k} \Delta \boldsymbol{\xi}_{j_{l'}^k}, \text{ for } l = 1, 2 \tag{2.33}$$

TABLE 2.5. Element Stiffness Matrices in Global Reference Frame due to Preexisting Axial Force

Structure	$\mathbf{G}_{11}^{\grave{N}_o^k} = \mathbf{G}_{22}^{\grave{N}_o^k} = -\mathbf{G}_{12}^{\grave{N}_o^k} = -\mathbf{G}_{21}^{\grave{N}_o^k}$	e
Space frame	$\frac{\grave{N}_o^k}{L^k}\begin{bmatrix} [\mathbf{g}\mathbf{g}^T + \mathbf{h}\mathbf{h}^T] & \cdot\ \cdot\ \cdot \\ & \cdot\ \cdot\ \cdot \\ \cdot\ \cdot\ \cdot & \cdot\ \cdot\ \cdot \\ & \\ \cdot\ \cdot\ \cdot & \cdot\ \cdot\ \cdot \end{bmatrix}^k$	6
Planar frame	$\frac{\grave{N}_o^k}{L^k}\begin{bmatrix} g_X^2 & g_X g_Y & \cdot \\ & g_Y^2 & \cdot \\ sym. & & \cdot \end{bmatrix}^k$	3
Gridwork frame	$\frac{\grave{N}_o^k}{L^k}\begin{bmatrix} \cdot & \cdot & \cdot \\ \cdot & \cdot & \cdot \\ \cdot & \cdot & 1 \end{bmatrix}^k$	3
Space truss	$\frac{\grave{N}_o^k}{L^k}[\mathbf{g}^k\mathbf{g}^{k^T} + \mathbf{h}^k\mathbf{h}^{k^T}]$	3
Planar truss	$\frac{\grave{N}_o^k}{L^k}\mathbf{g}^k\mathbf{g}^{k^T}$ *where* $\mathbf{g}^{k^T} = [g_X, g_Y]$	2
X, Y, Z: *global axes*	$\mathbf{g}^k, \mathbf{h}^k$ *unit vectors of* y^k, z^k *axes*	
$\mathbf{g}^{k^T} = [g_X, g_Y, g_Z]^k$	$\mathbf{h}^{k^T} = [h_X, h_Y, h_Z]^k$	
L^k: *length of k'th element*	$\grave{N}_o^k$: *preexisting axial force*	

where

$$\mathbf{G}_{ll'}^{\grave{N}_o^k} = (-1)^{l+l'}\frac{\grave{N}_o^k}{L^k}\begin{bmatrix} \mathbf{h}^k\mathbf{h}^{k^T} + \mathbf{g}^k\mathbf{g}^{k^T} & \cdot \\ \cdot & \cdot \end{bmatrix} \tag{2.34}$$

It may be observed that $\mathbf{G}_{ll'}^{\grave{N}_o^k}$ matrices are symmetric. They are specialized in Table 2.5 for various truss and frame elements studied in this book.

Using the identity of $\Delta\boldsymbol{\xi}_{j_{l'}^k} = \mathbf{I}_{j_{l'}^k}^T\Delta\boldsymbol{\xi}$, where matrix $\mathbf{I}_{j_{l'}^k}^T$ is the transpose of the j_l^kth column partition of the Neth order identity matrix partitioned e columns and rows at a time, in Eq.(2.33) and substituting the latter into Eq.(2.16), one obtains

$$\grave{\mathbf{q}} = \grave{\mathbf{K}}^o\Delta\boldsymbol{\xi} \tag{2.35}$$

where

$$\grave{\mathbf{K}}^o = \sum_{k=1}^{M}\sum_{l=1}^{2}\sum_{l'=1}^{2}\mathbf{I}_{j_l^k}\mathbf{G}_{ll'}^{\grave{N}_o^k}\mathbf{I}_{j_{l'}^k}^T \tag{2.36}$$

By using $\grave{\mathbf{q}}$ from Eq.(2.35) in Eq.(2.18), one may obtain

$$\grave{\mathbf{K}}^o\Delta\boldsymbol{\xi} + \mathbf{B}^o\Delta\mathbf{s} = \Delta\mathbf{p} \tag{2.37}$$

as the nodal force equilibrium equations for a discrete parameter structure without deflection and element force constraints, i.e., $b = f = 0$.

By deleting the columns of $\mathbf{\dot{K}}^o$ that correspond to prescribed nodal deflections, and by deleting the columns of $\mathbf{B}^o$ that correspond to zero element force components, and removing the nodal force equilibrium equations in the directions of the prescribed deflection directions, one may end up with the nodal force equilibrium equations of the structure with b deflection constraints and f element force constraints as in Eq.(2.24):

$$\mathbf{\dot{K}}\Delta\boldsymbol{\xi} + \mathbf{B}\Delta\mathbf{s} = \Delta\mathbf{p} \tag{2.38}$$

These are $n = Ne - b$ linearized equations of nodal force equilibrium in terms of response quantities $\Delta\boldsymbol{\xi}$ and $\Delta\mathbf{s}$. Note that matrix $\mathbf{\dot{K}}$ is symmetric per Eq.(2.36), and $\mathbf{\dot{K}} \in \mathrm{R}^{n\times n}$, $\mathbf{B} \in \mathrm{R}^{n\times m}$, where $n = Ne - b$ and $m = Ma - f$.

2.3 Geometric Relations

Since the element vertices are attached to the nodes, it is always possible to obtain unique expressions for the element deformations $\Delta\mathbf{v}$ (see the paragraph above Eq.(2.5)) in terms of the nodal deflections $\Delta\boldsymbol{\xi}$, by the rules of Euclidean geometry. These relations are intrinsically nonlinear functions of nodal deflection components. However, when the nodal deflections are very small relative to the typical size of the structure, the linearized relations become acceptable approximations of the nonlinear relations. The most general linear form of the relations is of the type of

$$\Delta\mathbf{v} = \mathbf{B}'\Delta\boldsymbol{\xi} + \Delta\mathbf{v}_o \tag{2.39}$$

where matrices $\mathbf{B}'$ and $\Delta\mathbf{v}_o$ are independent of any of the response quantities. In the following subsections, expressions for these matrices are developed.

2.3.1 Case When $\Delta\mathbf{v}_o = \mathbf{o}$

In Eq.(2.39) when $\Delta\boldsymbol{\xi} = \mathbf{o}$, i.e., when the structure is in a kinematically determinate state, $\Delta\mathbf{v} = \Delta\mathbf{v}_o$ which means that $\Delta\mathbf{v}_o$ represents the element deformations in the kinematically determinate state. As discussed in Subsection 2.1.2, the element deformations in the kinematically determinate state are caused by deflection-type loads. In this subsection it is assumed that there are no deflection-type loads, i.e., $\Delta\mathbf{v}_o = \mathbf{o}$. With this, Eq.(2.39) becomes

$$\Delta\mathbf{v} = \mathbf{B}'\Delta\boldsymbol{\xi} \tag{2.40}$$

where matrix $\mathbf{B}' \in \mathrm{R}^{m \times n}$ is yet to be determined. Note that $m = Ma - f$ and $n = Ne - b$. There are three ways of determining matrix $\mathbf{B}'$. These are discussed in the following paragraphs.

In the first method, by the rules of the Euclidean geometry and using the definitions of element deformation components and the sketches of the structure's geometry in the reference state and the final state, one may first obtain expressions for the element deformation components as non-linear functions of components of $\Delta\boldsymbol{\xi}$, and then linearize these expressions to obtain Eq.(2.40). This method, although very straightforward, is the hardest.

In the second method, one first observes that the kth column of $\mathbf{B}'$, i.e., $\mathbf{b}'_k$, is nothing but the element deformations corresponding to the deflection state of $\Delta\boldsymbol{\xi} = \mathbf{i}_k$, where $\mathbf{i}_k$ is the kth column of the nth order identity matrix, since from Eq.(2.40), $\Delta\mathbf{v} = \mathbf{B}'\mathbf{i}_k = \mathbf{b}'_k$. Therefore, one can obtain expressions for the columns of matrix $\mathbf{B}'$, one by one, by imposing on the structure at the kinematically determinate state the unit values of the nodal deflections as support settlements, and then computing the resulting element deformations by the small angle approximations of trigonometric functions.

In the third method, one uses the virtual work principle. When the structure is slowly released from the kinematically determinate state, it will move to its final equilibrium state where the nodal deflections $\Delta\boldsymbol{\xi}$ and the element deformations $\Delta\mathbf{v}$ are geometrically compatible. If one applies a force system with internal components $\Delta\mathbf{s}$ and external components $\Delta\mathbf{p}$ that satisfy equilibrium, then, according to the virtual work principle, the internal work equals the external work, namely

$$\Delta\mathbf{s}^T\Delta\mathbf{v} = \Delta\mathbf{p}^T\Delta\boldsymbol{\xi} \tag{2.41}$$

By using $\Delta\mathbf{v}$ from Eq.(2.40) and $\Delta\mathbf{p}$ from Eq.(2.21) with $\dot{\mathbf{q}} = \mathbf{o}$, one may rewrite it as

$$\Delta\mathbf{s}^T(\mathbf{B}' - \mathbf{B}^T)\Delta\boldsymbol{\xi} = 0 \tag{2.42}$$

which holds for arbitrary $\Delta\mathbf{s}$ and $\Delta\boldsymbol{\xi}$. Therefore

$$\mathbf{B}' = \mathbf{B}^T \tag{2.43}$$

The reader may verify this result by using either of the first two methods on a simple structure. This method is remarkable in the sense that it provides the definition of $\mathbf{B}'$ matrix in terms of a matrix already defined, since matrix $\mathbf{B}^T$ is the transpose of the coefficient matrix of element forces $\Delta\mathbf{s}$ in the nodal force equilibrium equations given by Eq.(2.21).

2.3.2 Case When $\Delta\boldsymbol{\xi} = \mathbf{o}$

As stated earlier, Eq.(2.39) becomes $\Delta\mathbf{v} = \Delta\mathbf{v}_o$ when $\Delta\boldsymbol{\xi} = \mathbf{o}$, i.e., when the structure is in a kinematically determinate state. In other words, $\Delta\mathbf{v}_o$ is

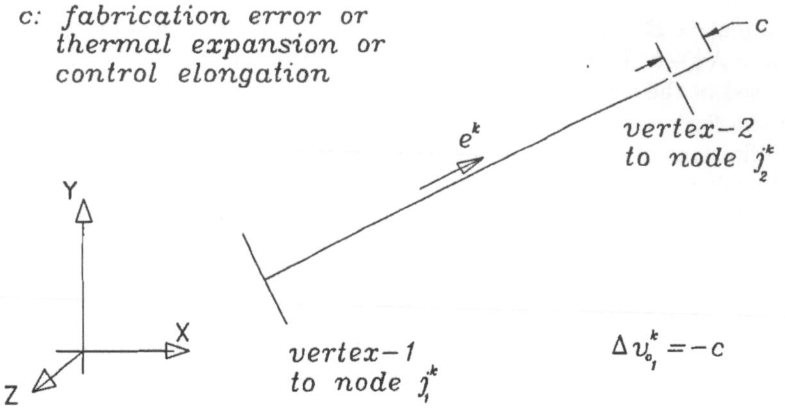

FIGURE 2.5. Structural element on statically determinate supports and evaluation of prescribed element deformations

the element deformations vector when the structure is in the kinematically determinate state. In Subsection 2.1.2, the two steps of an incremental static loading are explained. In the first step, nodes of the structure originally in reference state are put in the kinematically determinate state by locking them against any movement. Then the elements, with their loads, are attached to the nodes. For elements without loads or with force-type element loads, the attachment process requires no special attention, although the force-type element loads require the computation of the internal elemental forces and their fixed-end reactions as in the moment distribution method (see Fig. 2.1 and related discussions). The handling of elements subjected to deflection-type loads requires special attention and also provides the definition of $\Delta \mathbf{v}_o$. This is explained next.

Suppose a structural element, say, the kth, is subjected to deflection-type element loads. These loads may be due to shrinkage, fabrication error, thermal effects, and deformation inserting actuators of the element. The element is placed in a statically determinate support state, by fixing it at the first vertex as shown in Fig.2.5. Then its deflection-type loads are applied. The element in this state, in general, will not fit in the slot between the fixed nodes corresponding to its vertices, i.e., nodes j_1^k and j_2^k. By applying force and moment if necessary, the cross-section at the second vertex of the element is moved enough to fit its slot. The description, in the local reference frame, of the movements of the cross-section at the second vertex is called the prescribed element deformations of element k, and denoted by $\Delta \mathbf{v}_o^k$. This process is repeated for all elements to obtain $\Delta \mathbf{v}_o^k$, for $k = 1, \cdots, M$. If the structural element has an element force constraint, then the corresponding prescribed element deformation component is not considered. The complete list of *prescribed element deformations* defines

$\Delta\mathbf{v}_o$:

$$\Delta\mathbf{v}_o^T = [\Delta\mathbf{v}_o^{1^T}, \Delta\mathbf{v}_o^{2^T}, \cdots, \Delta\mathbf{v}_o^{M^T}] \quad (2.44)$$

Note that if an element is not subjected to deflection-type element loads, then its prescribed element deformation is a zero subvector in Eq.(2.44), hence $\Delta\mathbf{v}_o \in \mathrm{R}^{m\times 1}$ where $m = Ma - f$.

2.3.3 Final Form

By using the definitions given by Eqs.(2.43 and 2.44) in Eq.(2.39), the linearized relations between element deformations and nodal deflections may be obtained as

$$\Delta\mathbf{v} = \mathbf{B}^T\Delta\boldsymbol{\xi} + \Delta\mathbf{v}_o \quad (2.45)$$

where $\Delta\mathbf{v}, \Delta\mathbf{v}_o \in \mathrm{R}^{m\times 1}$, $\Delta\boldsymbol{\xi} \in \mathrm{R}^{n\times 1}$, and $\mathbf{B}^T \in \mathrm{R}^{m\times n}$, with $m = Ma - f$ and $n = Ne - b$. Eq.(2.45) consists of m number of scalar equations.

2.4 Stiffness Relations of Elements

In Subsection 2.1.4, during the discussion of the excitation-response relations given by Eq.(2.8), it is shown that the complete response $\Delta\psi$ requires $2m + n$ number of linearly independent scalar equations (recall that $m = Ma - f$ and $n = Ne - b$). The nodal force equilibrium equations of Eq.(2.21) provide n of the equations, and the relations between element deformations and nodal deflections given by Eq.(2.45) provide m of the equations. The remaining m equations are the stiffness relations between element deformations $\Delta\mathbf{v}$ and element forces $\Delta\mathbf{s}$, as a result of the constitutive relations of the structure's material. They are in the form of

$$\Delta\mathbf{s} = \vec{\mathbf{K}}\Delta\mathbf{v} \quad (2.46)$$

where $\vec{\mathbf{K}}$ is called the *matrix of element stiffnesses.* Since $\Delta\mathbf{s}, \Delta\mathbf{v} \in \mathrm{R}^{m\times 1}$, then $\vec{\mathbf{K}} \in \mathrm{R}^{m\times m}$ where $m = Ma - f$. Since the element forces of an element are related only to the element deformations of that element, Eq.(2.46) can be also stated as

$$\Delta\mathbf{s}^k = \mathbf{K}^k\Delta\mathbf{v}^k \text{ for } k = 1, \cdots, M \quad (2.47)$$

$\mathbf{K}^k$ is the stiffness matrix of the kth element. It follows from the comparison of Eq.(2.46) with Eq.(2.47) that

$$\vec{\mathbf{K}} = \begin{bmatrix} \mathbf{K}^1 & & & \\ & \mathbf{K}^2 & & \\ & & \ddots & \\ & & & \mathbf{K}^M \end{bmatrix} = diag(\mathbf{K}^1, \cdots, \mathbf{K}^M) = diag(\mathbf{K}^k) \quad (2.48)$$

Element stiffness matrices $\mathbf{K}^k$, $k = 1, \cdots, M$, are of $R^{a \times a}$ per Eq.(2.47). They should not be confused with the free-free stiffness matrices of structural elements, which are of order $2a$. Expressions are provided for the element stiffness matrices of truss and frame type structural elements with straight axes and overlapping centroids and shear centers.

2.4.1 Element Stiffness Matrices

From Eq.(2.47), one can write for the stiffness relations for the kth element

$$\Delta \mathbf{s}^k = \mathbf{K}^k \Delta \mathbf{v}^k \qquad (2.49)$$

where, assuming that there are no element force constraints, $\mathbf{K}^k \in R^{a \times a}$ and $\Delta \mathbf{s}^k, \Delta \mathbf{v}^k \in R^{a \times 1}$. In this equation if one uses $\Delta \mathbf{v}^k = \mathbf{i}_l$, where $\mathbf{i}_l$ is the lth column of the ath order identity matrix, one obtains $\Delta \mathbf{s}^k = \mathbf{K}^k \mathbf{i}_l = \mathbf{k}_l^k$ which is the lth column of $\mathbf{K}^k$.Then the lth column of $\mathbf{K}^k$ is the element forces in the kth element when it is subjected to the element deformations $\mathbf{i}_l$. One can compute the element forces by first fixing the vertices of the element as shown in Fig. 2.6 and then applying a unit support settlement at the second vertex in the direction of the lth deformation component. When

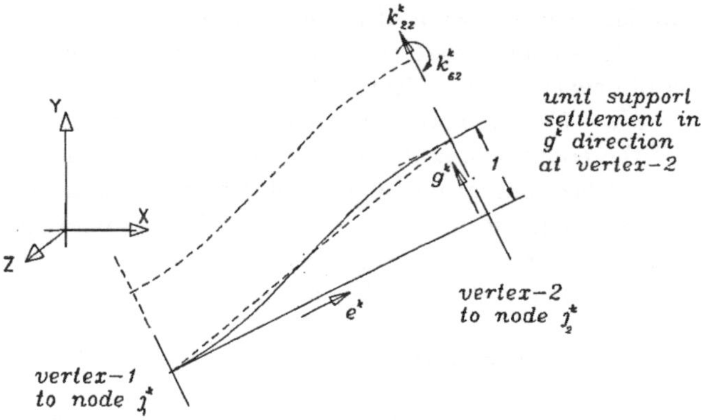

FIGURE 2.6. Element in a kinematically determinate state and subjected to unit support settlement in y-direction for computing (2,2) and (6,2) stiffness constants

both vertices are fixed, the element is in the kinematically determinate state and establishes a statically indeterminate structure. The methods of elementary structural analysis enable one to compute the forces developed at the second vertex due to the support settlement, and thus obtain the lth column of $\mathbf{K}^k$. By taking $l = 1, \cdots, a$, and repeating the computations, one can compute all of $\mathbf{K}^k$.

Because of Maxwell's reciprocity principle,

$$k^k_{i,j} = k^k_{j,i} \text{ for } i = 1, \cdots, a \text{ and } j = i, \cdots, a \qquad (2.50)$$

Namely, matrix $\mathbf{K}^k$ is symmetrical. The symmetry property may be used to reduce the number of computations mentioned in the preceding paragraph.

Due to the linear behavior of the material, the quantity ΔW, obtained by premultiplying both sides of Eq.(2.49) with $\Delta\mathbf{v}^{k^T}/2$, i.e.,

$$\Delta W = \frac{1}{2}\Delta\mathbf{v}^{k^T}\Delta\mathbf{s}^k = \frac{1}{2}\Delta\mathbf{v}^{k^T}\mathbf{K}^k\Delta\mathbf{v}^k \tag{2.51}$$

is the work done on the boundaries of the element during the slow support settlements $\Delta\mathbf{v}^k$. Alternately, it is the positive internal elastic energy stored in the element due to the support settlements. Since ΔW is always positive and zero only if $\Delta\mathbf{v}^k = \mathbf{o}$, then from the right side of Eq.(2.51), one observes that $\mathbf{K}^k$ is a positive definite matrix. As any other real, symmetric, and positive definite matrix, $\mathbf{K}^k$ possesses a unique inverse $\mathbf{F}^k$:

$$\mathbf{K}^{k^{-1}} = \mathbf{F}^k \tag{2.52}$$

which is also real, symmetric, and positive definite. The inverse of $\mathbf{K}^k$, i.e., $\mathbf{F}^k$, is called the *flexibility matrix* of the kth element. By inverting both sides of the last equation, the useful relation

$$\mathbf{K}^k = \mathbf{F}^{k^{-1}} \tag{2.53}$$

is obtained. As discussed in the next subsection, it is more convenient to compute $\mathbf{F}^k$ first and then use Eq.(2.53) to obtain $\mathbf{K}^k$.

2.4.2 Element Flexibility Matrices

The relation that one may obtain by inversion from Eq.(2.49), i.e.,

$$\Delta\mathbf{v}^k = \mathbf{F}^k\Delta\mathbf{s}^k \tag{2.54}$$

is called the flexibility relation of the kth element. As discussed above, the element flexibility matrix $\mathbf{F}^k \in R^{a\times a}$ is real, symmetric, and positive definite. The explicit expressions for its entries may be obtained column-by-column. If one chooses $\Delta\mathbf{s}^k = \mathbf{i}_l$, where $\mathbf{i}_l$ is the lth column of the ath order identity matrix, and observes that $\mathbf{F}^k\mathbf{i}_l = \mathbf{f}_l^k$, one obtains from Eq.(2.54) that $\Delta\mathbf{v}^k = \mathbf{f}_l^k$. In other words, the lth column of $\mathbf{F}^k$ lists the element deformations when the element is clamped at the first vertex and subjected to a unit load in the lth deformation direction. In Fig.(2.7) the deflections at the second vertex due to unit loads at the same vertex of a uniform space frame element clamped at the first vertex are shown. Considering that the loads and the deflections are measured in the directions defined by the local reference frame at the second vertex of the frame element clamped at the first vertex, the (i,j) entry of $\mathbf{F}^k$, that is, f_{ij}^k, is the deflection in the ith direction due to a unit load acting in the jth direction. Denoting the

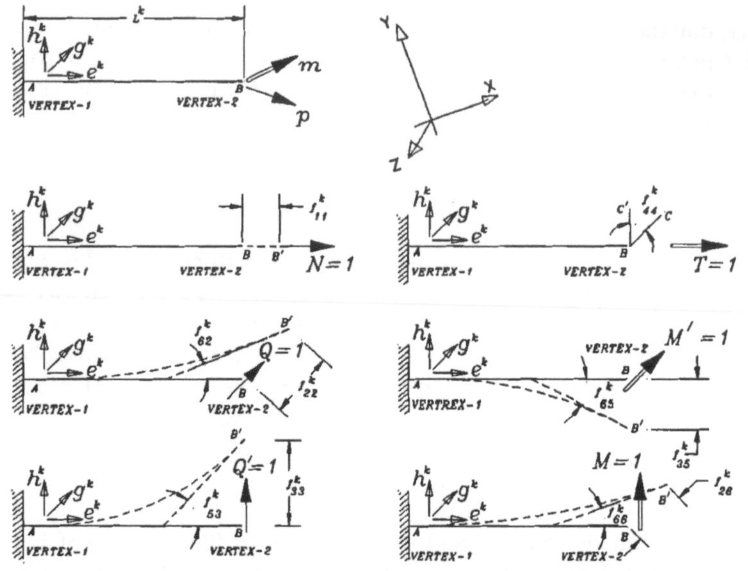

FIGURE 2.7. Flexibility constants as tip deflections of a cantilever due to unit loads at the tip

label of the element by k, Young's modulus by E, shear modulus by G, the cross-sectional area by A, the second area moments about local y and z axes by I_{yy} and I_{zz}, and the torsional constant by J, the f_{ij}^k constants may be computed by methods of the elementary structural analysis. Let $N_j(x)$, $Q_j(x)$, $Q_j'(x)$, $T_j(x)$, M_j', and $M_j(x)$, for $0 \leq x \leq L^k$, denote the internal force diagrams for normal force, y-shear force, z-shear force, torsion, y-moment, and z-moment, respectively, due to a unit load in the jth direction. Then, by the virtual work principle, one may write:

$$f_{ij}^k = \int_0^{L^k} \left(\frac{N_i N_j}{EA} + \frac{Q_i Q_j}{GA_y} + \frac{Q_i' Q_j'}{GA_z} + \frac{T_i T_j}{GJ} + \frac{M_i' M_j'}{EI_{yy}} + \frac{M_i M_j}{EI_{zz}}\right) dx \tag{2.55}$$

With the help of Fig.(2.7) and Eq.(2.55), and ignoring the shear force effect, the flexibility matrix of the uniform space frame element may be obtained as

$$\mathbf{F}^k = \begin{bmatrix} \frac{L}{EA} & \cdot & \cdot & \cdot & \cdot & \cdot \\ \cdot & \frac{L^3}{3EI_{zz}} & \cdot & \cdot & \cdot & \frac{L^2}{2EI_{zz}} \\ \cdot & \cdot & \frac{L^3}{3EI_{yy}} & \cdot & \frac{-L^2}{2EI_{yy}} & \cdot \\ \cdot & \cdot & \cdot & \frac{L}{GJ} & \cdot & \cdot \\ \cdot & \cdot & \frac{-L^2}{2EI_{yy}} & \cdot & \frac{L}{EI_{yy}} & \cdot \\ \cdot & \frac{L^2}{2EI_{zz}} & \cdot & \cdot & \cdot & \frac{L}{EI_{zz}} \end{bmatrix}^k \tag{2.56}$$

TABLE 2.6. Element Flexibility Matrices

Structure	**Element Flexibility Matrix**	a
Space frame	$\begin{bmatrix} \frac{L}{EA} & \cdot & \cdot & \cdot & \cdot & \cdot \\ \cdot & \frac{L^3}{3EI_{zz}} & \cdot & \cdot & \cdot & \frac{L^2}{2EI_{zz}} \\ \cdot & \cdot & \frac{L^3}{3EI_{yy}} & \cdot & \frac{-L^2}{2EI_{yy}} & \cdot \\ \cdot & \cdot & \cdot & \frac{L}{GJ} & \cdot & \cdot \\ \cdot & \cdot & \frac{-L^2}{2EI_{yy}} & \cdot & \frac{L}{EI_{yy}} & \cdot \\ \cdot & \frac{L^2}{2EI_{zz}} & \cdot & \cdot & \cdot & \frac{L}{EI_{zz}} \end{bmatrix}^k$	6
Planar frame	$\begin{bmatrix} \frac{L}{EA} & \cdot & \cdot \\ \cdot & \frac{L^3}{3EI_{zz}} & \frac{L^2}{2EI_{zz}} \\ \cdot & \frac{L^2}{2EI_{zz}} & \frac{L}{EI_{zz}} \end{bmatrix}^k$	3
Gridwork frm.	$\begin{bmatrix} \frac{L}{GJ} & \cdot & \cdot \\ \cdot & \frac{L}{EI_{yy}} & \frac{-L^2}{2EI_{yy}} \\ \cdot & \frac{-L^2}{2EI_{yy}} & \frac{L^3}{3EI_{yy}} \end{bmatrix}^k$	3
Spc/Pl truss	$\left[\frac{L}{EA}\right]^k$	1

where all matrix entries should be evaluated for the kth element. The flexibility matrix of other types of structural elements considered in this book can be obtained from this by retaining only the rows and columns associated with the element forces of the element. For example, for planar frame elements, columns/rows 1,2,6; for gridwork frame elements, columns/rows 4,5,3; for trusses, column/row 1 should be retained in the prescribed order. The results are summarized in Table 2.6.

2.4.3 Obtaining $\mathbf{K}^k$ *from* $\mathbf{F}^k$

If the flexibility matrix $\mathbf{F}^k$ is available, one may obtain the corresponding stiffness matrix $\mathbf{K}^k$ by inversion as suggested by Eq.(2.53). In the previous subsection, the flexibility matrices of straight axis uniform frame and truss elements are obtained by ignoring the shear force effects. In this subsection, the stiffness matrices of these elements are obtained by first inverting the space frame element flexibility matrix given by Eq.(2.56) and then selecting appropriate columns/rows.

By the use of permutation matrices the rows and columns of $\mathbf{F}^k$ given by Eq.(2.56) can be reordered to put it into block diagonal form for easy inversion. A permutation matrix is obtained by reordering the columns of the identity matrix. From the sixth order identity matrix $\mathbf{I} = [\mathbf{i}_1, \mathbf{i}_2, \mathbf{i}_3, \mathbf{i}_4, \mathbf{i}_5, \mathbf{i}_6]$, one may obtain the permutation matrix

$$\mathbf{I}' = [\mathbf{i}_1, \mathbf{i}_4, \mathbf{i}_2, \mathbf{i}_6, \mathbf{i}_3, \mathbf{i}_5]$$

Since postmultiplication with $\mathbf{I}'$ reorders columns, and premultiplication with $\mathbf{I}'^T$ reorders rows of $\mathbf{F}^k$, and since the permutation matrices are orthogonal, that is, $\mathbf{I}'^T\mathbf{I}' = \mathbf{I}$, one may rewrite Eq.(2.53) as

$$\left[\mathbf{I}'\mathbf{I}'^T\mathbf{F}^k\mathbf{I}'\mathbf{I}'^T\right]^{-1} = \mathbf{K}^k$$

This may be restated as

$$\mathbf{I}'\left[\mathbf{I}'^T\mathbf{F}^k\mathbf{I}'\right]^{-1}\mathbf{I}'^T = \mathbf{K}^k \tag{2.57}$$

It may be verified that

$$\left[\mathbf{I}'^T\mathbf{F}^k\mathbf{I}'\right] = \begin{bmatrix} \frac{L}{EA} & \cdot & \cdot & \cdot & \cdot & \cdot \\ \cdot & \frac{L}{GJ} & \cdot & \cdot & \cdot & \cdot \\ \cdot & \cdot & \frac{L^3}{3EI_{zz}} & \frac{L^2}{2EI_{zz}} & \cdot & \cdot \\ \cdot & \cdot & \frac{L^2}{2EI_{zz}} & \frac{L}{EI_{zz}} & \cdot & \cdot \\ \cdot & \cdot & \cdot & \cdot & \frac{L^3}{3EI_{yy}} & \frac{-L^2}{2EI_{yy}} \\ \cdot & \cdot & \cdot & \cdot & \frac{-L^2}{2EI_{yy}} & \frac{L}{EI_{yy}} \end{bmatrix}$$

Since this matrix is block diagonal, its inverse may be obtained as

$$\left[\mathbf{I}'^T\mathbf{F}^k\mathbf{I}'\right]^{-1} = \begin{bmatrix} \frac{AE}{L} & \cdot & \cdot & \cdot & \cdot & \cdot \\ \cdot & \frac{GJ}{L} & \cdot & \cdot & \cdot & \cdot \\ \cdot & \cdot & \frac{12EI_{zz}}{L^3} & -\frac{6EI_{zz}}{L^2} & \cdot & \cdot \\ \cdot & \cdot & -\frac{6EI_{zz}}{L^2} & \frac{4EI_{zz}}{L} & \cdot & \cdot \\ \cdot & \cdot & \cdot & \cdot & \frac{12EI_{yy}}{L^3} & \frac{6EI_{yy}}{L^2} \\ \cdot & \cdot & \cdot & \cdot & \frac{6EI_{yy}}{L^2} & \frac{4EI_{yy}}{L} \end{bmatrix}$$

Using this in Eq.(2.57), one may obtain $\mathbf{K}^k$ as

$$\mathbf{K}^k = \begin{bmatrix} \frac{AE}{L} & \cdot & \cdot & \cdot & \cdot & \cdot \\ \cdot & \frac{12EI_{zz}}{L^3} & \cdot & \cdot & \cdot & -\frac{6EI_{zz}}{L^2} \\ \cdot & \cdot & \frac{12EI_{yy}}{L^3} & \cdot & \frac{6EI_{yy}}{L^2} & \cdot \\ \cdot & \cdot & \cdot & \frac{GJ}{L} & \cdot & \cdot \\ \cdot & \cdot & \frac{6EI_{yy}}{L^2} & \cdot & \frac{4EI_{yy}}{L} & \cdot \\ \cdot & -\frac{6EI_{zz}}{L^2} & \cdot & \cdot & \cdot & \frac{4EI_{zz}}{L} \end{bmatrix}^k \tag{2.58}$$

The stiffness matrices of other types of structural elements considered in this book can be obtained from this by retaining only the rows and columns associated with the element force components listed in $\Delta\mathbf{s}^k$. For example, for planar frame elements, columns/rows 1,2,6; for gridwork frame elements, columns/rows 4,5,3; for trusses, column/row 1 should be retained, in the prescribed order, per Table 1.2. The results are summarized in Table 2.7.

TABLE 2.7. Element Stiffness Matrices

Structure	**Element Stiffness Matrix**	a
Space frame	$\begin{bmatrix} \frac{AE}{L} & \cdot & \cdot & \cdot & \cdot & \cdot \\ \cdot & \frac{12EI_{zz}}{L^3} & \cdot & \cdot & \cdot & -\frac{6EI_{zz}}{L^2} \\ \cdot & \cdot & \frac{12EI_{yy}}{L^3} & \cdot & \frac{6EI_{yy}}{L^2} & \cdot \\ \cdot & \cdot & \cdot & \frac{GJ}{L} & \cdot & \cdot \\ \cdot & \cdot & \frac{6EI_{yy}}{L^2} & \cdot & \frac{4EI_{yy}}{L} & \cdot \\ \cdot & -\frac{6EI_{zz}}{L^2} & \cdot & \cdot & \cdot & \frac{4EI_{zz}}{L} \end{bmatrix}^k$	6
Planar frame	$\begin{bmatrix} \frac{AE}{L} & \cdot & \cdot \\ \cdot & \frac{12EI_{zz}}{L^3} & -\frac{6EI_{zz}}{L^2} \\ \cdot & -\frac{6EI_{zz}}{L^2} & \frac{4EI_{zz}}{L} \end{bmatrix}^k$	3
Gridwork frame	$\begin{bmatrix} \frac{GJ}{L} & \cdot & \cdot \\ \cdot & \frac{4EI_{yy}}{L} & \frac{6EI_{yy}}{L^2} \\ \cdot & \frac{6EI_{yy}}{L^2} & \frac{12EI_{yy}}{L^3} \end{bmatrix}^k$	3
Space truss	$[\frac{AE}{L}]^k$	1
Planar truss	$[\frac{AE}{L}]^k$	1

2.5 Incremental Excitation-Response Relations

By combining the ($n = Ne - b$) number of scalar nodal force equilibrium equations given by $\grave{\mathbf{K}}\Delta\boldsymbol{\xi} + \mathbf{B}\Delta\mathbf{s} = \Delta\mathbf{p}$ in Eq.(2.24), the ($m = Ma - f$) number of scalar relations between element deformations and nodal deflections given by $\Delta\mathbf{v} = \mathbf{B}^T\Delta\boldsymbol{\xi} + \Delta\mathbf{v}_o$ in Eq.(2.45), and the m number of scalar stiffness relations given by $\Delta\mathbf{s} = \vec{\mathbf{K}}\Delta\mathbf{v}$ in Eq.(2.46), the excitation response relations may be rewritten compactly as

$$\begin{matrix} \text{Equilibrium:} \\ \text{Geometry:} \\ \text{Stiffness:} \end{matrix} \quad \begin{bmatrix} \grave{\mathbf{K}} & \mathbf{B} & \cdot \\ \mathbf{B}^T & \cdot & -\mathbf{I} \\ \cdot & -\mathbf{I} & \vec{\mathbf{K}} \end{bmatrix} \begin{Bmatrix} \Delta\boldsymbol{\xi} \\ \Delta\mathbf{s} \\ \Delta\mathbf{v} \end{Bmatrix} = \begin{Bmatrix} \Delta\mathbf{p} \\ -\Delta\mathbf{v}_o \\ \cdot \end{Bmatrix} \tag{2.59}$$

where the physical meanings of the equations are marked on the left, the dots represent the zero blocks, and $\mathbf{I}$ is the identity matrix of order m. These are $2m+n$ number of scalar relations in the same number of scalar response quantities listed in $\Delta\boldsymbol{\psi}^T = [\Delta\boldsymbol{\xi}^T, \Delta\mathbf{s}^T, \Delta\mathbf{v}^T]$. The excitation quantities are listed on the right-hand side where $\Delta\mathbf{p}$ represent force-type nodal loads, and $\Delta\mathbf{v}_o$ represent deflection type element loads.

The coefficient matrix of the response quantities is a real symmetric matrix of order $2m+n$. Since an engineering structure is not a mechanism, matrix $\mathbf{B}$ is not rank deficient, implying that rank($\mathbf{B}$) $= n$, for $m \geq n$. As discussed earlier, $\vec{\mathbf{K}}$ is positive definite, then rank($\vec{\mathbf{K}}$) $= m$. If one further assumes that $\det(\grave{\mathbf{K}} + \mathbf{B}\vec{\mathbf{K}}\mathbf{B}^T) \neq 0$ (see Subsection 3.1.1), then as a consequence of these, the coefficient matrix is not singular; therefore, for a given loading there is a unique response.

The coefficient matrix is a very sparse matrix, since $\mathbf{B}$ itself is sparse, $\vec{\mathbf{K}}$ is block diagonal symmetric, and $\grave{\mathbf{K}}$ is banded and symmetric. As discussed in this chapter, for the definitions of $\mathbf{B}$, $\grave{\mathbf{K}}$, $\vec{\mathbf{K}}$, one needs

$$\begin{aligned}
&(\mathbf{T}_l^k \in \mathrm{R}^{e\times a}, l = 1,2), k = 1,\cdots, M \text{ and } \mathbf{J} \in (\mathrm{Int})^{2\times M} \\
&(\mathbf{G}_{ll'}^{\bar{N}_o^k} \in \mathrm{R}^{e\times e}, l = 1,2; l' = 1,2), k = 1,\cdots, M \text{ and } \mathbf{J} \in (\mathrm{Int})^{2\times M} \\
&(\mathbf{K}^k \in \mathrm{R}^{a\times a}), k = 1,\cdots, M
\end{aligned} \tag{2.60}$$

respectively, plus the data defining element force and nodal deflection constraints. For large N and M, the storage for the sparse coefficient matrix of response quantities is about M times larger than the storage for the items listed in display (2.60).

Note that the incremental form of the excitation-response relations

$$\begin{bmatrix} \grave{\mathbf{K}} & \mathbf{B} & \cdot \\ \mathbf{B}^T & \cdot & -\mathbf{I} \\ \cdot & -\mathbf{I} & \vec{\mathbf{K}} \end{bmatrix} \begin{Bmatrix} \Delta\boldsymbol{\xi} \\ \Delta\mathbf{s} \\ \Delta\mathbf{v} \end{Bmatrix} = \begin{Bmatrix} \Delta\mathbf{p} \\ -\Delta\mathbf{v}_o \\ \cdot \end{Bmatrix} \tag{2.61}$$

can be used in nonlinear and time-dependent problems where updating the coefficient matrix at the end of each incremental state may be required.

3

Active Control of Response, Static Case

In this chapter we are mainly concerned with actively controlling *some* of the response components, without creating accelerations in the structure during the process. Because of earlier loadings, these response components may be unacceptably away from their nominal values. The sensors can tell us which components of the response exceeded how much of their nominal values. Taking advantage of the linear behavior of the structure in the current incremental step, we look for a convenient loading which will produce the negatives of the sensor measured responses so that at the end of the loading step these response components will revert back to their nominal values.

In Chapter 2, the linearized incremental excitation-response relations of discrete parameter structures are developed in the form of

$$\grave{\mathbf{A}}\Delta\boldsymbol{\psi} = \Delta\mathbf{f} \tag{3.1}$$

where $\grave{\mathbf{A}} \in \mathrm{R}^{(2m+n)\times(2m+n)}$ is a constant nonsingular matrix representing the structure

$$\grave{\mathbf{A}} = \begin{bmatrix} \cdot & \mathbf{B} & \cdot \\ \mathbf{B}^T & \cdot & -\mathbf{I} \\ \cdot & -\mathbf{I} & \vec{\mathbf{K}} \end{bmatrix} \tag{3.2}$$

$\Delta\mathbf{f} \in \mathrm{R}^{(2m+n)\times 1}$ is the *slowly* applied excitations which are

$$\Delta\mathbf{f} = [\Delta\mathbf{p}^T - \grave{\mathbf{q}}^T, -\Delta\mathbf{v}_o^T, \cdot]^T \tag{3.3}$$

and $\Delta\boldsymbol{\psi} \in \mathrm{R}^{(2m+n)\times 1}$ is the resulting response that can be displayed

$$\Delta\boldsymbol{\psi} = [\Delta\boldsymbol{\xi}^T, \Delta\mathbf{s}^T, \Delta\mathbf{v}^T]^T \tag{3.4}$$

In matrix $\grave{\mathbf{A}}$, submatrices $\mathbf{B}$ and $\vec{\mathbf{K}}$ are as studied in Chapter 2, and $\mathbf{I}$ is the identity matrix of order m. For discrete parameter structures studied in this book, as defined earlier, $m = Ma - f$, and $n = Ne - b$, where N is the number of nodes, M is the number of elements, a is the number of element forces per element, e is the deflection degrees of freedom at a node, f is the number of zeroed element forces, and b is the number of deflection constraints. In excitations, $\Delta\mathbf{p} \in \mathrm{R}^{n\times 1}$ is incremental nodal loads; $\Delta\mathbf{v}_o \in \mathrm{R}^{m\times 1}$ is incremental prescribed element deformations. In response, $\Delta\boldsymbol{\xi} \in \mathrm{R}^{n\times 1}$is incremental nodal deflections, $\Delta\mathbf{s} \in \mathrm{R}^{m\times 1}$is incremental element forces, and $\Delta\mathbf{v} \in \mathrm{R}^{m\times 1}$is incremental element deformations. As studied in Subsection 2.2.3, $\grave{\mathbf{q}} \in \mathrm{R}^{n\times 1}$ is the effect of prescribed element forces on nodal equilibrium, and defined as $\grave{\mathbf{q}} = \grave{\mathbf{K}}\Delta\boldsymbol{\xi}$. It is shown that structural elements, and the way they are brought together and supported, determine the matrix operator $\grave{\mathbf{A}}$ which involves many parameters representing geometry, material, element-to-element connections, and supports of the structure.

Suppose, out of possible $2m + n$ response components, sensors are continuously monitoring p of them, and we have q number of actuators to create a loading on the structure in order to compensate the measured deviations from the nominal state. Usually $p << 2m + n$, and $q \geq p$. Given the sensor output, the problem here is the identification of the q number of loading components and the determination of their values so that the incremental loading thus defined is (a) convenient to implement, (b) capable to produce the desired incremental response, and (c) minimally upsetting the unobserved response components. This is the subject of this chapter.

3.1 Inverse Relations

In order to assess the response of the structure to a given loading, one needs to invert the excitation-response relations of $\grave{\mathbf{A}}\Delta\boldsymbol{\psi} = \Delta\mathbf{f}$ into the inverse relations of $\Delta\boldsymbol{\psi} = \grave{\mathbf{A}}^{-1}\Delta\mathbf{f}$. There are two engineered methods for obtaining the inverse relations from these equations: (a)displacement method and (b) force method. These are discussed below.

3.1.1 By Displacement Method

In this method one may not defer the effect of preexisting element forces on nodal equilibrium. Using $\grave{\mathbf{q}} = \grave{\mathbf{K}}\Delta\boldsymbol{\xi}$ in the nodal equilibrium equations, the excitation-response relations may be written as in Eq.(2.61):

$$\begin{bmatrix} \grave{\mathbf{K}} & \mathbf{B} & \cdot \\ \mathbf{B}^T & \cdot & -\mathbf{I} \\ \cdot & -\mathbf{I} & \vec{\mathbf{K}} \end{bmatrix} \begin{Bmatrix} \Delta\boldsymbol{\xi} \\ \Delta\mathbf{s} \\ \Delta\mathbf{v} \end{Bmatrix} = \begin{Bmatrix} \Delta\mathbf{p} \\ -\Delta\mathbf{v}_o \\ \cdot \end{Bmatrix} \tag{3.5}$$

By substituting $\Delta\mathbf{v}$ from the geometric relations given by the second row partition of Eq.(3.5) into the third one and thus expressing $\Delta\mathbf{s}$ as

$$\Delta\mathbf{s} = \vec{\mathbf{K}}(\mathbf{B}^{T}\Delta\boldsymbol{\xi}+\Delta\mathbf{v}_o) \tag{3.6}$$

and then using it in the equilibrium equations given by the first partition of Eq.(3.5), one may obtain

$$\mathbf{K}'\Delta\boldsymbol{\xi} = \Delta\mathbf{p} - \mathbf{B}\vec{\mathbf{K}}\Delta\mathbf{v}_o \tag{3.7}$$

where

$$\mathbf{K}' = \mathbf{B}\vec{\mathbf{K}}\mathbf{B}^{T}+\grave{\mathbf{K}} \tag{3.8}$$

From this form of the nodal force equilibrium equations, one can solve for $\Delta\boldsymbol{\xi}$, since its coefficient matrix $\mathbf{K}'\in R^{n\times n}$, called the *global stiffness matrix of the structure*, is positive definite provided that the reference state of the structure is a stable equilibrium state. The first term in the right-hand side of Eq.(3.8), $\mathbf{B}\vec{\mathbf{K}}\mathbf{B}^{T}$, is positive definite since it is the congruent transform of symmetric positive definite $\vec{\mathbf{K}}\in R^{m\times m}$ with matrix $\mathbf{B}\in R^{n\times m}$ which is not rank deficient, i.e., rank($\mathbf{B}$) $= n$ and since $m \geq n$. The second term in the right-hand side of Eq.(3.8), $\grave{\mathbf{K}}$, although symmetric, is indefinite, since it is a function of initial stresses $\grave{\mathbf{s}}_o$, with entries not all positive, as discussed in Subsection 2.2.3. However, so long as the reference state of the structure is a stable equilibrium state, as explained with a simple example in Subsection 2.2.3, usually one has $\left\|\grave{\mathbf{K}}\right\| << \left\|\mathbf{B}\vec{\mathbf{K}}\mathbf{B}^{T}\right\|$. In fact in most problems, one may assume $\grave{\mathbf{K}} = 0$. Hence $\mathbf{K}'$ is symmetric and positive definite, and therefore it has a unique inverse $\mathbf{F}'$

$$\grave{\mathbf{F}}' = [\grave{\mathbf{K}} + \mathbf{B}\vec{\mathbf{K}}\mathbf{B}^{T}]^{-1} \tag{3.9}$$

implying that $\det(\mathbf{K}') = \det(\mathbf{B}\vec{\mathbf{K}}\mathbf{B}^{T}+\grave{\mathbf{K}}) \neq 0$. With this in mind, from Eq.(3.7), one can write

$$\Delta\boldsymbol{\xi} = \grave{\mathbf{F}}'(\Delta\mathbf{p} - \mathbf{B}\vec{\mathbf{K}}\Delta\mathbf{v}_o) \tag{3.10}$$

Substituting $\Delta\boldsymbol{\xi}$ from this equation into the middle partition of Eq.(3.5), one may obtain

$$\Delta\mathbf{v} = \mathbf{B}^{T}\{\grave{\mathbf{F}}'(\Delta\mathbf{p} - \mathbf{B}\vec{\mathbf{K}}\Delta\mathbf{v}_o)\}+\Delta\mathbf{v}_o \tag{3.11}$$

Finally, using $\Delta\boldsymbol{\xi}$ from Eq.(3.10) in Eq.(3.6)

$$\Delta\mathbf{s} = \vec{\mathbf{K}}[\mathbf{B}^{T}\{\grave{\mathbf{F}}'(\Delta\mathbf{p} - \mathbf{B}\vec{\mathbf{K}}\Delta\mathbf{v}_o)\}+\Delta\mathbf{v}_o] \tag{3.12}$$

is obtained. By rearranging these, the inverse relationship may be obtained as

$$\begin{Bmatrix} \Delta\boldsymbol{\xi} \\ \Delta\mathbf{s} \\ \Delta\mathbf{v} \end{Bmatrix} = \begin{bmatrix} \grave{\mathbf{F}}' & \grave{\mathbf{F}}'\mathbf{B}\vec{\mathbf{K}} & \grave{\mathbf{F}}'\mathbf{B} \\ \vec{\mathbf{K}}\mathbf{B}^T\grave{\mathbf{F}}' & \vec{\mathbf{K}}\mathbf{B}^T\grave{\mathbf{F}}'\mathbf{B}\vec{\mathbf{K}} - \vec{\mathbf{K}} & \vec{\mathbf{K}}\mathbf{B}^T\grave{\mathbf{F}}'\mathbf{B} - \mathbf{I} \\ \mathbf{B}^T\grave{\mathbf{F}}' & \mathbf{B}^T\grave{\mathbf{F}}'\mathbf{B}\vec{\mathbf{K}} - \mathbf{I} & \mathbf{B}^T\grave{\mathbf{F}}'\mathbf{B} \end{bmatrix} \begin{Bmatrix} \Delta\mathbf{p} \\ -\Delta\mathbf{v}_o \\ \cdot \end{Bmatrix} \tag{3.13}$$

where the facts that $\grave{\mathbf{A}} = \grave{\mathbf{A}}^T$ and $\grave{\mathbf{A}}\grave{\mathbf{A}}^{-1} = \mathbf{I}$ are also used. One may observe that the coefficient matrix of the loads, $\grave{\mathbf{A}}^{-1}$, although symmetric, is not sparse at all.

If one were to postpone consideration of the effect of preexisting element forces on nodal equilibrium, then Eq.(3.13) would become

$$\begin{Bmatrix} \Delta\boldsymbol{\xi} \\ \Delta\mathbf{s} \\ \Delta\mathbf{v} \end{Bmatrix} = \begin{bmatrix} \mathbf{F}' & \mathbf{F}'\mathbf{B}\vec{\mathbf{K}} & \mathbf{F}'\mathbf{B} \\ \vec{\mathbf{K}}\mathbf{B}^T\mathbf{F}' & \vec{\mathbf{K}}\mathbf{B}^T\mathbf{F}'\mathbf{B}\vec{\mathbf{K}} - \vec{\mathbf{K}} & \vec{\mathbf{K}}\mathbf{B}^T\mathbf{F}'\mathbf{B} - \mathbf{I} \\ \mathbf{B}^T\mathbf{F}' & \mathbf{B}^T\mathbf{F}'\mathbf{B}\vec{\mathbf{K}} - \mathbf{I} & \mathbf{B}^T\mathbf{F}'\mathbf{B} \end{bmatrix} \begin{Bmatrix} \Delta\mathbf{p} - \grave{\mathbf{q}} \\ -\Delta\mathbf{v}_o \\ \cdot \end{Bmatrix} \tag{3.14}$$

where

$$\mathbf{F}' = (\mathbf{B}\vec{\mathbf{K}}\mathbf{B}^T)^{-1} \tag{3.15}$$

and

$$\grave{\mathbf{q}} = \grave{\mathbf{K}}\Delta\boldsymbol{\xi} \tag{3.16}$$

If the objective of this analysis were the numerical computation of the response, one would never use Eq.(3.13) due to unwarranted computational cost. Instead, one would first compute $\Delta\boldsymbol{\xi}$ from Eq.(3.10), right to left and using the Cholesky factors of $\mathbf{B}\vec{\mathbf{K}}\mathbf{B}^T$, then use $\Delta\boldsymbol{\xi}$ in $\Delta\mathbf{v} = \mathbf{B}^T\Delta\boldsymbol{\xi} + \Delta\mathbf{v}_o$ and compute $\Delta\mathbf{v}$, and finally use $\Delta\mathbf{v}$ in $\Delta\mathbf{s} = \vec{\mathbf{K}}\Delta\mathbf{v}$ and compute $\Delta\mathbf{s}$. If corrections due to ignored $\grave{\mathbf{q}}$ are required, one may evaluate it from $\grave{\mathbf{q}} = \grave{\mathbf{K}}\Delta\boldsymbol{\xi}$ and then one may compute the corrections to the response by repeating the analysis with $\grave{\mathbf{q}}$ playing the role of $\Delta\mathbf{p}$. However, the objective here is to expose the structure of the coefficient matrix $\grave{\mathbf{A}}^{-1}$for future study.

3.1.2 By Force Method

In this method, consideration of the effect of preexisting element forces on nodal equilibrium is postponed, i.e., Eq.(3.5) is rewritten as

$$\begin{bmatrix} \cdot & \mathbf{B} & \cdot \\ \mathbf{B}^T & \cdot & -\mathbf{I} \\ \cdot & -\mathbf{I} & \vec{\mathbf{K}} \end{bmatrix} \begin{Bmatrix} \Delta\boldsymbol{\xi} \\ \Delta\mathbf{s} \\ \Delta\mathbf{v} \end{Bmatrix} = \begin{Bmatrix} \Delta\mathbf{p} - \grave{\mathbf{q}} \\ -\Delta\mathbf{v}_o \\ \cdot \end{Bmatrix} \tag{3.17}$$

where $\grave{\mathbf{q}} = \grave{\mathbf{K}}\Delta\boldsymbol{\xi}$. In many problems, one may take $\grave{\mathbf{q}} = \mathbf{o}$, and compute the response. If one is willing to pay the cost, one may update the computed response with the correction of $\grave{\mathbf{A}}^{-1}[\grave{\mathbf{q}}^T, \mathbf{o}^T, \mathbf{o}^T]^T$ which is usually negligible.

The method starts by computing the element forces $\Delta\mathbf{s}$ from the equilibrium equations given by the first row partition of Eq.(3.17):

$$\mathbf{B}\Delta\mathbf{s} = \Delta\mathbf{p} - \dot{\mathbf{q}} \tag{3.18}$$

where, as discussed before, $\mathbf{B} \in R^{n\times m}$. Since $m \geq n$ and $\mathbf{B}$ is not rank deficient for engineered structures, the general solution of Eq.(3.18) may be obtained as

$$\Delta\mathbf{s} = \mathbf{C}\Delta\mathbf{x} + \mathbf{C}'(\Delta\mathbf{p} - \dot{\mathbf{q}}) \tag{3.19}$$

where $\mathbf{C} \in R^{m\times r}$, $r = m - n$, $\mathbf{C}' \in R^{m\times m}$, and $\Delta\mathbf{x} \in R^{r\times 1}$ which lists arbitrary parameters called *redundants.* Matrices $\mathbf{C}$ and $\mathbf{C}'$ can be obtained from $\mathbf{B}$ by Gauss elimination or generated column-by-column, by applying unit loads to the primary structure in the direction of the redundants for $\mathbf{C}$ and in the direction of unrestrained nodal deflections for $\mathbf{C}'$.

In obtaining matrices $\mathbf{C}$ and $\mathbf{C}'$ by Gauss elimination, one first identifies n number of columns of $\mathbf{B}$, with labels $\beta_1, \beta_2, \cdots, \beta_n$, from the pool of m columns, such that matrix $\mathbf{B}_1$ defined by these columns:

$$\mathbf{B}_1 = [\mathbf{b}_{\beta_1}, \mathbf{b}_{\beta_2}, \cdots, \mathbf{b}_{\beta_n}] \tag{3.20}$$

has a determinant largest in magnitude among the $\frac{m!}{n!(m-n)!}$ possibilities. $\mathbf{B}_1$ is the best choice for numerical inversion. By applying the *algorithm of Gauss elimination with partial pivot search by interchanging columns* to matrix $\mathbf{B}$ with its column labels array:

$$\begin{bmatrix} \mathbf{B} \\ col.\ labels \end{bmatrix} = \left[\underbrace{\begin{matrix} \mathbf{b}_1 & \mathbf{b}_2 & \cdots & \mathbf{b}_n \\ 1 & 2 & \cdots & n \end{matrix}} \quad \underbrace{\begin{matrix} \mathbf{b}_{n+1} & \cdots & \mathbf{b}_m \\ n+1 & \cdots & m \end{matrix}} \right] \tag{3.21}$$

one may obtain

$$\begin{bmatrix} \cdots \\ c.\ lbls. \end{bmatrix} = \left[\underbrace{\begin{matrix} & \mathbf{B}_1^{-1} & \\ \beta_1 & \beta_2 & \cdots & \beta_n \end{matrix}} \quad \underbrace{\begin{matrix} & \mathbf{B}_1^{-1}\mathbf{B}_2 & \\ \beta_{n+1} & \cdots & \beta_m \end{matrix}} \right] \tag{3.22}$$

where $(\beta_1, \beta_2, \cdots, \beta_n)$ is the desired n-combination of $(1, 2, \cdots, m)$, and

$$\mathbf{B}_2 = [\mathbf{b}_{\beta_{n+1}}, \mathbf{b}_{\beta_{n+2}}, \cdots, \mathbf{b}_{\beta_m}] \tag{3.23}$$

It can be shown that[1]

$$\mathbf{C} = \begin{bmatrix} -\mathbf{B}_1^{-1}\mathbf{B}_2 \\ \mathbf{I} \end{bmatrix} \tag{3.24}$$

and

$$\mathbf{C}' = \begin{bmatrix} \mathbf{B}_1^{-1} \\ \mathbf{0} \end{bmatrix} \tag{3.25}$$

[1] See, for example, *Elementary Structural Analysis, 4th ed.*, Senol Utku, Charles Head Norris, and John Benson Wilbur, Part II, McGraw-Hill, Inc., 1991.

In obtaining matrices $\mathbf{C}$ and $\mathbf{C}'$ by unit loads, one first identifies a *primary structure* by introducing $r = m - n$ number of *cuts* in order to zero r of the element forces. The non-negative scalar r is called the *degree of statical indeterminacy.* The structure itself is called *statically determinate* when $r = 0$, and it is called *statically indeterminate to degree* r when $r > 0$. The primary structure is a statically determinate structure. It can be shown that the jth column of $\mathbf{C}$, i.e., $\mathbf{c}_j$, is the element forces computed for the primary structure when the primary structure is subjected to an *equal and opposite unit load pair* in the jth cut. By repeating this for $j = 1, 2, \cdots, r$ one may generate the complete $\mathbf{C}$ matrix. Similarly it can be shown that the jth column of $\mathbf{C}'$, i.e., $\mathbf{c}'_j$, is the element forces computed for the primary structure when it is subjected to a unit load in the direction of the jth unrestrained nodal deflection. By repeating this for $j = 1, 2, \cdots, n$, one may generate complete matrix $\mathbf{C}'$.

When the structure is statically determinate, it can be shown that $\mathbf{C}$ does not exit, and $\mathbf{C}'$ is the inverse of $\mathbf{B}$, namely

$$\mathbf{C} = 0, \text{ and } \mathbf{C}' = \mathbf{B}^{-1}, \text{ if } m = n \tag{3.26}$$

In this case, Eq.(3.19) becomes $\mathbf{s} = \mathbf{B}^{-1}(\mathbf{p} - \grave{\mathbf{q}})$. Moreover, if $\grave{\mathbf{q}} = \mathbf{o}$, then $\Delta\mathbf{s} = \mathbf{B}^{-1}\Delta\mathbf{p}$; namely, one can compute the internal forces without ever computing $\Delta\mathbf{v}$ and $\Delta\boldsymbol{\xi}$. In statically determinate structures the internal force response is uncoupled from the nodal deflection and element deformation responses, so long as the excitation-response relations are linear and preexisting element force effects are negligible.

When the structure is statically indeterminate, both $\mathbf{C}$ and $\mathbf{C}'$ exist. Since there are many primary structures associated with a given structure, they are not unique. However, they always satisfy the following:

$$\mathbf{BC} = 0 \tag{3.27}$$

$$\mathbf{BC}' = \mathbf{I} \tag{3.28}$$

According to Eq.(3.27), any linear combination of the columns of $\mathbf{C}$ is orthogonal to any linear combination of the rows of $\mathbf{B}$. Hence, the columns of $\mathbf{C}$ establish a *basis* in the *null space* of $\mathbf{B}$, which is unique. Matrix $\mathbf{C}'$ may be called a *generalized inverse* of $\mathbf{B}$.

By using $\mathbf{s}$ in the third row partition of Eq.(3.17) one may obtain

$$\Delta\mathbf{v} = \vec{\mathbf{F}}\{\mathbf{C}\Delta\mathbf{x} + \mathbf{C}'(\Delta\mathbf{p} - \grave{\mathbf{q}})\} \tag{3.29}$$

where $\vec{\mathbf{F}} = \vec{\mathbf{K}}^{-1}$ is used. Using $\Delta\mathbf{v}$ from this equation in the second row partition of Eq.(3.17) one may obtain

$$\Delta\mathbf{v} = \vec{\mathbf{F}}\{\mathbf{C}\Delta\mathbf{x} + \mathbf{C}'(\Delta\mathbf{p} - \grave{\mathbf{q}})\} = \mathbf{B}^T\Delta\boldsymbol{\xi} + \Delta\mathbf{v}_o \tag{3.30}$$

By premultiplying both sides of the last equation by $\mathbf{C}^T$ and using the transpose of Eq.(3.27), one may obtain an expression for the redundants $\Delta\mathbf{x}$ as

$$\Delta\mathbf{x} = [\mathbf{C}^T\vec{\mathbf{F}}\mathbf{C}]^{-1}\mathbf{C}^T\{\Delta\mathbf{v}_o - \vec{\mathbf{F}}\mathbf{C}'(\Delta\mathbf{p} - \grave{\mathbf{q}})\} \tag{3.31}$$

Premultiplying all sides of the last equation in Eq.(3.30) by $\mathbf{C}'^T$ and using the transpose of Eq.(3.28), one may obtain an expression for nodal deflections $\Delta\boldsymbol{\xi}$ as

$$\Delta\boldsymbol{\xi} = \mathbf{C}'^T\vec{\mathbf{F}}\{\mathbf{C}\Delta\mathbf{x} + \mathbf{C}'(\Delta\mathbf{p} - \grave{\mathbf{q}})\} - \mathbf{C}'^T\Delta\mathbf{v}_o \tag{3.32}$$

where $\Delta\mathbf{v}$ from Eq.(3.29) is also used. By substituting $\Delta\mathbf{x}$ from Eq.(3.31) into Eqs.(3.19, 3.30, and 3.32), and then rearranging, one may obtain the inverse relationship as

$$\left\{\begin{array}{c}\Delta\boldsymbol{\xi}\\ \Delta\mathbf{s}\\ \Delta\mathbf{v}\end{array}\right\} = \left[\begin{array}{ccc}\mathbf{C}'^T\Delta\mathbf{F}\,\mathbf{C}' & \mathbf{C}'^T\Delta\mathbf{F}\,\vec{\mathbf{K}} & \mathbf{C}'^T\Delta\mathbf{F}\\ \vec{\mathbf{K}}\,\Delta\mathbf{F}\,\mathbf{C}' & -\mathbf{K}_c & -\mathbf{K}_c\vec{\mathbf{F}}\\ \Delta\mathbf{F}\mathbf{C}' & -\vec{\mathbf{F}}\mathbf{K}_c & \Delta\mathbf{F}\end{array}\right]\left\{\begin{array}{c}\Delta\mathbf{p} - \grave{\mathbf{q}}\\ -\Delta\mathbf{v}_o\\ \cdot\end{array}\right\} \tag{3.33}$$

where

$$\mathbf{K}_c = \mathbf{C}[\mathbf{C}^T\vec{\mathbf{F}}\mathbf{C}]^{-1}\mathbf{C}^T \tag{3.34}$$

and

$$\left.\begin{array}{l}\mathbf{F}_c = \vec{\mathbf{F}}\mathbf{K}_c\vec{\mathbf{F}}\\ \Delta\mathbf{F} = \vec{\mathbf{F}} - \mathbf{F}_c\end{array}\right\} \tag{3.35}$$

and the facts that $\grave{\mathbf{A}} = \grave{\mathbf{A}}^T$ and $\grave{\mathbf{A}}\grave{\mathbf{A}}^{-1} = \mathbf{I}$ are also used. One may observe that the coefficient matrix of the loads, $\grave{\mathbf{A}}^{-1}$, although symmetric, is not sparse at all.

As stated for the displacement method, if the objective of this analysis were the numerical computation of the response, one would never use Eq.(3.33) due to unwarranted computational cost. Instead, one would first compute $\Delta\mathbf{x}$ from Eq.(3.31), right to left and using the Cholesky factors of $\mathbf{C}^T\vec{\mathbf{F}}\mathbf{C}$, then use $\Delta\mathbf{x}$ in $\Delta\mathbf{s} = \mathbf{C}\Delta\mathbf{x} + \mathbf{C}'(\Delta\mathbf{p} + \mathbf{o})$ and compute $\Delta\mathbf{s}$, use $\Delta\mathbf{s}$ in $\Delta\mathbf{v} = \vec{\mathbf{F}}\Delta\mathbf{s}$ and compute $\Delta\mathbf{v}$, and finally use $\Delta\mathbf{v}$ in $\Delta\boldsymbol{\xi} = \mathbf{C}'(\Delta\mathbf{v} - \Delta\mathbf{v}_o)$ and compute $\Delta\boldsymbol{\xi}$. If corrections due to ignored $\grave{\mathbf{q}}$ are required, one may evaluate it from $\grave{\mathbf{q}} = \grave{\mathbf{K}}\Delta\boldsymbol{\xi}$ and then one may compute the corrections to the response by repeating the analysis with $\grave{\mathbf{q}}$ playing the role of $\Delta\mathbf{p}$. However, the objective here is to expose the structure of the coefficient matrix $\grave{\mathbf{A}}^{-1}$ for future study.

3.1.3 Statically Determinate Case

The inverse relationships given by Eq.(3.14) simplify extensively in the statically determinate case where $m = n$ and $\mathbf{B}^{-1}$ exists. Noting that

$\mathbf{F}' = (\mathbf{B}\vec{\mathbf{K}}\mathbf{B}^T)^{-1} = \mathbf{B}^{-T}\vec{\mathbf{F}}\mathbf{B}^{-1}$, Eq.(3.14) becomes

$$\begin{Bmatrix} \Delta\boldsymbol{\xi} \\ \Delta\mathbf{s} \\ \Delta\mathbf{v} \end{Bmatrix} = \begin{bmatrix} \mathbf{B}^{-T}\vec{\mathbf{F}}\mathbf{B}^{-1} & \mathbf{B}^{-T} & \mathbf{B}^{-T}\vec{\mathbf{F}} \\ \mathbf{B}^{-1} & \cdot & \cdot \\ \vec{\mathbf{F}}\mathbf{B}^{-1} & \cdot & \vec{\mathbf{F}} \end{bmatrix} \begin{Bmatrix} \Delta\mathbf{p} - \grave{\mathbf{q}} \\ -\Delta\mathbf{v}_o \\ \cdot \end{Bmatrix}, \; if \; n = m \tag{3.36}$$

where $\vec{\mathbf{K}}^{-1} = \vec{\mathbf{F}}$ is used. The same result may be obtained from Eq.(3.33), since when $m = n$, Eq.(3.26) is applicable, i.e., $\mathbf{C}$ does not exist therefore $\mathbf{K}_c = \mathbf{F}_c = \mathbf{0}$, and $\mathbf{C}' = \mathbf{B}^{-1}$. The reader may verify that the inverse of the inverse relationship of Eq.(3.36) is Eq.(3.17) with all submatrices $\in \mathrm{R}^{n \times n}$.

As discussed in Subsection 2.2.3, $\grave{\mathbf{q}}$ appearing in Eq.(3.36) represents the effect of preexisting element forces that may be present before the application of the current loading. According to Eq.(3.16), $\grave{\mathbf{q}} = \grave{\mathbf{K}}\Delta\boldsymbol{\xi}$. As discussed in Subsection 2.2.3, $\grave{\mathbf{q}}$ is usually a negligible quantity, so long as the structure is not a cable structure.

3.2 Actuators of Adaptive Structures

As discussed in the beginning of this chapter, an important issue in adaptive structures is how to compensate deviations from the nominal values, in p number of controlled response components. Usually p number of sensors continuously monitor these response components. The number of controlled response components is usually much smaller than the total number of response components, i.e.,

$$p << 2m + n \tag{3.37}$$

Which p of the $2m + n$ response components are to be controlled is determined by the function of the structure. For example, if the adaptive structure is supporting an antenna surface, then the displacement components of nodes supporting the antenna surface, in the direction normal to the surface, should be controlled. If these displacement components become unacceptably large, they can be compensated by a control loading that would create the negatives of the measured displacements.

In structures, the control loading is created by q number of actuators. Usually, the number of actuators is much smaller than the number of possible loading components shown in the right-hand side of Eq.(3.17), i.e.,

$$q << m + n \tag{3.38}$$

If an actuator induces force-type loading at the nodes, then it is called a *force inducing actuator*. For example, the attitude control jets of a spacecraft are force inducing actuators. The physical location of a force inducing actuator determines which component of $\mathbf{p}$ in the right-hand side of Eq.(3.17) is to be created.

If an actuator induces deflection-type loading in the elements, then it is called a *deformation inducing actuator.* The muscles of a living organism are examples of deformation inducing actuators.[2] In this work, a controlled structure is an adaptive structure if the actuators are of deformation inducing type. The physical location of a deformation inducing actuator determines which component of $\mathbf{v}_o$ in the right-hand side of Eq.(3.17) is to be created.

One may observe from the inverse relationship given in Eq.(3.36) for statically determinate structures that, when $\grave{\mathbf{q}} = \mathbf{o}$, i.e., when there are no preexisting element forces or their effect on nodal equilibrium equations are negligible, the deformation inducing actuators create no stress response [since the (2,2) partition of coefficient matrix $\grave{\mathbf{A}}^{-1}$ in Eq.(3.36) is always zero], whereas the force inducing actuators always create stress response [since the (2,1) partition of coefficient matrix $\grave{\mathbf{A}}^{-1}$ in Eq.(3.36) is never zero]. This probably is the basic reason behind the use of the deformation inducing actuators in adaptive structures. It is quite appealing to be able to move the nodes of a structure without creating internal stresses. A structure is usually designed to be as fully stressed as possible in its nominal state. There is little allowance for increasing the magnitudes of existing stresses by actuator induced loadings. If the adaptive structure is statically determinate, one can change its geometry by deformation inducing actuators without changing its stress state.

If the structure is statically indeterminate, the inverse relationships given by Eq.(3.14) or Eq.(3.33) indicate that stresses are created even when the actuators are of deformation inducing type. This is because of the fact that the (2,2) partition of coefficient matrix $\grave{\mathbf{A}}^{-1}$, displayed in Eqs.(3.14) and (3.33) is not zero. However, as is discussed, in Chapter 5, by means of inducing geometrically compatible deformations, creation of internal stresses can be prevented even if the adaptive structure is statically indeterminate.

3.3 Basic Equations for Adaptive Structures

The excitation-response relations studied in the previous chapter may be specialized for adaptive structures by assuming that

- The effect of initial stresses on nodal equilibrium is negligible, i.e., $\Delta\grave{\mathbf{q}} = \mathbf{o}$.
- The incremental loading is caused by the deformation inducing actuators only, i.e., $\Delta\mathbf{v}_o$ exists, but $\Delta\mathbf{p} = \mathbf{o}$.

---

[2]For nature's reasons for choosing deformation inducing actuators in shape and motion control, see "Why the Wheels Won't Go" by Michael Barbera, *The American Naturalist*, v.121, n.3, pp.395-408, 1983.

- Some of the response components are prescribed as the negatives of the sensor measured values from the nominal.

With these, the excitation-response relations for adaptive structures may be obtained from Eq.(2.61) as

$$\begin{bmatrix} \cdot & \mathbf{B} & \cdot \\ \mathbf{B}^T & \cdot & -\mathbf{I} \\ \cdot & -\mathbf{I} & \vec{\mathbf{K}} \end{bmatrix} \begin{Bmatrix} \Delta\boldsymbol{\xi} \\ \Delta\mathbf{s} \\ \Delta\mathbf{v} \end{Bmatrix} = \begin{Bmatrix} \cdot \\ -\Delta\mathbf{v}_o \\ \cdot \end{Bmatrix} \tag{3.39}$$

such that the incremental response obtained from these equations, when added to the values before the loading, will yield the nominal values for the response components that are monitored by the sensors. The components that are not monitored may change. The incremental response will eliminate the excesses in the monitored response components. In these equations, $\mathbf{B} \in R^{n\times m}$, $(\vec{\mathbf{K}},\mathbf{I}) \in R^{m\times m}$, $(\Delta\mathbf{s},\Delta\mathbf{v},\Delta\mathbf{v}_o) \in R^{m\times 1}$, and $\Delta\boldsymbol{\xi} \in R^{n\times 1}$ where $n = Ne - b$ and $m = Ma - f$ for an N node, M element adaptive structure with e number of deflection degrees of freedom per node, a number of internal force components per element, b number of deflection constraints, and f number of zeroed element forces.

If the adaptive structure has q number of deformation inducing actuators, then only q number of components of $\Delta\mathbf{v}_o$ can be nonzero. If p number of response components are being controlled, then, when the response deviates from the nominal in these components, one can use basically Eq.(3.39) with the measured deviations to compute the required prescribed element deformations to be induced by the q number of actuators. The computed prescribed element deformations are the nonzero entries of $\Delta\mathbf{v}_o$ in Eq.(3.39). This $\Delta\mathbf{v}_o$ constitutes the control loading. The response corresponding to the control loading compensates the deviations in the observed components.

3.4 Actuator Locations and Controls

If one considers Eq.(3.33) in the form of Eq.(3.1), ignoring the effects of preexisting element forces, the control loading $\Delta\mathbf{f}$ becomes

$$\Delta\mathbf{f} = [\Delta\mathbf{p}^T, -\Delta\mathbf{v}_o^T, \cdot]^T \tag{3.40}$$

where $\Delta\mathbf{f} \in R^{(2m+n)\times 1}$, $\Delta\mathbf{p} \in R^{n\times 1}$ and $\Delta\mathbf{v}_o \in R^{m\times 1}$. Suppose the actuators may be of force and/or deformation inducing type. Since there are only q number of actuators, and $q << m+n$, many of the entries of $\Delta\mathbf{p}$ and $\Delta\mathbf{v}_o$ are zero. The physical locations of the actuators determine the locations of the nonzero entries of $\Delta\mathbf{f}$. The placement of actuators is one of the important problems of adaptive structures. Let $\beta_1, \beta_2, \cdots, \beta_q$ denote the row labels of the nonzero entries of $\Delta\mathbf{f}$. From Eq.(3.39) it may be seen that

$$n < \beta_j \leq n + m \text{ for } j = 1, \cdots, q \tag{3.41}$$

The actuator placement problem is to identify the labels $\beta_1, \beta_2, \cdots, \beta_q$.

Once the actuator locations are known, the next problem is to determine how much actuation is needed in each actuator. Let

$$\begin{array}{l} \Delta p_{\beta_j} = -\Delta u_j \text{ , if } \beta_j \leq n \\ \Delta v_o^{\beta_j} = \Delta u_j \text{ , if } \beta_j > n \end{array} \text{ , for } j = 1, \cdots, q \tag{3.42}$$

denote the amount of element deformation in each actuator. The determination of these is the basis of static control problem in adaptive structures, once the actuator placement problem is solved. The quantities Δu_j, $j = 1, \cdots, q$ are called *incremental controls.*

As a result of solving actuator placement and control computation problems, the control loading $\Delta\mathbf{f}$ may be defined as

$$\Delta\mathbf{f} = -\sum_{j=1}^{q} \mathbf{i}_{\beta_j} \Delta u_j \tag{3.43}$$

where $\mathbf{i}_{\beta_j}$ is the β_jth column of the $(2m+n)$th order identity matrix $\mathbf{I}$. The columns of $\mathbf{I}$, appearing in the summation sign, define the orthonormal matrix

$$\mathbf{I}_\beta = [\mathbf{i}_{\beta_1}, \mathbf{i}_{\beta_2}, \cdots, \mathbf{i}_{\beta_q}] \tag{3.44}$$

which is the *actuator location matrix*. Note that $\mathbf{I}_\beta \in (\text{Int})^{(2m+n)\times q}$. Using Eq.(3.42) and matrix $\mathbf{I}_\beta$, one may rewrite Eq.(3.43) as

$$\Delta\mathbf{f} = -\mathbf{I}_\beta \Delta\mathbf{u} \tag{3.45}$$

where

$$\Delta\mathbf{u} = [\Delta u_1, \Delta u_2, \cdots, \Delta u_q]^T \tag{3.46}$$

This is the definition of the control loading in terms of controls.

3.5 Observed Response Components and Output

If one considers Eq.(3.33) in the form of Eq.(3.1), the response $\Delta\psi$ of adaptive structure to control loading $\Delta\mathbf{f}$ may be displayed as

$$\Delta\psi = [\Delta\xi^T, \Delta\mathbf{s}^T, \Delta\mathbf{v}^T]^T \tag{3.47}$$

where $\Delta\psi \in \mathrm{R}^{(2m+n)\times 1}$. As discussed earlier, in Eq.(3.47) some of the response components are monitored continuously by sensors in order to control them. These controlled components are identified by their row labels in $\Delta\psi$. The determination of the number of controlled components p and their row labels in $\Delta\psi$ are important in adaptive structures. Let $\alpha_1, \alpha_2, \cdots, \alpha_p$

denote the row labels of the controlled entries of $\Delta\psi$. From Eq.(3.39) it may be seen that

$$1 \leq \alpha_j \leq 2m+n \ , \ \text{for } j = 1, \cdots, p \tag{3.48}$$

The sensor placement problem starts with the identification of the labels $\alpha_1, \alpha_2, \cdots, \alpha_p$. The actual locations of sensors are such that the controlled components can be observed with least error.

Let $\Delta\psi^d \in R^{(2m+n)\times 1}$ denote the deviations from the nominal values in all response components before the current control loading is applied. The deviations from the nominal values of the observed components are a subset of these, and they are measured in real time by the sensors. The measured values are called *disturbances* or *sensor output* or shortly *output*.[3] Let Δy_j denote the disturbance in the jth observed response component:

$$\Delta y_j = \Delta\psi^d_{\alpha_j}, \ j = 1, \cdots, p \tag{3.49}$$

Using the identity of

$$\Delta\psi^d_{\alpha_j} = \mathbf{i}^T_{\alpha_j} \Delta\psi^d \tag{3.50}$$

where $\mathbf{i}_{\alpha_j}$ is the α_jth column of the $(2m+n)$th order identity matrix $\mathbf{I}$, one can rewrite Eq.(3.49) as

$$\Delta \mathbf{y} = \mathbf{I}^T_\alpha \Delta\psi^d \tag{3.51}$$

where

$$\Delta \mathbf{y} = [\Delta y_1, \Delta y_2, \cdots, \Delta y_p]^T \tag{3.52}$$

and

$$\mathbf{I}_\alpha = [\mathbf{i}_{\alpha_1}, \mathbf{i}_{\alpha_2}, \cdots, \mathbf{i}_{\alpha_p}] \tag{3.53}$$

Note that $\mathbf{I}_\alpha \in (\text{Int})^{(2m+n)\times p}$ and it is orthonormal. Matrix $\mathbf{I}_\alpha$ is called the *observed component identification matrix.*

3.6 Determination of Controls

The inverse relations given by Eq.(3.33) may be rewritten for controlled structures under the control loading $\Delta \mathbf{f}$ as

$$\Delta\psi = \grave{\mathbf{A}}^{-1} \Delta \mathbf{f} \tag{3.54}$$

---

[3]The origin of this term may be found in dynamic control literature. As discussed later in Section 8.6; in dynamic control of vibrations of a linearly behaving time invariant *observable* system, knowledge of output $\Delta \mathbf{y}$ means the knowledge of current state $\Delta\psi$. In the static control of the same system, knowledge of output $\Delta \mathbf{y}$ never entails the knowledge of current state $\Delta\psi$. The structural system may not be *observable* when the control is static.

where $\Delta\boldsymbol{\psi}$ is the response to control loads as defined by Eq.(3.47). The control loading $\Delta\mathbf{f}$ is as in Eq.(3.40), and $\grave{\mathbf{A}}^{-1}$ is as shown below

$$\grave{\mathbf{A}}^{-1} = \begin{bmatrix} \mathbf{C}'^T\Delta\mathbf{F}\,\mathbf{C}' & \mathbf{C}'^T\Delta\mathbf{F}\,\vec{\mathbf{K}} & \mathbf{C}'^T\Delta\mathbf{F} \\ \vec{\mathbf{K}}\,\Delta\mathbf{F}\,\mathbf{C}' & -\mathbf{K}_c & -\mathbf{K}_c\vec{\mathbf{F}} \\ \Delta\mathbf{F}\mathbf{C}' & -\vec{\mathbf{F}}\mathbf{K}_c & \Delta\mathbf{F} \end{bmatrix} \tag{3.55}$$

where $\mathbf{K}_c$ and $\Delta\mathbf{F}$ are as defined in Eq.(3.34) and Eq.(3.35), respectively. By substituting $\Delta\mathbf{f}$ from Eq.(3.45), then premultiplying both sides of Eq.(3.54) by the transpose of $\mathbf{I}_\alpha$ of Eq.(3.53), and finally using this in the condition that the changes induced by the control loading in the observed response components, i.e., in $\mathbf{I}_\alpha^T\Delta\boldsymbol{\psi}$, equal the negatives of the *output*, that is, in

$$\mathbf{I}_\alpha^T\Delta\boldsymbol{\psi} = -\Delta\mathbf{y} \tag{3.56}$$

the following important relationship may be obtained:

$$\Delta\mathbf{y} = \mathbf{P}\Delta\mathbf{u} \tag{3.57}$$

where

$$\mathbf{P} = [\mathbf{I}_\alpha^T\grave{\mathbf{A}}^{-1}\mathbf{I}_\beta] \tag{3.58}$$

Note that $\mathbf{P} \in R^{p\times q}$ is the matrix obtained from $\grave{\mathbf{A}}^{-1}$ by the row labels of the observed response components and the knowledge of actuator locations. According to Eq.(3.58), $\mathbf{P}$ is obtained from $\grave{\mathbf{A}}^{-1}$ by retaining the rows corresponding to the observed response components and the columns corresponding to the actuator locations, and by deleting everything else. The column matrix $\Delta\mathbf{y} \in R^{p\times 1}$ is the *output*, and $\Delta\mathbf{u} \in R^{q\times 1}$ is the list of actuator induced element deformations, i.e., the *controls*. If one can solve $\Delta\mathbf{u}$ from Eq.(3.57), then the control loading defined by Eq.(3.45) with this $\Delta\mathbf{u}$ will create the negatives of the disturbances; thus, for the observed response components the nominal values will be maintained.

For the solvability of $\Delta\mathbf{u}$ from Eq.(3.57), the following may be observed from Eq.(3.57)

- If $p = q$, and $\det(\mathbf{P}) \neq 0$, then the controls are unique and they can be computed from the output as

$$\Delta\mathbf{u} = \mathbf{P}^{-1}\Delta\mathbf{y} \tag{3.59}$$

- If $p < q$, and $\mathbf{P}$ is not rank deficient, then there are infinitely many control possibilities corresponding to a given output.

- If $p > q$, and output $\Delta\mathbf{y}$ is in the column space of $\mathbf{P}$, then the controls are unique and they can be computed from

$$\Delta\mathbf{u} = [\mathbf{P}^T\mathbf{P}]^{-1}\mathbf{P}^T\Delta\mathbf{y} \tag{3.60}$$

- If $p > q$, and output $\Delta\mathbf{y}$ is not in the column space of $\mathbf{P}$, then there are no exact controls; however, one may use $\Delta\mathbf{u}$ from Eq.(3.60) as approximate controls.

In all other cases, the observed response components, hence the adaptive structure, are called *uncontrollable*,[4] unless one is willing to change the locations of actuators or the observed response components or both.

Suppose the adaptive structure is controllable, and there is at least one $\mathbf{u}$ in the form of

$$\Delta\mathbf{u} = \mathbf{G}\Delta\mathbf{y} \tag{3.61}$$

that satisfies Eq.(3.57). This is called the *control law*.[5] Note that the substitution of $\Delta\mathbf{u}$ from Eq.(3.61) into Eq.(3.57) yields

$$\mathbf{PG} = \mathbf{I} \tag{3.62}$$

and $\mathbf{G} \in R^{q\times p}$. Then, as mentioned earlier, the control loading defined by Eq.(3.45) with $\Delta\mathbf{u}$ from Eq.(3.61) will create the negatives of disturbances in the observed response components; thus, for these response components the nominal values will be maintained. The fate of the unobserved response components is discussed in the next section.

The short discussion above indicates the importance of the identification of observed response components and the placement of actuators. In the following chapters, these problems are further discussed from the standpoint of energy required by the actuators for controlling the structural response.

3.7 Fate of Unobserved Response Components

When the incremental control loading $\Delta\mathbf{f}$ corresponding to controls $\Delta\mathbf{u}$, i.e., $-\mathbf{I}_\beta\Delta\mathbf{u}$ according to Eq.(3.45), is applied on the structure, the response

$$\Delta\boldsymbol{\psi} = -\grave{\mathbf{A}}^{-1}\mathbf{I}_\beta\Delta\mathbf{u} \tag{3.63}$$

is induced per Eq.(3.54). Since this will superimpose on deviations $\Delta\boldsymbol{\psi}^d$ that were present before the current control loading, the combined response is $\Delta\boldsymbol{\psi}^d + \Delta\boldsymbol{\psi}$. Since premultiplication of $(\Delta\boldsymbol{\psi}^d + \Delta\boldsymbol{\psi})$ with $\mathbf{I}_a^T$ yields compactly the observed response components, and premultiplication of $\mathbf{I}_a^T(\Delta\boldsymbol{\psi}^d + \Delta\boldsymbol{\psi})$ with $\mathbf{I}_a$ enlarges the compact list back to an order of

---

[4] Controllable dynamic systems are discussed later Section 8.5.2 where it is shown that the vibrations of a time-invariant linearly behaving structure are always controllable. However, note that static control of the same structure may not be controllable.

[5] The reader may want to compare this control law with the ones discussed for dynamic control later in the book, for example, in Sections 8.4 and 8.5.

$2m+n$ by placing zero values at rows corresponding to unobserved components, one may define

$$\Delta\boldsymbol{\psi}^c = \mathbf{I}_\alpha \mathbf{I}_\alpha^T(\Delta\boldsymbol{\psi}^d + \Delta\boldsymbol{\psi}) \tag{3.64}$$

for the combined response in the observed components with zeros in the unobserved components. By using $\Delta\mathbf{y}$ from Eq.(3.51) in Eq.(3.61) and $\Delta\mathbf{u}$ from the latter in Eq.(3.63), one obtains

$$\Delta\boldsymbol{\psi} = -\grave{\mathbf{A}}^{-1}\mathbf{I}_\beta \mathbf{G}\mathbf{I}_\alpha^T \Delta\boldsymbol{\psi}^d \tag{3.65}$$

The use of this in Eq.(3.64) gives the observed response as

$$\Delta\boldsymbol{\psi}^c = \mathbf{I}_\alpha \mathbf{I}_\alpha^T \Delta\boldsymbol{\psi}^d - \mathbf{I}_\alpha[\mathbf{I}_\alpha^T \grave{\mathbf{A}}^{-1}\mathbf{I}_\beta \mathbf{G}]\mathbf{I}_\alpha^T \Delta\boldsymbol{\psi}^d \tag{3.66}$$

According to Eqs.(3.58 and 3.62), the bracketed factor is an identity matrix of order p. With this, the observed response in Eq.(3.66) becomes

$$\Delta\boldsymbol{\psi}^c = \mathbf{o} \tag{3.67}$$

as expected. Clearly, by using controls that satisfy Eq.(3.57), the disturbances in the observed components can be eliminated. However, nothing so decisive can be said for the unobserved response components, as shown below.

One may define $\Delta\boldsymbol{\psi}^u$ as the part of the combined response $(\Delta\boldsymbol{\psi}^d + \Delta\boldsymbol{\psi})$ which contains zeros in the observed response $\Delta\boldsymbol{\psi}^c$, and write

$$\Delta\boldsymbol{\psi}^u = (\Delta\boldsymbol{\psi}^d + \Delta\boldsymbol{\psi}) - \Delta\boldsymbol{\psi}^c \tag{3.68}$$

Using $\Delta\boldsymbol{\psi}^c$ from Eq.(3.67), this leads to

$$\Delta\boldsymbol{\psi}^u = \Delta\boldsymbol{\psi}^d + \Delta\boldsymbol{\psi} \tag{3.69}$$

which shows, as expected, that the unobserved combined response is the sum of the deviations that were present before the current control loading and the response caused by the current control loading. Although the current control loading ensures the cancellation of the deviations in the observed response components, there is no such assurance for the unobserved response components.

In order not to create undesired situations in the unobserved response components, it is imperative to make sure that the actuator induced control loading would create changes as small as possible in the unobserved response components.

For example, regardless of whether the structure is statically determinate or indeterminate, it is clear from the inverse relationships given by Eqs.(3.33) and (3.36) that one *should not use force inducing actuators for the control of nodal deflections.* This is because of the fact that the force inducing actuators create control loads in the first block, i.e., in the $\Delta\mathbf{p} - \grave{\mathbf{q}}$ block of the excitations; in turn, these cause not only an element force response but also element deformation and nodal deflection responses. This is due to the fact that the first column partitions of $\grave{\mathbf{A}}^{-1}$, i.e., submatrices $\mathbf{B}^{-T}\vec{\mathbf{F}}\overline{\mathbf{B}}^{-1}$, $\mathbf{B}^{-1}$, and $\vec{\mathbf{F}}\overline{\mathbf{B}}^{-1}$ are full submatrices [see Eqs.(3.33) and (3.36)].

Deformation inducing actuators, on the other hand, *should be always preferred for nodal deflection control in all discrete parameter structures* since they cause only nodal deflection response, but no element force and no element deformation responses in

- statically determinate structures for any values of the inserted deformations [see Eq.(3.39)], and in
- statically indeterminate structures for geometrically compatible values of the inserted deformations (see Section 3.9).

For this reason, in adaptive structures the actuators are always of deformation inducing type.

3.8 Control Energy

This is the energy one has to provide to run perfectly efficient actuators, in order to implement the control loading of the current step. Let ΔE denote this energy. In adaptive structures the deformation inducing actuators are used to create the control loading; hence, ΔE is the energy necessary to insert the prescribed element deformations $\Delta\mathbf{v}_o$ into the structure. The insertion of $\Delta\mathbf{v}_o$ into the structure may be resisted by two agents: (a) stiffness of the structure in the directions associated with the nonzero components of $\Delta\mathbf{v}_o$, and (b) preexisting element forces $\bar{\mathbf{s}}_o$. Taking $\Delta\mathbf{p} - \grave{\mathbf{q}} = \mathbf{o}$, from Eq.(3.33), one may obtain

$$\Delta\mathbf{s} = \mathbf{K}_c\Delta\mathbf{v}_o \tag{3.70}$$

where $\mathbf{K}_c \in \mathrm{R}^{m\times m}$. $\mathbf{K}_c$ is the stiffness of the structure in directions associated with $\Delta\mathbf{v}_o$. This matrix is defined by Eq.(3.34) as

$$\mathbf{K}_c = \mathbf{C}\overline{[\mathbf{C}^T\vec{\mathbf{F}}\mathbf{C}]}^{-1}\mathbf{C}^T \tag{3.71}$$

where $\vec{\mathbf{F}}$ is the matrix of element flexibilities, and matrix $\mathbf{C}$ is such that its columns define a basis in the null space of matrix $\mathbf{B}$, as discussed in Subsection 3.1.2. With these, the control energy ΔE may be expressed as

$$\Delta E = \int_{\mathbf{o}}^{\Delta\mathbf{v}_o} (\grave{\mathbf{s}}_o + \Delta\mathbf{s})^T d(\Delta\mathbf{v}_o) \tag{3.72}$$

By using $\Delta\mathbf{s}$ from Eq.(3.70) and observing from Eq.(3.71) that $\mathbf{K}_c$ is symmetrical, one may obtain from Eq.(3.72):

$$\Delta E = \frac{1}{2}\Delta\mathbf{v}_o^T\mathbf{K}_c\Delta\mathbf{v}_o + \grave{\mathbf{s}}_o^T\Delta\mathbf{v}_o \tag{3.73}$$

As discussed before, $\vec{\mathbf{F}}$ is positive definite; hence, the first term on the right of Eq.(3.73), i.e., the energy to overcome the stiffness of the structure, is always nonnegative. The sign of the second term depends on the signs of the corresponding entries of $\grave{\mathbf{s}}_o$ and $\Delta\mathbf{v}_o$. If the second term is positive, the insertion of $\Delta\mathbf{v}_o$ will be resisted; otherwise, it will be assisted by the preexisting element forces $\grave{\mathbf{s}}_o$.

For the effect of preexisting element forces on control energy, not much can be said without knowing $\grave{\mathbf{s}}_o$ and $\Delta\mathbf{v}_o$ explicitly. If the corresponding entries of $\grave{\mathbf{s}}_o$ and $\Delta\mathbf{v}_o$ are of the same sign, then one needs a positive energy to drive the actuators. However, if they are of different signs, one may gain energy by inserting the element deformation. For example, if one tries to insert an axial element deformation of $\Delta v_o < 0$, which corresponds to elongating the element per its definition in Subsection 2.3.2, to a column under compression by $\grave{s}_o < 0$, one needs positive energy to drive the actuator. On the other hand if one inserts the same deformation in a cable under tension, one may derive energy from the insertion. In an intelligent structure this may be sensed in real time, and procedures may be developed to take advantage of the sign parity in order to minimize the positive energy needs of the control.

There is a special situation where the preexisting element forces $\grave{\mathbf{s}}_o$ will not cause additional positive energy consumption for inserting $\Delta\mathbf{v}_o$. This situation arises when the preexisting element forces are self-equilibrating. Such element forces arise as a result of deflection-type element loads.[6] The preexisting element forces $\grave{\mathbf{s}}_o$ are called self-equilibrating, if

$$\mathbf{B}\grave{\mathbf{s}}_o = \mathbf{p}_o = \mathbf{o} \tag{3.74}$$

i.e., no nodal forces are required for the equilibrium of nodes in the unconstrained deflection directions. By the virtual work principle, when $\grave{\mathbf{s}}_o$ satisfies Eq.(3.74) and $\Delta\mathbf{v}_o$ is geometrically compatible as discussed in the next section and also in Chapter 5, then the internal work done by $\grave{\mathbf{s}}_o$ under $\Delta\mathbf{v}_o$, i.e., $\grave{\mathbf{s}}_o^T\Delta\mathbf{v}_o$, is zero, since, as in Eq.(2.41)

$$\grave{\mathbf{s}}_o^T\Delta\mathbf{v}_o = \mathbf{p}_o^T\Delta\boldsymbol{\xi} = \mathbf{o}^T\Delta\boldsymbol{\xi} = 0 \tag{3.75}$$

where $\Delta\boldsymbol{\xi}$ is the nodal deflections caused by $\Delta\mathbf{v}_o$.

[6] See "Geometry Control in Prestressed Adaptive Space Trusses," M. Sener, S. Utku, and B. K. Wada, *Journal of Smart Materials and Structures*, vol. 3, no. 2, pp. 219-225, 1994.

The energy to overcome the stiffness of the structure may be zero, if the adaptive structure is statically determinate. For statically determinate structures $m = n$, hence, $r = 0$ and the null space of matrix $\mathbf{B}$ does not exist as discussed in Subsection 3.1.2. Then, according to Eq.(3.71),

$$\mathbf{K}_c = 0, \text{ if } m = n \tag{3.76}$$

This may be seen also from Eq.(3.36) which is the inverse relation for statically determinate structures. Clearly, aside from the energy needed for causes not modeled in this work, *one may run the actuators of a statically determinate adaptive structure with no energy if there are no preexisting element forces.*

In statically indeterminate structures, positive matrix $\mathbf{K}_c$ always exists, and one may need nonnegative energy to drive the actuators. However, as is discussed in the next section, if the actuator induced element deformations are geometrically compatible, then one may not need any energy to drive the actuators even in statically indeterminate adaptive structures, provided that preexisting element forces are zero.

3.9 Compatibility of Controls in Adaptive Structures

The prescribed element deformations $\Delta\mathbf{v}_o$ induced by the actuators constitute the control loading in adaptive structures. As discussed in the previous section, if the adaptive structure is statically determinate, no element forces $\Delta\mathbf{s}$ develop in the structure because of $\Delta\mathbf{v}_o$. If the adaptive structure is statically indeterminate, according to Eqs.(3.70 and 3.71) it appears that element forces $\Delta\mathbf{s}$ develop proportional to $\Delta\mathbf{v}_o$:

$$\Delta\mathbf{s} = \mathbf{C}[\mathbf{C}^T\vec{\mathbf{F}}\mathbf{C}]^{-1}\mathbf{C}^T\Delta\mathbf{v}_o \tag{3.77}$$

However, if the product of the last two factors in the right-hand side of this equation vanishes, i.e., if

$$\mathbf{C}^T\Delta\mathbf{v}_o = \mathbf{o} \tag{3.78}$$

then

$$\Delta\mathbf{s} = \mathbf{C}[\mathbf{C}^T\vec{\mathbf{F}}\mathbf{C}]^{-1}\mathbf{o} = \mathbf{o} \tag{3.79}$$

Eq.(3.78) is called the *compatibility equation* for the actuator induced element deformations $\Delta\mathbf{v}_o$. As studied in Subsection 3.1.2, the rows of matrix $\mathbf{C}^T \in R^{r\times m}$ define a basis in the null space of matrix $\mathbf{B}$. If actuator induced element deformations $\Delta\mathbf{v}_o$ satisfy Eq.(3.78), i.e., if $\Delta\mathbf{v}_o$ is in the null space of $\mathbf{B}$, then they are called *geometrically compatible.* According to Eq.(3.77), the geometrically compatible prescribed element deformations $\Delta\mathbf{v}_o$ create no stresses in the structure.

Since, according to Eq.(3.27), $\mathbf{C}^T\mathbf{B}^T = \mathbf{0}$, then from Eq.(3.78) it follows that $\Delta\mathbf{v}_o$ *is geometrically compatible so long as it is in the row space of* $\mathbf{B}$. The number of scalar compatibility equations in Eq.(3.78) is $r = m - n$, i.e., the degree of statical indeterminacy of the adaptive structure.

Note that in statically determinate adaptive structures $r = 0$; therefore there is no compatibility equation to satisfy. Hence any $\Delta\mathbf{v}_o$ in these structures is geometrically compatible (see, for example, Fig. 1.6a). Then one can state that *when actuator induced element deformations are used as control loading and they are geometrically compatible, no element forces develop as a result of the control loading.* This, of course, cannot be said for the force inducing actuators.

In the absence of preexisting element forces, i.e., when $\mathbf{\mathring{s}}_o = \mathbf{o}$, with the help of Eqs.(3.71 and 3.73), the control energy ΔE may be expressed as

$$\Delta E = \frac{1}{2}\Delta\mathbf{v}_o^T \mathbf{C}[\mathbf{C}^T\vec{\mathbf{F}}\mathbf{C}]^{-1}\mathbf{C}^T\Delta\mathbf{v}_o \tag{3.80}$$

Clearly, *when the actuator induced element deformations* $\Delta\mathbf{v}_o$ *are geometrically compatible, aside from the energy needed for causes not modeled in this work, no energy is required to control an adaptive structure*, namely,

$$\Delta E = 0, \quad \text{if} \quad \mathbf{C}^T\Delta\mathbf{v}_o = \mathbf{o} \tag{3.81}$$

This important conclusion is exploited extensively during the discussions of statically indeterminate adaptive structures.

3.10 Recapitulation

In this chapter the problems associated with static feedback control of p of the $2m + n$ response quantities by means of q deformation inducing actuators are studied. It is shown that in structural control, deformation inducing actuators should be preferred over the force inducing actuators when such options exist. Also discussed are the problems of control energy, robustness, and actuator placement.

4
Statically Determinate Adaptive Structures

Discussions in the previous chapter show that if the adaptive structure is statically determinate and the control loading is created by deformation inducing actuators, then (a) aside from the energy needed for causes not modeled in this work, no energy is needed to run the actuators when pre-existing element forces are zero, and (b) aside from the unobserved nodal deflection components, none of the other response components are altered by the control loading. These two observations warrant further study of the statically determinate adaptive structures.

4.1 Excitation-Response Relations

The linearized incremental excitation-response relations for N node, M element structure with a number of element forces per element, e number of deflection degrees of freedom per node, b number of nodal deflection constraints, and f number of zeroed element forces are studied in Chapter 3 and may be specialized for the statically determinate adaptive structures. Copying from Eq.(3.17), these relations are:

$$\begin{bmatrix} \cdot & \mathbf{B} & \cdot \\ \mathbf{B}^T & \cdot & -\mathbf{I} \\ \cdot & -\mathbf{I} & \vec{\mathbf{K}} \end{bmatrix} \begin{Bmatrix} \Delta\boldsymbol{\xi} \\ \Delta\mathbf{s} \\ \Delta\mathbf{v} \end{Bmatrix} = \begin{Bmatrix} \cdot \\ -\Delta\mathbf{v}_o \\ \cdot \end{Bmatrix} \tag{4.1}$$

where the effect of preexisting element forces $\grave{\mathbf{s}}_o$ on nodal equilibrium equations, i.e., $\grave{\mathbf{q}} = \grave{\mathbf{K}}\Delta\boldsymbol{\xi}$ studied in Subsection 2.2.3, is ignored, that is:

$$\grave{\mathbf{q}} = \mathbf{o} \tag{4.2}$$

is assumed. Since for statically determinate structures

$$m = n \tag{4.3}$$

where $m = Ma - f$ and $n = Ne - b$, all submatrices in the coefficient matrix on the left-hand side of Eq.(4.1) are square matrices, i.e., $\mathbf{B}, \vec{\mathbf{K}}, \mathbf{I} \in R^{n\times n}$ and $\Delta\boldsymbol{\xi}, \Delta\mathbf{s}, \Delta\mathbf{v}, \Delta\mathbf{v}_o \in R^{n\times 1}$. As studied in Subsection 2.2.1, in an engineering structure, $\mathbf{B}$ is not rank deficient; hence, one may readily obtain the inverse relation from Eq.(4.1) as in Eq.(3.36):

$$\begin{Bmatrix} \Delta\boldsymbol{\xi} \\ \Delta\mathbf{s} \\ \Delta\mathbf{v} \end{Bmatrix} = \begin{bmatrix} \mathbf{B}^{-T}\vec{\mathbf{F}}\mathbf{B}^{-1} & \mathbf{B}^{-T} & \mathbf{B}^{-T}\vec{\mathbf{F}} \\ \mathbf{B}^{-1} & \cdot & \cdot \\ \vec{\mathbf{F}}\mathbf{B}^{-1} & \cdot & \vec{\mathbf{F}} \end{bmatrix} \begin{Bmatrix} \cdot \\ -\Delta\mathbf{v}_o \\ \cdot \end{Bmatrix} \tag{4.4}$$

where $\vec{\mathbf{F}} \in R^{m\times m}$ is the block diagonal matrix of element flexibilities studied in Subsection 2.4.2. One may rewrite Eq.(4.4) as

$$\begin{Bmatrix} \Delta\boldsymbol{\xi} \\ \Delta\mathbf{s} \\ \Delta\mathbf{v} \end{Bmatrix} = \begin{bmatrix} -\mathbf{B}^{-T} \\ \cdot \\ \cdot \end{bmatrix} \Delta\mathbf{v}_o \tag{4.5}$$

which shows that in statically determinate adaptive structures, one cannot control element forces $\Delta\mathbf{s}$ and element deformations $\Delta\mathbf{v}$, but one can control nodal deflections $\Delta\boldsymbol{\xi}$. In many applications this may be a blessing, since controls cannot alter the stress state of the structure, in which, especially in earthbound structures, the alteration of the stress state may not be allowed.

In statically determinate adaptive structures, as shown by Eqs.(4.4 and 4.5), the only excitation response relation is the geometric relation of

$$\mathbf{B}^T\Delta\boldsymbol{\xi} = -\,\Delta\mathbf{v}_o \tag{4.6}$$

and its inverse

$$\Delta\boldsymbol{\xi} = -\,\mathbf{B}^{-T}\Delta\mathbf{v}_o \tag{4.7}$$

where $\mathbf{B} \subset R^{n\times m}$ and $m = n$.

Eq.(4.6) is the special case for $\Delta\mathbf{v} = \mathbf{o}$ of the geometric relation

$$\Delta\mathbf{v} = \mathbf{B}^T\Delta\boldsymbol{\xi} + \Delta\mathbf{v}_o \tag{4.8}$$

given by Eq.(2.45) and studied in Section 2.3. By differentiating both sides with respect to the components of $\Delta\boldsymbol{\xi}$, one component at a time, and then writing compactly, one may obtain

$$\mathbf{v}_{,\xi} = \mathbf{B}^T \tag{4.9}$$

where the notation $\mathbf{v}_{,\boldsymbol{\xi}}$ is used to denote the *Jacobian matrix of element deformations* $\mathbf{v}$ with respect to nodal deflections $\boldsymbol{\xi}$, at the state before the control loading is applied:

$$\mathbf{v}_{,\boldsymbol{\xi}} = \sum_{j=1}^{m} \sum_{k=1}^{n} \mathbf{i}_j \frac{\partial v^j}{\partial \xi_k} \mathbf{i}_k'^T \tag{4.10}$$

where $\mathbf{i}_j$ is the jth column of the mth order identity matrix, and $\mathbf{i}_k'$ is the kth column of the nth order identity matrix. Eq.(4.9) shows that matrix $\mathbf{B}^T$ is the Jacobian matrix of element deformations $\mathbf{v}$ with respect to nodal deflections $\boldsymbol{\xi}$, at the state before the controls are applied.

Since $m = n$ and $\mathbf{B}$ is not rank deficient, from Eq.(4.8), one may obtain:

$$\Delta\boldsymbol{\xi} = \mathbf{B}^{-T}\Delta\mathbf{v} - \mathbf{B}^{-T}\Delta\mathbf{v}_o \tag{4.11}$$

By differentiating both sides with respect to the components of $\Delta\mathbf{v}$, one component at a time, and then writing compactly, one may obtain

$$\boldsymbol{\xi}_{,\mathbf{v}} = \mathbf{B}^{-T} \tag{4.12}$$

where the notation $\boldsymbol{\xi}_{,\mathbf{v}}$ is used to denote the *Jacobian matrix of nodal deflections* $\boldsymbol{\xi}$ with respect to the element deformations $\mathbf{v}$, at the state before the control loading is applied:

$$\boldsymbol{\xi}_{,\mathbf{v}} = \sum_{j=1}^{n} \sum_{k=1}^{m} \mathbf{i}_j' \frac{\partial \xi^j}{\partial v^k} \mathbf{i}_k^T \tag{4.13}$$

where $\mathbf{i}_j'$ is the jth column of the nth order identity matrix, and $\mathbf{i}_k$ is the kth column of the mth order identity matrix. Eq.(4.12) shows that matrix $\mathbf{B}^{-T}$ is the Jacobian matrix of nodal deflections $\boldsymbol{\xi}$ with respect to element deformations $\mathbf{v}$, at the state before the control loads are applied.

As discussed in previous chapter, from Eqs.(3.72 and 3.73), the control energy ΔE to run the actuators may be expressed as

$$\Delta E = \mathbf{s}_o^T \Delta\mathbf{v}_o \tag{4.14}$$

where $\mathbf{s}_o$ is the preexisting element forces. Later in the chapter the issue of control energy will be discussed further.

4.2 Observed Response Components and Disturbances

In statically determinate adaptive structures, since only some of the nodal deflection components can be controlled, the sensors are used to monitor the changes in these components. The observed response components are identified by their labels in the list of nodal deflections $\boldsymbol{\xi}$. The determination of the number of observed deflection components p and their labels is very important. The determination of these depend on the objective in controlling the adaptive structure. Here, it is assumed that p number of deflection components are being observed, and their labels in the list of nodal deflections $\boldsymbol{\xi} \in \mathrm{R}^{n\times 1}$, i.e., $\alpha_1, \alpha_2, \cdots, \alpha_p$, are known. Usually

$$p << n \tag{4.15}$$

Since the observed response components are p of those listed in $\boldsymbol{\xi}$, the following also hold:

$$1 \leq \alpha_j \leq n \text{ , for } j = 1, 2, \cdots, p \tag{4.16}$$

Labels $\alpha_1, \alpha_2, \cdots, \alpha_p$ are positive integers.

Let $\Delta\boldsymbol{\xi}^d \in \mathrm{R}^{n\times 1}$ denote the nodal deflection response deviations from the nominal before the application of the current control loading. The deviations in the observed response components are measured by the sensors. These measurements are called disturbances. Let

$$\Delta y_j = \Delta\xi^d_{\alpha_j}, \text{ for } j = 1, 2, \cdots, p \tag{4.17}$$

denote the disturbances. Using the identity of

$$\Delta\xi^d_{\alpha_j} = \mathbf{i}^T_{\alpha_j}\Delta\boldsymbol{\xi}^d \tag{4.18}$$

where $\mathbf{i}_{\alpha_j}$ is the α_jth column of the nth order identity matrix, one can rewrite Eq.(4.17) as

$$\Delta\mathbf{y} = \mathbf{I}^T_{\alpha}\Delta\boldsymbol{\xi}^d \tag{4.19}$$

where

$$\Delta\mathbf{y} = [\Delta y_1, \Delta y_2, \cdots, \Delta y_p]^T \tag{4.20}$$

and

$$\mathbf{I}_{\alpha} = [\mathbf{i}_{\alpha_1}, \mathbf{i}_{\alpha_2}, \cdots, \mathbf{i}_{\alpha_p}] \tag{4.21}$$

Matrix $\mathbf{I}_{\alpha}$ is orthonormal and it identifies the observed response components. Note that $\mathbf{I}_{\alpha} \in (\mathrm{Int})^{n\times p}$. Thanks to the sensors, $\Delta\mathbf{y}$ is available before the control loading is applied.

4.3 Actuators and Controls

To control the p number of observed nodal deflection components, one may use q number of deformation inducing actuators. Since the adaptive structure is statically determinate, $m = n$. In general, $q << m$. The element deformations induced by these actuators are the controls, and they define the nonzero entries of $\Delta\mathbf{v}_o \in R^{m\times 1}$. Let $\beta_1, \beta_2, \cdots, \beta_q$ denote the labels of the nonzero entries of $\Delta\mathbf{v}_o$. One may observe that

$$1 \leq \beta_j \leq m \text{ , for } j = 1, 2, \cdots, q \tag{4.22}$$

The actuator placement problem is to identify the labels $\beta_1, \beta_2, \cdots, \beta_q$.

Once the actuator locations are known, the next problem is to determine how much actuation is needed in each actuator. Let

$$\Delta v_o^{\beta_j} = \Delta u_j \text{ , for } j = 1, 2, \cdots, q \tag{4.23}$$

denote the amount of element deformation to be induced by the jth actuator. The determination of these is the basis of the control problem in statically determinate adaptive structures. The quantities Δu_j , $j = 1, 2, \cdots, q$, are called controls. The complete list of prescribed element deformations $\Delta\mathbf{v}_o$, i.e., the control loading, may be expressed in terms of the controls as

$$\Delta\mathbf{v}_o = \sum_{j=1}^{q} \mathbf{i}_{\beta_j} \Delta u_j \tag{4.24}$$

Defining

$$\Delta\mathbf{u} = [\Delta u_1, \Delta u_2, \cdots, \Delta u_q]^T \tag{4.25}$$

and

$$\mathbf{I}_\beta = [\mathbf{i}_{\beta_1}, \mathbf{i}_{\beta_2}, \cdots, \mathbf{i}_{\beta_q}] \tag{4.26}$$

Eq.(4.24) may rewritten compactly as

$$\Delta\mathbf{v}_o = \mathbf{I}_\beta \Delta\mathbf{u} \tag{4.27}$$

which is the definition of the control loading in terms of controls. Matrix $\mathbf{I}_\beta \in R^{m\times q}$ is orthonormal, and is called the actuator locations matrix. Note that $\Delta\mathbf{v}_o$ is defined by the locations of the actuators, and also by the numerical values of the controls. The determination of these will be discussed in following sections.

4.4 Actuator Placement and Control Problems

It is assumed that p number of deflection components are already identified as the observed deflection components, i.e., matrix $\mathbf{I}_\alpha \in (\text{Int})^{n\times p}$, as defined by Eq.(4.21), is available. Also available is the numerical values of the current disturbances measured by the sensors, i.e., $\Delta\mathbf{y} \in R^{p\times 1}$.

The number of actuators q may be equal to or greater than the number of unobserved deflection components p. If $q \geq p$, then one may have flexibility in controlling the observed deflections. The cases of $q = p$ and $q > p$ are discussed separately in this chapter. In either case, the first problem is the determination of locations of the deformation inducing actuators, i.e., the identification of the actuator location matrix $\mathbf{I}_\beta \in (\mathrm{Int})^{m\times q}$. The second problem is the determination of $\Delta\mathbf{u} \in \mathrm{R}^{q\times 1}$, i.e., the amounts of actuation in each actuator to compensate the disturbance $\Delta\mathbf{y}$.

One may start with the observation that in the current control step, the actuator induced control loading create nodal deflections $\mathbf{I}_\alpha^T\Delta\boldsymbol{\xi}$ in the observed deflection directions that cancel out the measured disturbances $\Delta\mathbf{y}$, i.e.:

$$\mathbf{I}_\alpha^T\Delta\boldsymbol{\xi}+\Delta\mathbf{y} = \mathbf{o} \tag{4.28}$$

By premultiplying both sides of the inverse relation given in Eq.(4.7) by $\mathbf{I}_\alpha^T$ then using $\mathbf{I}_\alpha^T\Delta\boldsymbol{\xi}$ from Eq.(4.28), and $\Delta\mathbf{v}_o$ from Eq.(4.27), one may obtain from Eq.(4.7):

$$\left.\begin{aligned} \Delta\mathbf{y} &= \mathbf{H}\Delta\mathbf{u} \\ \mathbf{H} &= [\mathbf{I}_\alpha^T\mathbf{B}^{-T}\mathbf{I}_\beta] \end{aligned}\right\} \tag{4.29}$$

where $\mathbf{H} \in \mathrm{R}^{p\times q}$. One may observe that matrix $\mathbf{H}$ is obtained by retaining rows $\alpha_1, \alpha_2, \cdots, \alpha_p$ and columns $\beta_1, \beta_2, \cdots, \beta_q$, and deleting everything else in matrix $\mathbf{B}^{-T}$. Although we know $\alpha_1, \alpha_2, \cdots, \alpha_p$, hence $\mathbf{I}_\alpha$ from Eq.(4.21), we are yet to find $\beta_1, \beta_2, \cdots, \beta_q$ for the definition of $\mathbf{I}_\beta$ in Eq.(4.26).

For the moment, suppose $\beta_1, \beta_2, \cdots, \beta_q$ are known, and investigate the solvability of $\Delta\mathbf{u}$, from Eq.(4.29). As discussed in Section 3.6, the following cases can be identified:

1. If $q = p$ and $\det(\mathbf{H}) \neq 0$, then the controls are unique.

2. If $q > p$ and $\mathbf{H}$ is not rank deficient, then there are infinitely many control possibilities.

3. If $p > q$ but $\Delta\mathbf{y}$ is in the column space of $\mathbf{H}$, then the controls are unique.

4. In all other cases, there are no controls to satisfy Eq.(4.28) exactly, i.e., the adaptive structure is *not controllable.*

Case #3 is hypothetical, since it relies on the numerical value of current $\Delta\mathbf{y}$. This implies that each time $\Delta\mathbf{y}$ changes, the location of the actuators will also change. Since the actuators are part of the hardware of an adaptive structure, it is highly unlikely that all the random disturbances $\Delta\mathbf{y}$ affecting the adaptive structure can be in the fixed column space defined by the locations of the actuators already placed in the hardware. Case #4 is not a concern in this book. The book aims for controllable adaptive structures. The scenarios for approximate control are too many to fit in this short work. Therefore only Case #1 and Case #2 are discussed further in the following two sections.

4.5 Actuator Placement and Control Computation When $q = p < n$

In this case, the location of actuators should be such that for any disturbance $\Delta\mathbf{y}$, the control should be as effective as possible. Because there is no prior knowledge about $\Delta\mathbf{y}$, the placement should be done by expecting the worst. One criterion for the placement problem may be the minimum deformation inducement by the actuators to compensate the disturbances. This will minimize the energy to drive the actuators regardless of the values of preexisting element forces $\grave{\mathbf{s}}_o$.

The effect of preexisting element forces $\grave{\mathbf{s}}_o$ on the actuator placement problem is not discussed here. The arguments above are valid for the cases when (a) $\mathbf{s}_o = \mathbf{o}$, and (b) $\grave{\mathbf{s}}_o \neq \mathbf{o}$, but may change at each control step. The case where $\grave{\mathbf{s}}_o$ is nonzero but constant during the control steps is an interesting one; however, it is not considered here.

4.5.1 Actuator Placement Problem

By taking the norms of both sides of Eq.(4.29) and then using the Cauchy-Schwarz inequality, one may write

$$||\mathbf{H}|| \; ||\Delta\mathbf{u}|| > ||\Delta\mathbf{y}|| \tag{4.30}$$

which states that $||\Delta\mathbf{u}||$ is minimum, if $||\mathbf{H}||$ is as large as possible. Instead of directly maximizing $||\mathbf{H}||$, one may take advantage of the identity[1]

$$||\mathbf{H}|| \geq |\det(\mathbf{H})|^{1/p} \tag{4.31}$$

and maximize $d = |\det(\mathbf{H})|$ for identifying $\mathbf{I}_\beta$. In this way, the robustness of the control is also guaranteed, since the solution of $\Delta\mathbf{u}$ from Eq.(4.29) is most robust, when d is the largest; i.e., d is a measure of controllability. Such $\mathbf{H}$ may be obtained by applying the *algorithm of Gauss elimination with partial pivot search by interchanging columns*[2] to matrix

$$\grave{\mathbf{H}} = \mathbf{I}_\alpha^T \mathbf{B}^{-T} \tag{4.32}$$

The algorithm transforms, by elementary row operations,

$$\begin{bmatrix} \grave{\mathbf{H}} \\ labels \end{bmatrix} = \begin{bmatrix} \underbrace{\begin{matrix} \mathbf{h}_1 & \mathbf{h}_2 & \cdots & \mathbf{h}_p \\ 1 & 2 & \cdots & p \end{matrix}} & \underbrace{\begin{matrix} \mathbf{h}_{p+1} & \cdots & \mathbf{h}_n \\ p+1 & \cdots & n \end{matrix}} \end{bmatrix} \tag{4.33}$$

---

[1]Since $|\det(\mathbf{H})| = \Pi_{i=1}^{p} |\lambda_i|$ where λ_i is the ith eigenvalue of $\mathbf{H}$, and since $||\mathbf{H}|| \geq |\lambda_i|$ for any i, the identity follows.

[2]See lines 1-13 of Algorithm 14.113 on pp. 507 of *Elementary Structural Analysis, 4th ed.*, Utku, Norris, Wilbur, McGraw-Hill Inc., 1991, where n is used for p, and m is used for n.

into its upper Hessenberg form

$$\left[\begin{array}{c}\grave{\mathbf{H}}' \\ labels\end{array}\right] = \left[\begin{array}{cccc} \underbrace{\begin{array}{cccc}\mathbf{h}'_{\beta_1} & \mathbf{h}'_{\beta_2} & \cdots & \mathbf{h}'_{\beta_p} \\ \beta_1 & \beta_2 & \cdots & \beta_p\end{array}} & \underbrace{\begin{array}{ccc}\mathbf{h}'_{\beta'_{p+1}} & \cdots & \mathbf{h}'_{\beta'_n} \\ \beta'_{p+1} & \cdots & \beta'_n\end{array}}\end{array}\right] \tag{4.34}$$

where $[\mathbf{h}'_{\beta_1}, \mathbf{h}'_{\beta_2}, \cdots, \mathbf{h}'_{\beta_p}]$ is upper triangular, and

$$\left|\det([\mathbf{h}_{\beta_1}, \mathbf{h}_{\beta_2}, \cdots, \mathbf{h}_{\beta_p}])\right| = \left|\det([\mathbf{h}'_{\beta_1}, \mathbf{h}'_{\beta_2}, \cdots, \mathbf{h}'_{\beta_p}])\right| = \Pi_{i=1}^{p} \left|h'_{i\beta_i}\right| \tag{4.35}$$

is not smaller than any of those corresponding to other $(\beta_1, \beta_2, \cdots, \beta_p)$ column combinations. Namely, the process identifies the labels β_1, β_2, $\cdots$, β_p which are required for the definition of matrix $\mathbf{I}_\beta$ in Eq.(4.26). The computational effort for this is proportional to $O(p^2 n)$. Since there are $\frac{n!}{p!(n-p)!}$ possibilities in selecting p columns out of the n columns of $\grave{\mathbf{H}} \in R^{p\times n}$ and the cost of computing the determinant of each is $O(p^3)$, the Gaussian algorithm with partial pivot search is a remarkable success.[3] Once the locations of the actuators are identified by labels $\beta_1, \beta_2, \cdots, \beta_p$, then matrix $\mathbf{I}_\beta$ is available, and the actuators can be physically placed.

4.5.2 Computation of Controls

For the computation of controls $\Delta\mathbf{u}$, corresponding to the current disturbance $\Delta\mathbf{y}$, Eq.(4.29) may be used. Since the controlled deflection components are identified and the locations of actuators are known, with the knowledge of $\mathbf{B}^{-T}$, matrix

$$\mathbf{H} = \mathbf{I}_\alpha^T \mathbf{B}^{-T} \mathbf{I}_\beta^T \tag{4.36}$$

can be computed. Using $\mathbf{H} \in R^{p\times p}$, Eq.(4.29) may be reproduced

$$\mathbf{H}\Delta\mathbf{u} = \Delta\mathbf{y} \tag{4.37}$$

The solution of this equation for $\Delta\mathbf{u}$ may be displayed as

$$\Delta\mathbf{u} = \mathbf{G}\Delta\mathbf{y} \tag{4.38}$$

where

$$\mathbf{G} = \mathbf{H}^{-1} \tag{4.39}$$

Note that in this case $\mathbf{G} \in R^{p\times p}, \mathbf{H} \in R^{p\times p}$, $\Delta\mathbf{y} \in R^{p\times 1}$, $\Delta\mathbf{u} \in R^{p\times 1}$, and the actuator placement is achieved by a criterion of robustness in control.

Since $q = p$, there is no actuator selection problem, i.e., the problem of deciding which p of the q actuators should be used for compensating the current $\Delta\mathbf{y}$ does not exist.

---

[3] Instead of partial pivot search one may use complete pivot search. See pp. 212-214, the *Algebraic Eigenvalue Problem*, J. H. Wilkinson, Oxford University Press, 1965.

4.6 Actuator Selection and Control Computation When $p < q = n$

Actuator placement problem in this case does not exist, since there is already an actuator in each of the possible n locations. However, there is the problem of selecting which p of the n actuators needs to be activated in order to compensate the current $\Delta\mathbf{y}$. In general, this is a computationally very intensive operation. One may activate appropriate p of the available n actuators, in order to eliminate the current disturbance $\Delta\mathbf{y}$ with least energy. Which p of the n actuators should be used in creating the current control loading is the *actuator selection problem*. Since, according to Eq.(4.14), the energy needed to run the actuators depends on the preexisting element forces $\grave{\mathbf{s}}_o$, the actuator selection problem should take them into account. Here only the $\grave{\mathbf{s}}_o = \mathbf{o}$ case is considered.

4.6.1 Actuator Selection Problem

A possible way of handling this problem is as follows.[4] If $q = n$, then one may order the actuators with their effectiveness in compensating current $\Delta\mathbf{y}$, assuming that the actuators are activated one at a time. From Eq.(4.29) one may write

$$\Delta\mathbf{y} = \grave{\mathbf{H}}\Delta\mathbf{u} \tag{4.40}$$

where

$$\grave{\mathbf{H}} = \mathbf{I}_\alpha^T\mathbf{B}^{-T} \tag{4.41}$$

Note that $\grave{\mathbf{H}} \in R^{p\times n}$. One may rewrite Eq.(4.40) as

$$\Delta\mathbf{y} = \sum_{j=1}^{n}\mathbf{h}_j\Delta u_j \tag{4.42}$$

If only the jth actuator is active, its control Δu_j may be computed from Eq.(4.42), by premultiplying both sides by $\mathbf{h}_j^T$ and then solving, as $\Delta u_j = c_j$ where

$$c_j = \mathbf{h}_j^T\Delta\mathbf{y}/\mathbf{h}_j^T\mathbf{h}_j \tag{4.43}$$

By computing c_j for each of the n actuators, and then ordering them with their magnitudes as

$$\left|c_{\beta_1'}\right| \geq \left|c_{\beta_2'}\right| \geq \cdots \geq \left|c_{\beta_n'}\right| \tag{4.44}$$

one may find the location of the most effective first actuator for the current $\Delta\mathbf{y}$ as β_1'. Then one may repeat the operation using $\Delta\mathbf{y} - c_{\beta_1'}\mathbf{h}_{\beta_1'}$ in

[4]The solution described here is equivalent to identifying those basis vectors from the n columns of $\grave{\mathbf{H}} \in R^{p\times n}$ that can span current $\Delta\mathbf{y} \in R^{p\times 1}$.

TABLE 4.1. Algorithm for Identifying p Actuators out of Possible n

No	Statement
1	$\mathbf{a} \leftarrow \Delta\mathbf{y};\ k \leftarrow 0$
2	*for* $i = 1$ to n
3	$\quad b = \mathbf{h}_i^T \mathbf{h}_i$
4	$\quad$ *if* $b < \varepsilon$ *then next* i
5	$\quad c \leftarrow \mathbf{h}_i^T \mathbf{a}/b$
6	$\quad \beta_i \leftarrow i;\ k \leftarrow k+1$
7	$\quad$ *for* $j = i+1$ to n
8	$\qquad b = \mathbf{h}_j^T \mathbf{h}_j$
9	$\qquad$ *if* $b < \varepsilon$ *then next* j
10	$\qquad c' \leftarrow \mathbf{h}_j^T \mathbf{a}/b$
11	$\qquad$ *if* $\lvert c' \rvert > \lvert c \rvert$ *then* $(c \leftarrow c';\ \beta_i \leftarrow j)$
12	$\quad$ *if* $k \geq p$ *then exit*
13	$\quad \mathbf{a} \leftarrow \mathbf{a} - c\mathbf{h}_{\beta_i};\ \mathbf{h}_{\beta_i} \leftarrow \mathbf{o}$
	ε *is zero − number criterion.* **H** *is destroyed.*

Eq.(4.43) for $\Delta\mathbf{y}$, and thus one may obtain the location of the most effective second actuator for the current $\Delta\mathbf{y}$, similarly. Doing this operation p times would yield which of the p actuators should be used for the current $\Delta\mathbf{y}$. In algorithmic form, the process may be stated as in Table 4.1. Such a scheme would yield $\beta_1, \beta_2, \cdots, \beta_p$, as the locations of the preferred actuators for compensating the current $\Delta\mathbf{y}$.

4.6.2 Computation of Controls in Selected Actuators

The computation of the controls of the step may be performed by using Eq.(4.36) which requires the computation of **H**, according to Eq.(4.34), and its inverse. Note that at every control step a new matrix **H** must be generated. Since the computations are done in real time, the parallelization of the algorithm described in Section 4.10 may prove useful for this purpose.

4.7 Actuator Placement and Control When $p < q < n$

If $q < n$, then one cannot use the algorithm of the previous section, since the actuator locations $\beta_1, \beta_2, \cdots, \beta_p$, determined by the algorithm of Table 4.1 may not coincide with the locations of the q actuators already in the structure. In this case there are two problems: (a) How to determine the locations of the q actuators, and (b) which p of the q actuators should be used for the current $\Delta\mathbf{y}$. These two problems are discussed in the following subsections, assuming that the preexisting element forces are either zero or they are nonzero but varying randomly from step to step.

4.7.1 Actuator Placement Problem

Since $q < n$, the actuator locations cannot be determined as in the previous section. However, the method used earlier for $p = q < n$ case may be extended for the present case. From Eq.(4.29), as before, one may write

$$\Delta \mathbf{y} = \mathbf{H}\Delta \mathbf{u} \tag{4.45}$$

where

$$\mathbf{H} = \mathbf{I}_\alpha^T \mathbf{B}^{-T} \mathbf{I}_\beta \tag{4.46}$$

In the present case $\mathbf{H} \in R^{p\times q}$. The solution of $\Delta\mathbf{u}$ from Eq.(4.29) is most robust when the magnitude of the pth order minor of $\mathbf{H}$, used for the *special solution* of the general solution of Eq.(4.45), is the largest.[5] After determining the p columns of $\mathbf{H}$ associated with the largest determinant, as before by the algorithm of Gauss elimination with partial pivot search by interchanging columns, one may choose additional $q-p$ columns as the replacements of the pth column such that the determinant will be as large as possible. The algorithm will transform matrix $\grave{\mathbf{H}}$ and its column labels as shown in Eq.(4.33) in p steps into its upper Hessenberg form and its column labels as shown in Eq.(4.34) and reproduced below

$$\begin{bmatrix} \grave{\mathbf{H}}' \\ labels \end{bmatrix} = \begin{bmatrix} \underbrace{\begin{matrix} \mathbf{h}'_{\beta_1} & \mathbf{h}'_{\beta_2} & \cdots & \mathbf{h}'_{\beta_{p-1}} & \mathbf{h}'_{\beta_p} \\ \beta_1 & \beta_2 & \cdots & \beta_{p-1} & \beta_p \end{matrix}} & \underbrace{\begin{matrix} \mathbf{h}'_{\beta'_{p+1}} & \cdots & \mathbf{h}'_{\beta'_n} \\ \beta'_{p+1} & \cdots & \beta'_n \end{matrix}} \end{bmatrix} \tag{4.47}$$

As before $[\mathbf{h}'_{\beta_1}, \mathbf{h}'_{\beta_2}, \cdots \mathbf{h}'_{\beta_{p-1}}, \mathbf{h}'_{\beta_p}]$ is upper triangular, and its determinant is the largest in magnitude. In order to determine the locations of the remaining $q-p$ actuators, one may reorder the last $n-p$ columns of matrix $\grave{\mathbf{H}}'$ by the magnitudes of the entries in the pth row. Such a process will transform the right-hand side of Eq.(4.47) into

$$\begin{bmatrix} \underbrace{\begin{matrix} \mathbf{h}'_{\beta_1} & \mathbf{h}'_{\beta_2} & \cdots & \mathbf{h}'_{\beta_{p-1}} & \mathbf{h}'_{\beta_p} \\ \beta_1 & \beta_2 & \cdots & \beta_{p-1} & \beta_p \end{matrix}} & \underbrace{\begin{matrix} \mathbf{h}'_{\beta_{p+1}} & \cdots & \mathbf{h}'_{\beta_q} & \cdots & \mathbf{h}'_{\beta_n} \\ \beta_{p+1} & \cdots & \beta_q & \cdots & \beta_n \end{matrix}} \end{bmatrix} \tag{4.48}$$

---

[5] General solution of $\Delta\mathbf{u}$ from Eq.(4.45) may be expressed as

$$\Delta\mathbf{u} = \mathbf{C_H}\Delta\mathbf{x} + \mathbf{C}'_{\mathbf{H}}\Delta\mathbf{y}$$

where the second term on the right-hand side is the *special solution*, $\Delta\mathbf{x} \in R^{(q-p)\times 1}$ is the vector of arbitrary constants, columns of $\mathbf{C_H}$ constitute a basis in the null space of $\mathbf{H}$, and $\mathbf{C}'_{\mathbf{H}}$ is a generalized inverse of $\mathbf{H}$ (see, for example, Art. 14.1.3 of *Elementary Structural Analysis, 4th ed*, Utku, Norris, Wilbur, McGraw-Hill, Inc., 1991).

Let $\mathbf{H}_p, \mathbf{H}_{p+1}, \cdots, \mathbf{H}_q$ denote the pth order upper triangular matrices defined by augmenting matrix $[\mathbf{h}'_{\beta_1}, \mathbf{h}'_{\beta_2}, \cdots \mathbf{h}'_{\beta_{p-1}}] \in R^{p\times(p-1)}$ by columns $\mathbf{h}'_{\beta_p}, \mathbf{h}'_{\beta_{p+1}}, \cdots, \mathbf{h}'_{\beta_q}$, respectively. Clearly, because of the column reordering, the following holds

$$\det(\mathbf{H}_p) \geq \det(\mathbf{H}_{p+1}) \geq \cdots \geq \det(\mathbf{H}_q) \tag{4.49}$$

Thus, labels $(\beta_1, \beta_2, \cdots, \beta_{p-1}, \beta_p)$, $(\beta_1, \beta_2, \cdots, \beta_{p-1}, \beta_{p+1}), \cdots$ $(\beta_1, \beta_2, \cdots, \beta_{p-1}, \beta_q)$ will refer to the pth order submatrices of $\grave{\mathbf{H}}$ which are associated with the largest determinants in magnitude (with the limitations of partial pivoting). The label sequence $\beta_1, \beta_2, \cdots, \beta_p, \cdots, \beta_q$ define matrix $\mathbf{I}_\beta$ and thus the physical locations of the actuators.

4.7.2 Selection of Actuators and Computation of Controls

There are q actuators; which p of them should be activated in order to compensate the current $\Delta\mathbf{y}$? This problem is studied in the previous section for $q = n$. The solution algorithm developed there can also be used for $q < n$. By replacing n with q, the algorithm in Table 4.1 would yield the label sequence $\beta_1, \beta_2, \cdots, \beta_p$ which identifies matrix $\mathbf{I}_\beta$, and, therefore, the best p actuators to be used to compensate the current $\Delta\mathbf{y}$.

The computation of the controls may be performed by using Eq.(4.38) which requires the computation of $\mathbf{H}$ according to Eq.(4.36), and its inverse. Note that at every control step a new matrix $\mathbf{H}$ must be generated. This may require the parallelization of the algorithm in Section 4.10.

4.8 Precision Control

Especially in space applications, adaptive structures may provide supports to sensitive instruments, such as optical interferometers, or optical collectors. The deflections of the nodes that are used as supports for such instruments must be kept at the close proximity of their nominal values at all times when the instruments are operating. For this purpose, suppose p number of deflection components are being observed continuously and controlled by q number of actuators. From the mathematical description of the statically determinate adaptive structure, matrix $\mathbf{B}$ and, therefore, $\mathbf{B}^{-T}$ are available. By the procedures of the previous three sections, one can find the best locations for the necessary q actuators, and the best p actuators to compensate current disturbance $\Delta\mathbf{y}$. With these, matrix $\mathbf{H}$ of Eq.(4.36) and its inverse $\mathbf{G}$ may be computed. The controls $\Delta\mathbf{u}$ necessary to compensate the disturbance $\Delta\mathbf{y}$ may be computed from Eq.(4.38) as $\Delta\mathbf{u} = \mathbf{G}\Delta\mathbf{y}$.

The measurement of the disturbances, and the computation of controls to compensate them, may be placed in a control loop as shown in Fig.(4.1).

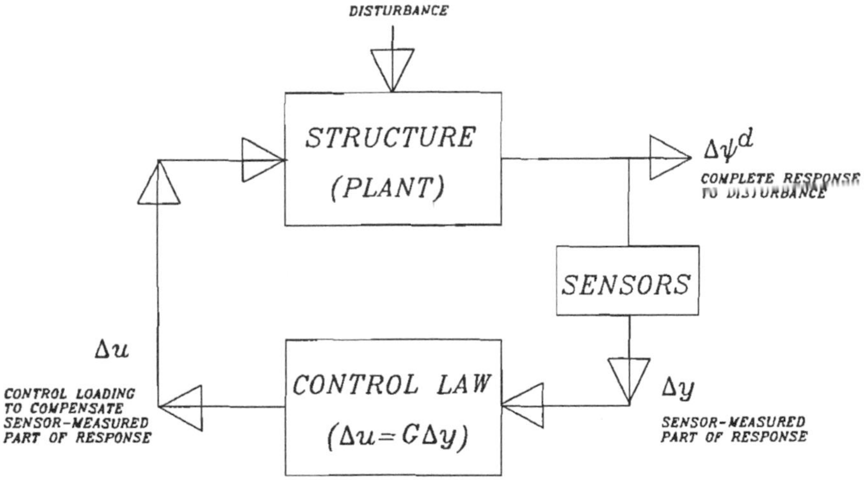

FIGURE 4.1. Static direct output feedback control for precision

Because of the requirement that no accelerations should be present in the structure, not only the disturbances but also the controls must take place very slowly. In other words, the control frequency should be very small, usually a small multiple of the fundamental vibration frequency of the structure. If the disturbances are also cyclic, their period must also be sufficiently large relative to the fundamental period of the structure itself. In practice, there are many disturbances which take place with very large periods relative to the fundamental period of the structure. For example, the thermal loads on an earth satellite are of low frequency type, relative to the fundamental frequency of the satellite. For such phenomena, the static control described here may be applicable.

4.9 Adaptive Trusses as Slow Moving Mechanical Manipulators

An adaptive structure can become a mechanical manipulator, if vector $\Delta\mathbf{y} \in \mathrm{R}^{p\times 1}$, discussed earlier in this chapter as sensor measured disturbances, is the desired values for p number of the nodal deflection components, and the controls $\Delta\mathbf{u} \in \mathrm{R}^{q\times 1}$ are the means of achieving such response. The relationships between $\Delta\mathbf{y}$ and $\Delta\mathbf{u}$, developed in the earlier sections, are usable for the current problem, if both the desired nodal deflections and the controls to induce them are to cause only negligible inertial forces in the structure. In mechanical manipulators this can be achieved by completing the desired maneuver in a time which is a multiple of the fundamental period of the structure. Such manipulators may be called slow moving. By decreasing the fundamental period of the structure through the use of stiff materials, less mass, and appropriate geometry, the maneuver time may be increased to levels that may be perceived psychologically as fast, although the manipulator is in the slow moving category. In fact many modern industrial robot arms, construction cranes, and construction machinery fall in the slow moving category. They are basically slow moving statically determinate adaptive structures.

The relationships developed in the earlier sections of this chapter for disturbance control can be used here for motion control provided that one is careful to use $-\Delta\mathbf{u}$ for $\Delta\mathbf{u}$, due to the fact that, in the earlier developments, the controls are to eliminate $\Delta\mathbf{y}$, whereas in the current problem they are to induce $\Delta\mathbf{y}$.

The inverse of the excitation-response relationship of statically determinate adaptive structures may be reproduced from Eq.(4.7) as

$$\Delta\boldsymbol{\xi} = -\mathbf{B}^{-T}\Delta\mathbf{v}_o \tag{4.50}$$

where $\Delta\boldsymbol{\xi} \in R^{n\times 1}$ is incremental nodal deflections caused by the prescribed element deformations $\Delta\mathbf{v}_o \in R^{m\times 1}$. Since the adaptive structure is statically determinate, $m = n$. It is discussed in the beginning of this chapter that $\mathbf{B}^{-T} \in R^{n\times m}$ is the Jacobian matrix of the nodal deflections $\boldsymbol{\xi}$ with respect to the element deformations $\mathbf{v}$, i.e., $\boldsymbol{\xi}_{,\mathbf{v}}$, as stated by Eq.(4.12). In the absence of preexisting element forces $\mathring{\mathbf{s}}_o$, according to Eq.(4.50), the p components of $\Delta\boldsymbol{\xi}$ can take desired values if q components of $\Delta\mathbf{v}_o$ are assigned appropriate values by means of q actuators. In the absence of preexisting element forces $\mathring{\mathbf{s}}_o$, e.g., as in space trusses, no energy is required to run the actuators. It may be possible to compute the appropriate values of the q components of $\Delta\mathbf{v}_o$, corresponding to the desired values of the p components of $\Delta\boldsymbol{\xi}$, when $p \leq q$. In slow moving mechanical manipulators, the *trajectory planning* provides the number p and the desired values of p components of $\Delta\boldsymbol{\xi}$. Since $\mathbf{B}^T$ is not rank deficient, one needs $q = m = n$ actuators to maneuver an adaptive structure as a mechanical manipulator when $p = n$. The quantity p is the measure of *dexterity* of the mechanical manipulator. Adaptive structures can be very dexterous. They are not limited to the cantilevers, i.e., the open loop structures favored by the modern robot arms.[6,7,8]

---

[6] "Control of a Slow-Moving Space Crane as an Adaptive Structure," S. Utku, A. V. Ramesh, S. K. Das, B. K. Wada, and G. S. Chen, *AIAA Journal*, vol.29, no.6, pp.961-967, June 1991 (presented as paper 89-1286 at the 30th Structures, Structural Dynamics, and Materials Conference, Mobile, Alabama, April 1989).

[7] "Real-time Control of Geometry and Stiffness in Adaptive Structures," A. V. Ramesh, S. Utku, B. K. Wada, *Computer Methods in Applied Mechanics and Engineering 90* (1991) pp.761-779, North-Holland.

[8] *Geometry Control in Adaptive Truss Structures*, Anapathur Viswanathan Ramesh, Doctoral Dissertation, Jan. 1991, Department of Civil and Environmental Engineering, Duke University, Durham, NC.

The function of the mechanical manipulator determines which p components of $\Delta\boldsymbol{\xi}$ are to be controlled. Let $\alpha_1, \alpha_2, \cdots, \alpha_p$ denote the labels of the components of $\Delta\boldsymbol{\xi}$ which are to take, at the end of the current control step, the values, determined by the trajectory planners. Let $\Delta\mathbf{y} \in R^{p\times 1}$ denote these components. Then one may write

$$\Delta\mathbf{y} = \mathbf{I}_\alpha^T \Delta\boldsymbol{\xi} \tag{4.51}$$

where

$$\mathbf{I}_\alpha = [\mathbf{i}_{\alpha_1}, \mathbf{i}_{\alpha_2}, \cdots, \mathbf{i}_{\alpha_p}] \tag{4.52}$$

In the last equation, $\mathbf{i}_{\alpha_j}$ is the α_jth column of the nth order identity matrix.

In the current control step, in order to induce the nodal deflections $\Delta\mathbf{y}$, one may use q number of deformation inducing actuators. These actuators are to define the q components of the prescribed element deformations $\Delta\mathbf{v}_o$. The problem of determining which q components of $\Delta\mathbf{v}_o$ are to be created depends on the best way to induce $\Delta\mathbf{y}$. This is the actuator placement problem. This problem will be dealt with later in this section. Presently, suppose that components $\beta_1, \beta_2, \cdots, \beta_q$ are chosen for actuator locations, and the controls Δu_j, $j = 1, 2, \cdots, q$ are

$$\Delta v_o^{\beta_j} = \Delta u_j \text{ , for } j = 1, 2, \cdots, q \tag{4.53}$$

and all other $\Delta\mathbf{v}_o$ components are zero. Then, one may write

$$\Delta\mathbf{v}_o = \mathbf{I}_\beta \Delta\mathbf{u} \tag{4.54}$$

where $\Delta\mathbf{u} = [\Delta u_1, \Delta u_2, \cdots, \Delta u_q]$ and

$$\mathbf{I}_\beta = [\mathbf{i}_{\beta_1}, \mathbf{i}_{\beta_2}, \cdots, \mathbf{i}_{\beta_q}] \tag{4.55}$$

In the last equation, $\mathbf{i}_{\beta_j}$ is the β_jth column of the nth order identity matrix.

By premultiplying both sides of Eq.(4.50) by $\mathbf{I}_\alpha^T$, then using $\mathbf{I}_\alpha^T \Delta\boldsymbol{\xi}$ from Eq.(4.51) and $\Delta\mathbf{v}_o$ from Eq.(4.54), one may obtain

$$\Delta\mathbf{y} = -\mathbf{H}\Delta\mathbf{u} \tag{4.56}$$

where

$$\mathbf{H} = \mathbf{I}_\alpha^T \mathbf{B}^{-T} \mathbf{I}_\beta \tag{4.57}$$

Note that $\mathbf{H} \in R^{p\times q}$. When the trajectory planners prescribe $\Delta\mathbf{y}$, Eq.(4.56) may be used to compute the necessary controls $\Delta\mathbf{u}$, provided that the location of the actuators are known, i.e., matrix $\mathbf{I}_\beta$ is available for the computation of $\mathbf{H}$. In other words, one has to solve the actuator placement problem first.

The actuator placement problem is closely related with the energy requirement to drive the actuators. In statically determinate adaptive structures, as stated in Eq.(4.14), this energy is the work against the preexisting element forces $\grave{\mathbf{s}}_o$. Actuator placement problems can be solved without involving the preexisting element forces only in the cases where $\grave{\mathbf{s}}_o = \mathbf{o}$ or it is variable without a definite trend during the control steps. In all other cases one should take $\grave{\mathbf{s}}_o$ into account. In the following brief discussions of the actuator placement problem, $\grave{\mathbf{s}}_o$ is not considered.

4.9.1 Actuator Placement When $p = q < n$

This is the case studied in Section 4.5 for disturbance control. The location of the actuators is determined by applying the algorithm of Gauss elimination with partial pivot search by interchanging columns on matrix

$$\grave{\mathbf{H}} = \mathbf{I}_\alpha^T \mathbf{B}^{-T} \tag{4.58}$$

For a given $\mathbf{B}^{-T}$, this would give the locations of the actuators for a robust control. In disturbance control problems, matrix $\mathbf{B}^{-T}$ is a constant matrix which remains the same in all control steps. However, in the mechanical manipulator problem, $\mathbf{B}^{-T}$ changes from step to step. Unless there are reasons to use a particular $\mathbf{B}^{-T}$ corresponding to a particular configuration of the structure, the method of Section 4.5 cannot be used directly. Here, the actuator placement problem may be solved by numerical optimization using simulation of all expected configurations of the structure and their preexisting element force states. Although interesting, the numerical optimization methods are not considered in this book.

4.9.2 Actuator Selection When $p < q = n$

This is the case studied in Section 4.6 for disturbance control. Since there is an actuator in each of the n possible locations, the actuator placement problem does not exist. However, at each control step which p of the n actuators is to be used, i.e., the *actuator selection problem* exists. For this, in the absence of preexisting element forces $\grave{\mathbf{s}}_o$, one may use the algorithm given in Table 4.1, which determines the actuators to be used in a step by identifying the ones that would contribute best towards the realization of the prescribed deflections $\Delta\mathbf{y}$ of that step.

4.9.3 Actuator Placement When $p < q < n$

This is the case studied in Section 4.7 for disturbance control. The location of the actuators is determined by applying the algorithm of Gauss elimination with partial pivot search by interchanging columns on matrix $\grave{\mathbf{H}}$. For a given $\mathbf{B}^{-T}$, this would give the locations of the actuators for a robust control. In disturbance control problems, matrix $\mathbf{B}^{-T}$ is a constant matrix which remains the same in all control steps. However, in the mechanical manipulator problem, $\mathbf{B}^{-T}$ changes from step to step. Unless there are reasons to use a particular $\mathbf{B}^{-T}$ corresponding to a particular configuration of the structure, the method of Section 4.5 cannot be used directly. Here, the actuator placement problem may be solved by numerical optimization using simulation of all expected configurations of the structure and their preexisting element force states. Although interesting, the numerical optimization methods are not considered.

Once the locations of the q actuators are determined, the actuator selection problem, i.e., which p of the available q actuators is to be used for the current $\Delta\mathbf{y}$, may be handled as in the previous subsection. Here, too, one may use the algorithm of Table 4.1 by substituting q for n.

4.9.4 Computation of Controls

Once one determines which p of the q actuators is to be used for inducing the current desired nodal deflections $\Delta\mathbf{y}$, there is the problem of computing the necessary controls $\Delta\mathbf{u}$ to do the job. The solution of the actuator selection problem yields the labels $\beta_1, \beta_2, \cdots, \beta_p$ of the actuators to be used in the current step, i.e., $\mathbf{I}_\beta$ is available. With this, one may compute matrix $\mathbf{H} \in R^{p\times p}$ using Eq.(4.57), and obtain $\Delta\mathbf{u}$ from Eq.(4.56) as

$$\Delta\mathbf{u} = \mathbf{G}\Delta\mathbf{y} \tag{4.59}$$

where

$$\mathbf{G} = -\mathbf{H}^{-1} \tag{4.60}$$

Matrix $\mathbf{G}$ always exists, since the labels $\beta_1, \beta_2, \cdots, \beta_p$ are selected to ensure the nonsingularity of $\mathbf{H}$.

4.10 Generation of Output-Control Matrix

In Subsection 2.2.2, the generation of matrix $\mathbf{B}$ is discussed. Because of its importance in the actuator placement, actuator selection, and control computation problems, one needs to discuss this subject further. In precision control problems, one needs to generate matrix $\mathbf{B}$ and its inverse once, in order to place the actuators correctly in the structure, and to select them and compute the controls at every control cycle during the operations. In mechanical manipulator problems, one needs the generation of matrix $\mathbf{B}$ and its inverse for a multitude of configurations of the structure in the *work space* of the manipulator. This is required in order to place the actuators correctly during the design phase, and in order to select the active actuators and compute their controls during the operation phase. When the generation of matrix $\mathbf{B}$ and its inverse is needed more than once, either in batch mode during the design, or in real time during the operations, it needs to be engineered.

In actuator placement, actuator selection, and control computation problems, one needs *output-control* matrix $\grave{\mathbf{H}}$, not matrix $\mathbf{B}$. This matrix is defined in Eq.(4.41) as the p number of rows of matrix $\mathbf{B}^{-T}$, i.e., the rows associated with the controlled components of nodal deflections (or observed components in precision control problems):

$$\grave{\mathbf{H}} = \mathbf{I}_\alpha^T \mathbf{B}^{-T} \tag{4.61}$$

Matrix $\mathbf{I}_\alpha$ identifies the observed deflection components in precision control problems, and the controlled deflection components in the manipulator problems. For statically determinate adaptive structures $m = n$, and matrix $\mathbf{B} \in \mathrm{R}^{n\times m}$. For an N node, M element structure with e degrees of deflection freedom per node, a number element forces per element, b number of deflection constraints, and f number of zeroed element force components, one may observe that $n = Ne - b$, and $m = Ma - f$.

As discussed in Subsection 2.2.2, matrix $\mathbf{B}$ is very sparse, and may be easily generated from the global descriptions of the position vectors of nodes $\grave{\mathbf{X}} \in \mathrm{R}^{3\times N}$, connectivity matrix $\mathbf{J} \in \mathrm{R}^{2\times M}$, and the data defining element force and nodal deflection constraints. These data need a storage of the order $O(5M)$, whereas matrix $\mathbf{B}$ needs a storage of the order of $O(e^2M^2)$. The computational effort to generate the matrix from the data is of the order of $O(M)$.

With the above observations, one may look for algorithms that can generate matrix $\grave{\mathbf{H}} \in \mathrm{R}^{p\times n}$ directly from the *basic data* available in an $O(5M)$ storage. Such a method may be devised by taking advantage of the *extended method of joints method*,[9] and the fact that matrix $\mathbf{B}$ is the coefficient matrix of element forces in the nodal equilibrium equations

$$\mathbf{Bs} = \mathbf{p} \tag{4.62}$$

where $\mathbf{B} \in \mathrm{R}^{n\times n}$ and it is not rank deficient.

Starting with the identity, $\mathbf{BB}^{-1} = \mathbf{I}$ and postmultiplying both sides with matrix $\mathbf{I}_\alpha$, one may write

$$\mathbf{BH}' = \mathbf{I}_\alpha \tag{4.63}$$

where

$$\mathbf{H}' = \mathbf{B}^{-1}\mathbf{I}_\alpha \tag{4.64}$$

From Eqs.(4.61) and (4.64), one may observe that

$$\mathbf{H}' = \grave{\mathbf{H}}^T \tag{4.65}$$

Since

$$\mathbf{I}_\alpha = [\mathbf{i}_{\alpha_1}, \mathbf{i}_{\alpha_2}, \cdots, \mathbf{i}_{\alpha_p}] \tag{4.66}$$

---

[9] This is a generalized version of Henneberg's method mentioned on p.139 of *Elementary Structural Analysis, 4th ed.*, Utku, Norris, Wilbur, McGraw-Hill, Inc., 1991.

where $\mathbf{i}_{\alpha_j}$ is the α_jth column of the nth order identity matrix, and matrix $\mathbf{H}'$ may be displayed by its columns as:

$$\mathbf{H}' = [\mathbf{h}'_1, \mathbf{h}'_2, \cdots, \mathbf{h}'_p] \tag{4.67}$$

one may rewrite Eq.(4.63) as

$$\mathbf{B}\mathbf{h}'_j = \mathbf{i}_{\alpha_j}, \quad \text{for } j = 1, 2, \cdots, p \tag{4.68}$$

The comparison of this with Eq.(4.62) reveals that the jth column of $\mathbf{H}'$, or the jth row of $\grave{\mathbf{H}}$ according to Eq.(4.65), is the element forces when the statically determinate adaptive structure is subjected as unit load in the direction of the α_jth nodal deflection component. By repeating this unit load analysis for α_j, $j = 1, 2, \cdots, p$, one may obtain matrix $\grave{\mathbf{H}}$, row by row.

For the computation of element forces for the unit loading described in the previous paragraph, one may use the *extended method of joints*. By proper use of node and element labels, matrix $\mathbf{B}$ may be made block lower triangular, as in *simple* trusses, or *almost* block lower triangular with occasional nonzero entries in upper off-diagonal blocks, as in *compounded* or *complex* trusses.[10] Blocks are of order $e \times e$ or smaller. If matrix $\mathbf{B}$ can be made block lower triangular, one can solve $\mathbf{s}^1$ from the first block of equations, and $\mathbf{s}^2$ from the second block of equations after substituting $\mathbf{s}^1$. In this manner, all of the element forces may be computed with an effort proportional to $O(Me^2)$, and a storage requirement of $O(Me^2)$, without ever actually generating $\mathbf{B}$, but instead using its basic blocks of $\mathbf{T}^k_l$, $l = 1, 2$; $k = 1, \cdots, M$ (discussed in Subsection 2.2.2, and defined in Tables 2.2 and 2.3). The process requires also the *basic data*. By scanning the connectivity matrix $\mathbf{J}$ and the constraint data, one may find the order in which the element forces can be computed *serially*, one block at a time. This is the usual *method of joints*. The extension comes into play when there are nonzero entries in the upper off-diagonal blocks. It is basically the *Henneberg method*.[11] Its generalization is explained below.

Suppose the usual method of joints cannot be continued at some stage, due to the presence of g number of nonzero entries in upper off-diagonal blocks of the current block of equations. The scanning of the connectivity matrix $\mathbf{J}$, the basic data, and the bookkeeping of already computed element forces enable one to find the block of equations with smallest g. Let $\mathbf{s}^g \in \mathrm{R}^{g\times 1}$ denote the true values of the element forces where g is the number of nonzero entries in upper off-diagonal blocks of the current block of equations. In the *generalized Henneberg algorithm*, one makes $g+1$ linearly independent guesses ($\mathbf{s}^g_j$, $j = 1, 2, \cdots, g+1$) $\in \mathrm{R}^{g\times 1}$, for element forces $\mathbf{s}^g$, and for each guess one obtains the equation residues of

---

[10] See, for example, Chapter 4, pp.105-164, *Elementary Structural Analysis, 4th ed.*, Utku, Norris, Wilbur, McGraw-Hill, Inc., 1991.

[11] L. Henneberg, *Static der starren Systeme*, Darmstadt, 1886.

$(\mathbf{r}_j,\ j = 1,2,\cdots,g+1) \in R^{g\times 1}$, in order to establish equilibrium in nodal deflection directions associated with the final g number of equations. Since the guesses and the residues are related linearly, one may write

$$\mathbf{A}^g(\mathbf{s}^g - \mathbf{s}^g_j) = \mathbf{r}_j, \quad for \quad j = 1,2,\cdots,g+1 \tag{4.69}$$

where $\mathbf{A}^g \in R^{g\times g}$ are nonsingular so long as $\mathbf{B}$ is not rank deficient. The aim here is to find the true values $\mathbf{s}^g$. If by chance $\mathbf{r}_j = \mathbf{o}$ for some j, then the corresponding guess $\mathbf{s}^g_j$ is the true value. This can be seen from Eq.(4.69). In general, one may have to use the following procedure for the true value. By subtracting side-by-side the last of $g+1$ equations from the first g equations, one may obtain from Eq.(4.69)

$$\mathbf{A}^g[\mathbf{s}^g_{g+1} - \mathbf{s}^g_1, \cdots, \mathbf{s}^g_{g+1} - \mathbf{s}^g_g] = [\mathbf{r}_{g+1} - \mathbf{r}_1, \cdots, \mathbf{r}_{g+1} - \mathbf{r}_g] \tag{4.70}$$

Then using Eqs.(4.69) and (4.70), one may eliminate $\mathbf{A}^g$, and obtain

$$\mathbf{s}^g = \mathbf{s}^g_1 + [\mathbf{r}_{g+1} - \mathbf{r}_1, \cdots, \mathbf{r}_{g+1} - \mathbf{r}_g]^{-1}[\mathbf{s}^g_{g+1} - \mathbf{s}^g_1, \cdots, \mathbf{s}^g_{g+1} - \mathbf{s}^g_g]\mathbf{r}_1 \tag{4.71}$$

If the inversion operation in this equation fails, as an indication of the usage of linearly dependent guesses, one may simply repeat the process by changing a few of the guesses. In statically determinate structures, g can be of $O(e)$; therefore, the method is practical.[12]

Considering the fact that the actuator selection problem needs to be done at each control cycle, in real time, the method explained above is quite helpful especially for large n and large p. Moreover it is quite amenable for parallelization.[13]

4.11 Recapitulation

Studied in this chapter are the problems associated with static feedback control of p of the n nodal deflection quantities by means of q deformation inducing actuators in statically determinate discrete parameter adaptive structures. The problems include robustness of control, actuator placement, actuator selection, and computation of controls. These problems are studied also when the static control is of the open loop type as in slow moving mechanical manipulators.

---

[12] See, for example, "DETRANS: A Fast and Storage Efficient Algorithm for Static Analysis of Determinate Trusses," A. V. Ramesh, S. Utku, L. Y. Lu, *JPL D-6194*, Jet Propulsion Laboratory, Pasadena, CA, Feb. 1989(also *Journal of Aerospace Engineering, American Society of Civil Engineers*, vol.4, no.3, pp.274-285, July 1991).

[13] See "Parallel Computation of Geometry Control in Adaptive Truss Structures," A. V. Ramesh, S. Utku, and B. K. Wada, *Proceedings of First US National Congress on Computational Mechanics*, Chicago, IL, July 1991.

5

Statically Indeterminate Adaptive Structures

Although statically determinate adaptive structures are preferable for precision control and advantageous in mechanical manipulators, many of the actual structures, because of stiffness and diverse loading requirements, are not statically determinate. Especially in vibration control problems, a structure is rarely statically determinate. Because of these reasons, the statically indeterminate adaptive structures are discussed in this chapter.

5.1 Excitation-Response Relations

These relationships are studied in Chapter 2 for N node, M element structure with a number of element forces per element, e number of deflection degrees of freedom per node, b number of nodal deflection constraints, and f number of zeroed element forces. The linearized incremental equations are summarized in Eq.(2.61). The form which is suitable for the adaptive structures is given in Eqs.(3.1-3.3). They can be restated here as

$$\begin{bmatrix} \cdot & \mathbf{B} & \cdot \\ \mathbf{B}^T & \cdot & -\mathbf{I} \\ \cdot & -\mathbf{I} & \vec{\mathbf{K}} \end{bmatrix} \begin{Bmatrix} \Delta\boldsymbol{\xi} \\ \Delta\mathbf{s} \\ \Delta\mathbf{v} \end{Bmatrix} = \begin{Bmatrix} \Delta\mathbf{p} - \dot{\mathbf{q}} \\ -\Delta\mathbf{v}_o \\ \cdot \end{Bmatrix} \tag{5.1}$$

where the partitioned coefficient matrix on the left represents the structure, $[\Delta\boldsymbol{\xi}^T, \Delta\mathbf{s}^T, \Delta\mathbf{v}^T]^T$ represents the incremental response consisting of the incremental nodal deflection response $\Delta\boldsymbol{\xi} \in R^{n\times 1}$, the incremental element force response $\Delta\mathbf{s} \in R^{m\times 1}$, and the incremental element deformations response $\Delta\mathbf{v} \in R^{m\times 1}$, and the right-hand side represents the incremental excitations consisting of the incremental force-type nodal loads $\Delta\mathbf{p} \in R^{n\times 1}$, deflection-type element loads $\Delta\mathbf{v}_o \in R^{m\times 1}$, and the incremental effect $\grave{\mathbf{q}} \in R^{n\times 1}$ of preexisting element forces $\grave{\mathbf{s}}_o$ on nodal force equilibrium. Submatrix $\mathbf{B} \in R^{n\times m}$ is studied in Subsection 2.2.2, block diagonal submatrix $\vec{\mathbf{K}} \in R^{m\times m}$ is studied in Section 2.4, and $\mathbf{I}$ is the mth order identity matrix. The matrix orders are defined as $n = Ne - b$ and $m = Ma - f$. As studied in Chapter 2, the $(2m + n)$th order coefficient matrix of the left-hand side is nonsingular and symmetric, due to the fact that $\mathbf{B}$ is not rank deficient and $\vec{\mathbf{K}}$ is positive definite.

5.1.1 Relations for Adaptive Structures

It is discussed in Subsection 2.2.3 that, in non-cable type structures, the incremental effect of preexisting element forces $\grave{\mathbf{s}}_o$ on nodal force equilibrium is negligible. In the discussions of this chapter

$$\grave{\mathbf{q}} = \mathbf{o} \tag{5.2}$$

is assumed.

For adaptive structures, one is interested in the response associated with actuator induced element deformations that define the prescribed element deformations $\Delta\mathbf{v}_o$. Using

$$\Delta\mathbf{p} = \mathbf{o} \tag{5.3}$$

and Eq.(5.2) in Eq.(5.1), the latter can be restated as

$$\begin{bmatrix} \cdot & \mathbf{B} & \cdot \\ \mathbf{B}^T & \cdot & -\mathbf{I} \\ \cdot & -\mathbf{I} & \vec{\mathbf{K}} \end{bmatrix} \begin{Bmatrix} \Delta\boldsymbol{\xi} \\ \Delta\mathbf{s} \\ \Delta\mathbf{v} \end{Bmatrix} = \begin{Bmatrix} \cdot \\ -\Delta\mathbf{v}_o \\ \cdot \end{Bmatrix} \tag{5.4}$$

which is the excitation-response relation used in this chapter.

5.1.2 Inverse Relations

The inverse of excitation-response relations is studied in Section 3.1. Using the results obtained by the force method in Eq.(3.33), the inverse relations corresponding to Eq.(5.4) may be written as

$$\begin{Bmatrix} \Delta\boldsymbol{\xi} \\ \Delta\mathbf{s} \\ \Delta\mathbf{v} \end{Bmatrix} = \begin{bmatrix} \mathbf{C}'^T\Delta\mathbf{F}\,\mathbf{C}' & \mathbf{C}'^T\Delta\mathbf{F}\,\vec{\mathbf{K}} & \mathbf{C}'^T\Delta\mathbf{F} \\ \vec{\mathbf{K}}\,\Delta\mathbf{F}\,\mathbf{C}' & -\mathbf{K}_c & -\mathbf{K}_c\,\vec{\mathbf{F}} \\ \Delta\mathbf{F}\,\mathbf{C}' & -\vec{\mathbf{F}}\,\mathbf{K}_c & \Delta\mathbf{F} \end{bmatrix} \begin{Bmatrix} \cdot \\ -\Delta\mathbf{v}_o \\ \cdot \end{Bmatrix} \tag{5.5}$$

where

$$\mathbf{K}_c = \mathbf{C}[\mathbf{C}^T\vec{\mathbf{F}}\,\mathbf{C}]^{-1}\mathbf{C}^T \tag{5.6}$$

$$\mathbf{F}_c = \vec{\mathbf{F}}\,\mathbf{K}_c\,\vec{\mathbf{F}} \tag{5.7}$$

and

$$\left.\begin{array}{l}\Delta\mathbf{F} = \vec{\mathbf{F}} - \mathbf{F}_c \\ \vec{\mathbf{F}} = \vec{\mathbf{K}}^{-1}\end{array}\right\} \tag{5.8}$$

In these relations $\vec{\mathbf{F}} \in \mathrm{R}^{m \times m}$ is the block diagonal matrix of element flexibilities:

$$\vec{\mathbf{F}} = \begin{bmatrix} \mathbf{F}^1 & \cdot & \cdots & \cdot \\ \cdot & \mathbf{F}^2 & \cdot & \cdot \\ \vdots & \vdots & \ddots & \vdots \\ \cdot & \cdot & \cdots & \mathbf{F}^M \end{bmatrix} = diag(\mathbf{F}^1, \cdots, \mathbf{F}^M) = diag(\mathbf{F}^k) \tag{5.9}$$

The diagonal blocks of $\vec{\mathbf{F}}$ are studied in Subsection 2.4.2.

The matrix $\mathbf{C} \in \mathrm{R}^{m \times r}$ appearing in the inverse relations above is defined in Eq.(3.24). Its columns establish a basis in the null space of **B**. It follows from its definition that

$$\mathbf{BC} = \mathbf{0} \tag{5.10}$$

The quantity r, i.e., the number of columns of **C**, is the degree of statical indeterminacy of the structure, which is defined by

$$r = m - n \tag{5.11}$$

The matrix $\mathbf{C}' \in \mathrm{R}^{m \times n}$, also appearing in the inverse relations above, is defined in Eq.(3.25). It follows from its definition that

$$\mathbf{BC}' = \mathbf{I} \tag{5.12}$$

It may be considered a generalized inverse of **B**.

5.1.3 Computation of Matrices C *and* C′

In Subsection 3.1.2, it is shown that a *robust primary structure* and the corresponding matrices **C** and **C′** may be obtained from matrix **B** by Gauss elimination. The term robust also refers, among other things, to the fact that the inverse appearing in the definitions of **C** and **C′** can be achieved with least error. The elimination method starts with matrix **B** and its column labels $(1, 2, \cdots, m)$ in row matrix $\boldsymbol{\gamma}$

$$\begin{bmatrix} \mathbf{B} \\ \boldsymbol{\gamma} \end{bmatrix} = \begin{bmatrix} \underbrace{\begin{matrix} \mathbf{b}_1 & \mathbf{b}_2 & \cdots & \mathbf{b}_n \\ 1 & 2 & \cdots & n \end{matrix}} & \underbrace{\begin{matrix} \mathbf{b}_{n+1} & \cdots & \mathbf{b}_m \\ n+1 & \cdots & m \end{matrix}} \end{bmatrix} \tag{5.13}$$

and obtains

$$\begin{bmatrix} \mathbf{B}' \\ \boldsymbol{\gamma}' \end{bmatrix} = \begin{bmatrix} \underbrace{\begin{matrix} \mathbf{B}_1^{-1} \\ \gamma_1 \quad \gamma_2 \quad \cdots \quad \gamma_n \end{matrix}} & \underbrace{\begin{matrix} \mathbf{B}_1^{-1}\mathbf{B}_2 \\ \gamma_{n+1} \quad \cdots \quad \gamma_m \end{matrix}} \end{bmatrix} \tag{5.14}$$

by elementary row operations with partial pivot search by interchanging columns[1] where the column labels $(\gamma_1, \gamma_2, \cdots, \gamma_n)$ identify the robust primary structure, the column labels $(\gamma_{n+1}, \cdots, \gamma_m)$ identify the redundants, and matrices $\mathbf{B}_1^{-1}\mathbf{B}_2$ and $\mathbf{B}_1^{-1}$ are the ones used in Eqs.(3.24) and (3.25), that is, in

$$\mathbf{C} = \begin{bmatrix} -\mathbf{B}_1^{-1}\mathbf{B}_2 \\ \mathbf{I} \end{bmatrix} \tag{5.15}$$

and

$$\mathbf{C}' = \begin{bmatrix} \mathbf{B}_1^{-1} \\ \mathbf{0} \end{bmatrix} \tag{5.16}$$

which are the definitions of matrices $\mathbf{C}$ and $\mathbf{C}'$. The *Gauss elimination with partial pivot search by interchanging columns* procedure is given in Table 5.1 in algorithmic form. It requires a computational effort proportional to $O(MN^2e^3)$, and storage of $O(MNe^2)$.

Alternately, after selecting a primary structure, matrices $\mathbf{C}$ and $\mathbf{C}'$ may be generated column-by-column, as element forces in the primary structure corresponding to unit loads in the direction of the redundants and the nodal deflection directions, respectively. The element forces may be computed by the *extended method of joints* using the *generalized Henneberg method*, explained in Section 4.10. It requires a computational effort of $O(MNe^2)$ and storage $O(Me)$. The alternate method requires n times less computational resource and is suited better to parallelization than Gaussian elimination. However, it may be more susceptible to numerical errors, and requires that the primary structure be identified beforehand by other means.

5.1.4 Response due to $\Delta\mathbf{v}_o$

Since the only interesting loading of an adaptive structure during a control cycle is the one induced by the prescribed element deformations $\Delta\mathbf{v}_o$, from Eqs.(5.5-5.8) one may obtain

$$\begin{Bmatrix} \Delta\boldsymbol{\xi} \\ \Delta\mathbf{s} \\ \Delta\mathbf{v} \end{Bmatrix} = \begin{bmatrix} -\mathbf{C}'^T(\mathbf{I} - \vec{\mathbf{F}}\,\mathbf{K}_c) \\ \mathbf{K}_c \\ \vec{\mathbf{F}}\,\mathbf{K}_c \end{bmatrix} \Delta\mathbf{v}_o \tag{5.17}$$

In adaptive structures, the incremental prescribed element deformations $\Delta\mathbf{v}_o$ are created by the actuators.

It is interesting to observe from Eq.(5.17) that, in statically indeterminate adaptive structures, the incremental response due to the actuator induced incremental prescribed element deformations $\Delta\mathbf{v}_o$ includes not only the nodal deflection response $\Delta\boldsymbol{\xi}$, but also the element force response $\Delta\mathbf{s}$

---

[1] Instead of partial pivot search one may use complete pivot search. See pp. 212-214, the *Algebraic Eigenvalue Problem*, J. H. Wilkinson, Oxford University Press, 1965.

TABLE 5.1. Algorithm to Identify the Robust Primary Structure

No	Statement
1	*for* $i = 1$ *to* n
2	*find* r $(i \leq r \leq m)$ *so that for* $k = i, \cdots, m$:
3	$\|b_{i,r}\| \geq \|b_{i,k}\|$ *is true*
4	*if* $r \neq i$ *then interchange complete*
5	*columns* i *and* r *in* $\mathbf{B}$ *and* $\boldsymbol{\gamma}$
6	*if* $i < n$ *then*
7	*for* $k = i + 1$ *to* n
8	$c \leftarrow -b_{k,i}/b_{i,i}$
9	*for* $j = 1$ *to* m
10	$b_{k,j} \leftarrow (b_{k,j} + cb_{i,j})$
11	$b_{k,i} \leftarrow c$
12	*for* $i = 1$ *to* n
13	$c \leftarrow b_{i,i}$
14	*for* $j = 1$ *to* m
15	$b_{i,j} \leftarrow b_{i,j}/c$
16	$b_{i,i} \leftarrow 1/c$
17	*for* $i = 1$ *to* $n - 1$
18	*for* $k = i + 1$ *to* n
19	$c \leftarrow -b_{i,k}$
20	$b_{i,k} \leftarrow 0$
21	*for* $j = 1$ *to* m
22	$b_{i,j} \leftarrow (b_{i,j} + cb_{k,j})$

and the element deformation response $\Delta\mathbf{v}$. This should be compared with the facts one may observe from Eq.(4.5) that, in statically determinate adaptive structures, one can control only the nodal deflections, and the actuator induced prescribed element deformations cannot alter internal stress and deformation states. The situation in statically indeterminate adaptive structures opens the opportunity of stress and deformation control, but also causes difficulties in nodal deflection control due to alterations in stress and deformation states.

The discussions of Sections 3.4-3.8 can be useful in stress and deformation control in statically indeterminate adaptive structures. Stress control by means of introducing element deformations is not new to engineers. The Freyssinet method[2] of inserting deformations at the crown section of fixed end arch bridges for controlling its stresses and the whole field of prestressed concrete are examples of such. In this chapter, only the nodal deflection control problem will be discussed further.

5.2 Observed Response Components and Disturbances

In adaptive structures, sensors are used to measure deviations from the nominal state of the structure for p of the $2m+n$ response components. The design objective of the adaptive structure determines the number p, and also which p of the response components is to be observed and controlled. As stated earlier, the nodal deflection control is the main aim of this chapter. It is not possible to infer changes in a deflection component through changes in other response quantities, unless the disturbance loading is known precisely. Moreover, it is not possible to infer the complete deflection state from observed $p < n$ deflection components. In other words, in static control, a structure is not *observable.*[3] For this reason, one observes the deviations from the nominal only for the nodal deflection components that one wants to control. Out of n components of nodal deflections $\boldsymbol{\xi}$, suppose components $\alpha_1, \alpha_2, \cdots, \alpha_p$ are to be controlled. It is assumed that there are appropriate sensors to observe the deviations from the nominal in p number of deflection components $\xi_{\alpha_1}, \xi_{\alpha_2}, \cdots, \xi_{\alpha_p}$. Let

$$\Delta\boldsymbol{\xi}^d = [\Delta\xi_1^d, \Delta\xi_2^d, \cdots, \Delta\xi_n^d]^T \tag{5.18}$$

denote the complete list of deviations from the nominal for nodal deflections, and let

$$\Delta\mathbf{y} = [\Delta\xi_{\alpha_1}^d, \Delta\xi_{\alpha_2}^d, \cdots, \Delta\xi_{\alpha_p}^d]^T \tag{5.19}$$

[2]See, for example, p. 9, *Design of Prestressed Concrete Structures*, T. Y. Lin, John Wiley & Sons, 1955.

[3]However, in vibration control it is *observable*. See Section 8.6.

where

$$\Delta \mathbf{y} = [\Delta y_1, \Delta y_2, \cdots, \Delta y_p]^T \tag{5.20}$$

denotes the ones measured by the sensors. Using the identity

$$\Delta \xi^d_{\alpha_j} = \mathbf{i}^T_{\alpha_j} \Delta \boldsymbol{\xi}^d \tag{5.21}$$

where $\mathbf{i}_{\alpha_j}$ is the α_jth column of the nth order identity matrix, one may rewrite Eq.(5.19) as

$$\Delta \mathbf{y} = \mathbf{I}^T_{\alpha} \Delta \boldsymbol{\xi}^d \tag{5.22}$$

where

$$\mathbf{I}_{\alpha} = [\mathbf{i}_{\alpha_1}, \mathbf{i}_{\alpha_2}, \cdots, \mathbf{i}_{\alpha_p}] \tag{5.23}$$

Note that the orthonormal matrix $\mathbf{I}_{\alpha} \in (\mathrm{Int})^{n \times p}$ identifies the observed deflection components, and $\Delta \mathbf{y} \in \mathrm{R}^{p \times 1}$ lists the deviations from nominal in these components as measured by the sensors.

5.3 Actuators and Controls

To control p number of deflection components, one may use q number of deformation inducing actuators. As discussed in Chapter 3, in order to control p deflection components exactly, the number of actuators should not be less than the number of controlled deflection components, i.e.,

$$q \geq p \tag{5.24}$$

The induced element deformations by these actuators define the nonzero entries of $\Delta \mathbf{v}_o$ in Eq.(5.17). Let $\beta_1, \beta_2, \cdots, \beta_q$ denote the labels of these nonzero entries, and let Δu_j denote the induced element deformation in the jth actuator. Then one may write

$$\Delta \mathbf{v}_o = \sum_{j=1}^{q} \mathbf{i}_{\beta_j} \Delta u_j \tag{5.25}$$

where $\mathbf{i}_j$ is the jth column of the mth order identity matrix. Defining

$$\Delta \mathbf{u} = [\Delta u_1, \Delta u_2, \cdots, \Delta u_q]^T \tag{5.26}$$

and

$$\mathbf{I}_{\beta} = [\mathbf{i}_{\beta_1}, \mathbf{i}_{\beta_2}, , \cdots, \mathbf{i}_{\beta_q}] \tag{5.27}$$

one may rewrite Eq.(5.25) as

$$\Delta \mathbf{v}_o = \mathbf{I}_{\beta} \Delta \mathbf{u} \tag{5.28}$$

which is the definition of prescribed element deformations in terms of the controls $\Delta \mathbf{u} \in \mathrm{R}^{q \times 1}$.

The design of adaptive structure requires the locations of actuators. This is determined by the knowledge of matrix $\mathbf{I}_\beta$. To compensate the disturbances $\Delta\mathbf{y}$ by deformation inserting actuators, one needs to compute $\Delta\mathbf{u}$. The determination of $\mathbf{I}_\beta$ is the actuator placement problem, and given $\Delta\mathbf{y}$, the computation of the corresponding $\Delta\mathbf{u}$ is the control computation problem.[4] These are discussed in the remainder of this chapter.

5.4 Prevention of Stress Build-up

One may obtain the response corresponding to the control loading $\Delta\mathbf{v}_o$ defined in Eq.(5.28) by using Eq.(5.17). It may be observed that the control loading creates not only nodal deflection response $\Delta\boldsymbol{\xi}$ but also element force response $\Delta\mathbf{s}$ and element deformation response $\Delta\mathbf{v}$, mainly because of the presence of matrix $\mathbf{K}_c \in R^{m\times m}$. In other words, the control loading, which is conceived to control some of the nodal deflections, causes stresses and strains in the structure. This is not desirable because of two reasons: (a) One needs additional energy to drive the actuators in order to overcome the resistance of the structure to deformation insertion as discussed in Section 3.8, and (b) The response spills over to all of the uncontrolled response components, including internal stresses and internal strains. As discussed in Section 3.7, as one tries to control p number of deflection components, one should not cause undesired situations in the remaining response components. As is discussed below, by the use of secondary actuators, the development of element force response $\Delta\mathbf{s}$ and element deformation response $\Delta\mathbf{v}$ can be prevented.[5]

By using the definition of $\mathbf{K}_c$ from Eq.(3.34) in the second row partition of Eq.(5.17), the latter may be rewritten as

$$\Delta\mathbf{s} = \mathbf{C}[\mathbf{C}^T\vec{\mathbf{F}}\mathbf{C}]^{-1}\mathbf{C}^T\Delta\mathbf{v}_o \tag{5.29}$$

Clearly, if the control loading $\Delta\mathbf{v}_o$ is such that

$$\mathbf{C}^T\Delta\mathbf{v}_o = \mathbf{o} \tag{5.30}$$

then it follows from Eq.(5.29) that

$$\Delta\mathbf{s} = \mathbf{o} \tag{5.31}$$

In fact, it can be shown that, if Eq.(5.30) is true, then Eq.(5.17) becomes

$$\left\{\begin{array}{c}\Delta\boldsymbol{\xi}\\ \Delta\mathbf{s}\\ \Delta\mathbf{v}\end{array}\right\} = \left[\begin{array}{c}-\mathbf{C}'^T\\ \cdot\\ \cdot\end{array}\right]\Delta\mathbf{v}_o \tag{5.32}$$

---

[4] This is the problem of identifying *control law*.

[5] "Real-Time Control of Geometry and Stiffness in Adaptive Structures," A. V. Ramesh, S. Utku, and B. K. Wada, *Proceedings of the Second World Congress of Computational Mechanics*, Aug 27-31,1990, Stuttgart, Germany.

Comparing this with Eq.(4.5), one may state that the control response of a statically indeterminate adaptive structure can be similar to that of a statically determinate adaptive structure so long as Eq.(5.30) is satisfied by the control loading $\Delta\mathbf{v}_o$, i.e., the prescribed element deformations induced by the actuators. Eq.(5.30) is called the *compatibility equation of inserted deformations* $\Delta\mathbf{v}_o$.

Matrix $\mathbf{C} \in R^{m\times r}$ is studied in Subsections 5.1.2 and 5.1.3. Its r columns establish a basis in the null space of $\mathbf{B}$. Noting that $r = m - n$ is the degree of statical indeterminacy of the structure, when the structure is statically determinate then $\mathbf{C}$ vanishes, and the compatibility equation displayed in Eq.(5.30) is satisfied for all control loadings $\Delta\mathbf{v}_o$. If, however, the structure is statically indeterminate to the degree r, then there are r scalar equations in Eq.(5.30) which need to be satisfied for geometry control without creating stresses.

By using the definition of control loading from Eq.(5.28) in Eq.(5.30), the latter may be rewritten as

$$\mathbf{C}^T\mathbf{I}_\beta\Delta\mathbf{u} = \mathbf{o} \tag{5.33}$$

where $\Delta\mathbf{u} \in R^{q\times 1}$ is the vector of controls, $\mathbf{I}_\beta \in R^{m\times q}$ is the actuator location matrix, and $\mathbf{C}^T \in R^{r\times m}$ is the compatibility matrix. Using the definition of $\mathbf{I}_\beta$ from Eq.(5.27), and the identity that $\mathbf{A}\mathbf{i}_{\beta_j} = \mathbf{a}_{\beta_j}$, this equation may be restated as

$$\sum_{j=1}^{q}(\mathbf{c}^T)_{\beta_j}\Delta u_j = \mathbf{o} \tag{5.34}$$

where $(\mathbf{c}^T)_i$ refers to the ith column of $\mathbf{C}^T$ [$(\mathbf{c}^T)_i$ should not be confused with $\mathbf{c}_i^T$ which is the transpose of the ith column of $\mathbf{C}$].[6]

Clearly, if the columns of $\mathbf{C}^T$ that multiply the controls $\Delta u_j, j = 1, \cdots, q$, are zero columns, then the compatibility requirement of Eq.(5.30) is identically satisfied; hence, the statically indeterminate adaptive structure would behave similar to a statically determinate one. This situation is sketched in Fig.5.1(a) where a statically indeterminate truss of degree $r = 3$ is equipped with $q = 2$ actuators. It may be observed from the figure that the actuators can insert elongations/shortenings without any resistance, an indication that the compatibility requirement is satisfied. In Fig.5.1(b) the actuator placement is such that a zero column multiplies the control of the top actuator, and a nonzero column multiplies that of the bottom actuator; therefore, the compatibility requirement is not satisfied. Likewise in Fig.5.1(c), the actuator placement is such that the controls of both actuators are multiplied by nonzero columns in Eq.(5.34) and the compatibility

[6] This point is usually missed by many authors. See, for example, pp.6-7, *Matrix Computations, 2ed.*, G. H. Golub and C. F. Van Loan, The Johns Hopkins University Press, 1989.

requirement is not satisfied. For each nonzero column associated with an actuator in Eq.(5.34), one needs an additional actuator, subservient to the actuator already in place, to satisfy the compatibility equation. These ad-

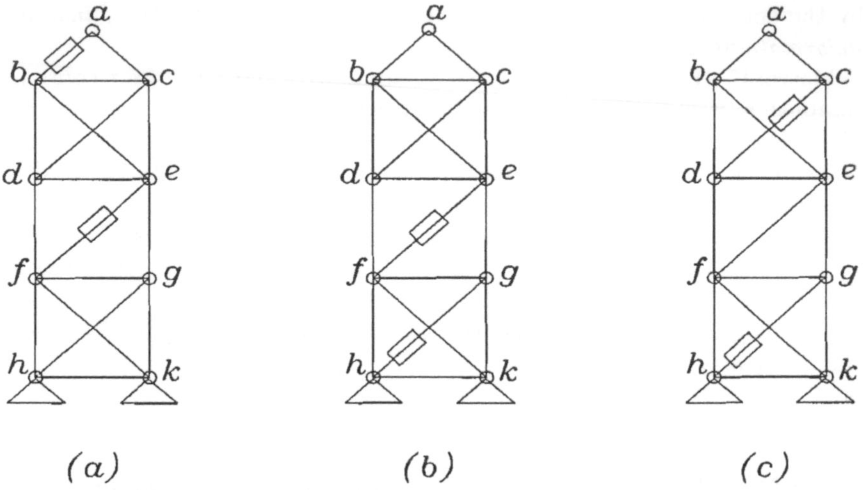

FIGURE 5.1. Third degree statically indeterminate truss with two actuators

ditional actuators are called *secondary or slave actuators*, since they induce just enough element deformation to satisfy the compatibility equation for the control vector $\Delta\mathbf{u}$ of the q actuators already in place.[7] The q actuators already in place are called *primary actuators*. The locations of the secondary actuators and their controls can be computed from the compatibility equation given by Eq.(5.30).

Let s denote the number of secondary actuators. Since there are r scalar compatibility equations, in the worst case, the number of secondary actuators s can be as high as the degree of indeterminacy r of the adaptive structure. However, as examples in Fig.5.1 show, a statically indeterminate adaptive structure, depending on element connectivity, may need (a) no secondary actuators, or (b) less than r secondary actuators, or (c) exactly r secondary actuators. In the next section, assuming that the locations of q number of primary actuators, i.e., matrix $\mathbf{I}_\beta$, is known, the determination of the number s and the locations $\mathbf{I}_{\beta'}$ of secondary actuators are explained. Also explained is the determination of the controls $\Delta\mathbf{u}' \in R^{s\times 1}$ of the secondary actuators corresponding to the controls $\Delta\mathbf{u} \in R^{q\times 1}$ of the primary actuators.

[7] "Energy Consumption in Active Vibration Control of Statically Indeterminate Structures," C. M. Baycan, S. Utku, Paper AIAA-95-1134, *Proceedings of 36th Structures, Structural Dynamics and Materials Conference*, New Orleans, LA, April 1995.

5.5 Secondary Actuators

If the actuator induced prescribed element deformations $\Delta\mathbf{v}_o$ satisfy the compatibility requirement of Eq.(5.30), then the response corresponding to $\Delta\mathbf{v}_o$ may be computed from Eq.(5.32), instead of Eq.(5.17). It is clear from Eq.(5.32) that when actuator induced deformations satisfy Eq.(5.30), i.e., when they are compatible, then only the nodal deflection response $\Delta\boldsymbol{\xi}$ will take place and the element force response $\Delta\mathbf{s}$ and the element deformation response $\Delta\mathbf{v}$ will take zero values.

When q number of primary actuators are in place, one may use at the most s number of secondary actuators in order to ensure the satisfaction of the compatibility requirement of Eq.(5.30). It is assumed that the location of primary actuators, i.e., the actuator locations matrix $\mathbf{I}_\beta \in(\text{Int})^{m\times q}$ is already known. In order to find the locations of the secondary actuators, one needs to operate on the compatibility equations as explained below.[8]

5.5.1 Transformation of Compatibility Equation

For an arbitrary $\Delta\mathbf{v}_o$, the compatibility equations in Eq.(5.30) will not be satisfied. When q components of $\Delta\mathbf{v}_o$ are prescribed by controls $\Delta\mathbf{u}$ of the primary actuators and the remaining $m-q$ components are zero, the use of such $\Delta\mathbf{v}_o$ in the compatibility equation will produce r number of equation residues. One may use the negatives of these residues as the controls of secondary actuators in order to ensure the satisfaction of the compatibility equations. Which zero components of $\Delta\mathbf{v}_o$ are to be replaced by the negatives of the residues is the question one has to answer for the secondary actuator placement problem. For this, the compatibility equations may first be transformed into a better form to identify the locations.

The r number of scalar compatibility equations may be rewritten by transposing both sides of Eq.(5.30) as

$$\Delta\mathbf{v}_o^T\mathbf{C} = \mathbf{0} \tag{5.35}$$

where $\mathbf{0} \in\text{R}^{1\times r}$ and $\mathbf{C}$ is as defined in Eq.(5.15). It may be observed from its definition that matrix $\mathbf{C}$ is not rank deficient.[9] Using the labels $\boldsymbol{\gamma}'$ displayed in Eq.(5.14) and the control loading $\Delta\mathbf{v}_o$ from Eq.(5.28), Eq.(5.35) may be

[8] See *Effect of the Compatibility of Inserted Strains on Energy Consumption in Active Control of Adaptive Structures*, Mehmet Can Baycan, Doctoral Dissertation, Civil and Environmental Engineering Department, Duke University, Durham, NC, Dec. 1996.

[9] Thanks to the identity matrix at the lower partition in the display of Eq.(5.15).

displayed in detailed form as

$$[\cdot,\cdots,\Delta v_{\beta_1},\cdots,\Delta v_{\beta_2},\cdots,\Delta v_{\beta q},\cdots]\begin{bmatrix} \longleftarrow \mathbf{C}_{\gamma_1} \longrightarrow \\ \longleftarrow \mathbf{C}_{\gamma 2} \longrightarrow \\ \vdots \\ \longleftarrow \mathbf{C}_{\gamma_n} \longrightarrow \\ 1 \quad \cdot \quad \cdots \quad \cdot \\ \cdot \quad 1 \quad \cdots \quad \cdot \\ \vdots \quad \cdot \quad \ddots \quad \vdots \\ \cdot \quad \cdot \quad \cdots \quad 1 \end{bmatrix} = [0,\cdots,0] \tag{5.36}$$

where $\mathbf{C}_k$ is the kth row of matrix $\mathbf{C} \in \mathrm{R}^{m \times r}$ and $n = m - r$. By reordering the rows of $\mathbf{C}$ so that the rows associated with the primary actuators are at the top, and using from Eq.(5.25) the fact that

$$\Delta u_j = \Delta v_{\beta_j}\ , \quad j = 1,\cdots,q \tag{5.37}$$

one may rewrite Eq.(5.36) as

$$[\Delta u_1,\Delta u_2,\cdots,\Delta u_q,0,\cdots,0]\begin{bmatrix} \leftarrow \mathbf{C}_{\beta_1} \rightarrow \\ \leftarrow \mathbf{C}_{\beta_2} \rightarrow \\ \vdots \\ \leftarrow \mathbf{C}_{\beta_q} \rightarrow \\ \leftarrow \mathbf{C}_{\beta'_{q+1}} \rightarrow \\ \vdots \\ \leftarrow \mathbf{C}_{\beta'_m} \rightarrow \end{bmatrix} = [0,\cdots,0] \tag{5.38}$$

Let $\mathbf{C}^q \in \mathrm{R}^{q \times r}$ denote the submatrix defined by the first q rows of $\mathbf{C}$ in Eq.(5.38), and let $\mathbf{C}'^s \in \mathrm{R}^{(m-q)\times r}$ denote the remainder, so that

$$\begin{bmatrix} \leftarrow \mathbf{C}_{\beta_1} \rightarrow \\ \leftarrow \mathbf{C}_{\beta_2} \rightarrow \\ \vdots \\ \leftarrow \mathbf{C}_{\beta_q} \rightarrow \\ \leftarrow \mathbf{C}_{\beta'_{q+1}} \rightarrow \\ \vdots \\ \leftarrow \mathbf{C}_{\beta'_m} \rightarrow \end{bmatrix} = \begin{bmatrix} \mathbf{C}^q \\ \mathbf{C}'^s \end{bmatrix} . \tag{5.39}$$

Using this we may rewrite Eq.(5.38) as

$$\begin{bmatrix} \Delta \mathbf{u}^T & \mathbf{0}' \end{bmatrix}\begin{bmatrix} \mathbf{C}^q \\ \mathbf{C}'^s \end{bmatrix} = \mathbf{0} \tag{5.40}$$

where zero row $\mathbf{0}' \in R^{(m-q)\times 1}$. Since the premultiplier of $\mathbf{C}'^s$ is a zero row, we may replace it by its unit upper triangular form which may be obtained from $\mathbf{C}'^s$ by the algorithm of Gauss elimination with partial pivot search by interchanging rows:

$$\mathbf{0}'\mathbf{C}'^s = \mathbf{0}'\mathbf{C}^s = [0,\cdots,0]\begin{bmatrix} 1 & \times & \cdots & \times \\ \cdot & 1 & \cdots & \times \\ \vdots & \vdots & \ddots & \times \\ \cdot & \cdot & \cdots & 1 \\ \vdots & \vdots & \vdots & \vdots \end{bmatrix} \tag{5.41}$$

where $\mathbf{C}^s$ is the upper unit triangular form of $\mathbf{C}'^s$. During the Gauss algorithm, if one encounters a zero pivot, one may skip the normalization by the pivot and continue on the next row as usual. Therefore some of the diagonals of $\mathbf{C}^s$ may be zeros instead of ones. By using $\mathbf{C}^s$ from Eq.(5.41) in Eq.(5.38), the latter may be displayed as

$$[\Delta u_1, \Delta u_2, \cdots, \Delta u_q, 0, \cdots, 0]\begin{bmatrix} \leftarrow & \mathbf{C}_{\beta_1} & \rightarrow \\ \leftarrow & \mathbf{C}_{\beta_2} & \rightarrow \\ & \vdots & \\ \leftarrow & \mathbf{C}^*_{\beta_q} & \rightarrow \\ \leftarrow & \mathbf{C}^*_{\beta_{q+1}} & \rightarrow \\ & \vdots & \\ \leftarrow & \mathbf{C}^*_{\beta_m} & \rightarrow \end{bmatrix} = [0,\cdots,0] \tag{5.42}$$

where the primes in the indices are dropped to indicate that the rows below row β_q are shuffled, and stars are introduced in lower $m-q$ rows to indicate that the values of the entries are changed.

Let $\mathbf{c}_k$ denote the kth column of matrix $\mathbf{C}$ in Eq.(5.42). Then, one may write the kth scalar compatibility equation as

$$\sum_{j=1}^{q} \Delta u_j \, c_{\beta_j,k} + \sum_{l=1}^{m-q} 0 \; c^*_{\beta_{q+l},k} = 0, \quad \text{for } k = 1,\cdots,r \tag{5.43}$$

5.5.2 Placement and Control of Secondary Actuators

From the scalar compatibility equations in Eq.(5.43), one may obtain the number s, the locations, and, when $\Delta\mathbf{u}$ is known, the expected element deformations $\Delta\mathbf{u}'$ of the secondary actuators. The following may be concluded from Eq.(5.43):

1. If all the coefficients of primary controls $\Delta\mathbf{u}$ in the kth scalar compatibility equation are zero, then the kth compatibility is satisfied

for any $\Delta\mathbf{u}$; hence, no secondary actuator is required for the kth compatibility requirement. If the coefficients of $\Delta\mathbf{u}$ in the kth scalar compatibility equation are zero, this means the kth column of submatrix $\mathbf{C}^q$ is a zero column. Let z denote the number of zero columns of submatrix $\mathbf{C}^q$. Then the number of secondary actuators s may be given as

$$s = r - z \tag{5.44}$$

2. If not all coefficients of primary controls $\Delta\mathbf{u}$ in the kth scalar compatibility equation are zero, i.e., when the kth column of submatrix $\mathbf{C}^q$ is not a zero column, then one needs a secondary actuator for the element deformation component β_{q+k}, i.e., the component corresponding to the unit diagonal element in the kth column of upper triangular matrix $\mathbf{C}^s$, to induce the deformation $\Delta u'_k$ which should be numerically the negative of the first sum on the left hand side of Eq.(5.43):

$$\Delta u'_k = -\sum_{j=1}^{q} \Delta u_j \; c_{\beta_j,k} \;, \quad \text{for } k = 1, \cdots, r \tag{5.45}$$

in order to satisfy the kth compatibility requirement. Since matrix $\mathbf{C}$ is not rank deficient according to its definition in Eq.(5.15), the diagonal position in the kth column of matrix $\mathbf{C}^s$ cannot be zero; namely, when one needs a secondary actuator for the deformation component β_{q+k}, then the constant $c_{\beta_{q+k},k}$ is not zero. Thus the secondary actuator placement matrix $\mathbf{I}_{\beta'}$ may be defined as

$$\mathbf{I}_{\beta'} = [\mathbf{i}_{\beta_{q+1}}, \mathbf{i}_{\beta_{q+2}}, \cdots, \mathbf{i}_{\beta_{q+r}}] \tag{5.46}$$

when $z = 0$. If $z \neq 0$ then for each zero column of $\mathbf{C}^q$, the corresponding column in $\mathbf{I}_{\beta'}$ should be deleted.

The process explained above for identifying the locations of secondary actuators is given in Table 5.6 in algorithmic form. Note that the labels $(\beta_1, \beta_2, \cdots, \beta_m)$ and $(\gamma_1, \gamma_2, \cdots, \gamma_m)$ refer to the reordered form of the original labels $(1, 2, \cdots, m)$ of the element deformation components.

5.6 Primary Actuators

Given that p number of deflection components with locations identified by matrix $\mathbf{I}_\alpha \in (\text{Int})^{m\times p}$ are to be controlled, one needs to answer the following questions:

1. How many primary actuators are needed, i.e., what is q?

TABLE 5.2. Secondary Actuator Placement Algorithm

Statement
Generate $\mathbf{C} \in \mathrm{R}^{m\times r}$ *per Eq.(5.15):*
$\mathbf{C} \leftarrow [\ -\mathbf{B}_2^T\mathbf{B}_1^{-T}\ \ \mathbf{I}\ ]^T$
Generate $\gamma' \in (\mathrm{Int})^{m\times 1}$ *per Eq.(5.14):*
$\gamma' \leftarrow [\gamma_1, \cdots, \gamma_n, \gamma_{n+1}, \cdots, \gamma_m]^T$
Input the location labels of q primary actuators
Reorder rows of γ' *and* $\mathbf{C}$ *such that the rows*
associated with q actuators are at top, hence
$\mathbf{C}$ *becomes* $[\ \mathbf{C}^{q^T}\ \ \mathbf{C}^{*s^T}\ ]^T$ *and*
γ' *becomes* $[\beta_1, \cdots, \beta_q, \beta'_{q+1}, \cdots, \beta'_m]$
Apply Gauss elimination with partial pivot
search by interchanging rows on $\mathbf{C}^{*s} \in \mathrm{R}^{(m-q)\times r}$
to make it unit upper triangular. Then γ' *becomes*
$[\beta_1, \cdots, \beta_q, \beta_{q+1}, \cdots, \beta_{q+r}, \beta_{q+r+1} \cdots, \beta_m]$
Tentative locations of secondary actuators are:
$\mathbf{t} \leftarrow [\beta_{q+1}, \cdots, \beta_{q+r}]^T$, $\mathbf{t} \in (\mathrm{Int})^{r\times 1}$
Set number of secondary actuators $s \leftarrow r$
for $k = 1$ *to* r
if ($\mathbf{c}_k^q$ *is a zero column) then*
$s \leftarrow s - 1$
delete k th entry of $\mathbf{t}$
Output the number s and the locations
$\mathbf{t} \in (\mathrm{Int})^{s\times 1}$ *of the secondary actuators*

2. What are the locations of q primary actuators, i.e., what is matrix $\mathbf{I}_\beta \in (\mathrm{Int})^{m\times q}$? This is the *actuator placement* problem.

3. If $q > p$, then which p of the primary actuators needs to be activated in the current control step? This is the *actuator selection* problem.

4. Given the current disturbance $\Delta\mathbf{y}$, what is the corresponding primary controls $\Delta\mathbf{u}$? This is the problem of *control computation.*

Each of the questions is discussed in one of the subsections below. The basic philosophy here is to determine the primary actuators first, and then take advantage of the discussions in the previous section to handle the compatibility problem.

5.6.1 Number of Primary Actuators q

The number of primary actuators that one needs to use in an adaptive structure depends on many things. Among them the most important one is the number of response components one would like to control. As discussed earlier, this chapter discusses mainly nodal deflection control. In precision structures, one is interested in controlling only those nodal deflection components which have a direct influence on the operations of a subsystem attached to the adaptive structure. In mechanical manipulators, one is interested in moving some of the nodes along prescribed trajectories. In both cases, as before, it is assumed that the number p and the locations of the controlled deflection components, as defined by matrix $\mathbf{I}_\alpha \in (\mathrm{Int})^{m\times p}$ of Eq.(5.23), are known. At a control step, as a result of control loading $\Delta\mathbf{v}_o$, one would like to have these deflection components to assume prescribed values $\Delta\mathbf{y}$.

For statically indeterminate adaptive structures, one may use the first row partition of Eq.(5.17) for the nodal deflection response $\Delta\boldsymbol{\xi}$ corresponding to control loading $\Delta\mathbf{v}_o$, i.e:

$$\Delta\boldsymbol{\xi} = -\mathbf{C}'^T(\mathbf{I} - \vec{\mathbf{F}}\mathbf{K}_c)\Delta\mathbf{v}_o \tag{5.47}$$

where $\mathbf{K}_c = \mathbf{C}[\mathbf{C}^T\vec{\mathbf{F}}\mathbf{C}]^{-1}\mathbf{C}^T$ per Eq.(5.6). One is interested in not creating element force response by the control loading. Then, if required, one should use appropriate secondary actuators in order to make sure that the compatibility requirement $\mathbf{C}^T\Delta\mathbf{v}_o = \mathbf{o}$ is satisfied. With this in mind, the relationship in Eq.(5.47) becomes

$$\Delta\boldsymbol{\xi} = -\,\mathbf{C}'^T\Delta\mathbf{v}_o \tag{5.48}$$

which is the first row partition of Eq.(5.32) discussed earlier.

Of the incremental nodal deflection response $\Delta\boldsymbol{\xi}$, appearing on the left hand side of Eq.(5.48), p number of components on rows $\alpha_1, \alpha_2, \cdots, \alpha_p$ are expected to have the prescribed values $\Delta\mathbf{y}$ which are the negatives of the sensor measured disturbances in precision structures or the values obtained from the prescribed trajectory in mechanical manipulators. By premultiplying both sides of Eq.(5.48) by $\mathbf{I}_\alpha^T$ and noting that

$$\Delta\mathbf{y} = \mathbf{I}_\alpha^T\Delta\boldsymbol{\xi} \tag{5.49}$$

one may obtain:

$$\Delta\mathbf{y} = -\,\mathbf{I}_\alpha^T\mathbf{C}'^T\Delta\mathbf{v}_o\ . \tag{5.50}$$

The control loading $\Delta\mathbf{v}_o \in \mathrm{R}^{m\times 1}$ contains nonzero entries only at rows $\beta_1, \beta_2, \cdots, \beta_q$ associated with q number of primary actuators. As before, denoting the element deformation induced in the jth primary actuator by Δu_j, and noting that

$$\Delta u_j = \Delta v_o^{\beta_j}\ ,\ \text{for}\quad j = 1, \cdots, q \tag{5.51}$$

one may write

$$\Delta\mathbf{v}_o = \mathbf{I}_\beta\Delta\mathbf{u} \tag{5.52}$$

where $\mathbf{I}_\beta \in(\text{Int})^{m\times q}$ is the actuator location matrix as defined in Eq.(5.27). Using $\Delta\mathbf{v}_o$ from Eq.(5.52) in Eq.(5.50), one may obtain

$$\Delta\mathbf{y} = -\,[\mathbf{I}_\alpha^T\mathbf{C}'^T\mathbf{I}_\beta]\Delta\mathbf{u} \tag{5.53}$$

where matrix $[\mathbf{I}_\alpha^T\mathbf{C}'^T\mathbf{I}_\beta] \in \mathrm{R}^{p\times q}$ consists of columns $\beta_1,\beta_2,\cdots,\beta_q$ of the coefficient matrix $\mathbf{I}_\alpha^T\mathbf{C}'^T \in \mathrm{R}^{p\times m}$ in Eq.(5.50). The questions to answer are what is q, i.e., how many primary actuators needed, and what are $\beta_1,\beta_2,\cdots,\beta_q$, i.e., which q columns of matrix $\mathbf{I}_\alpha^T\mathbf{C}'^T \in \mathrm{R}^{p\times m}$ should be selected.

Since Eq.(5.53) is to be used in obtaining the controls $\Delta\mathbf{u}$ when $\Delta\mathbf{y}$ is available, the coefficient matrix $\mathbf{I}_\alpha^T\mathbf{C}'^T\mathbf{I}_\beta$ should have at least one non-zero pth order minor. Namely, since

$$q \geq p \tag{5.54}$$

it should not be rank deficient. This means that the number of primary actuators cannot be less than the number of controlled deflection components.

If the number of primary actuators q exceeds the number of controlled deflection components, then, at a control step, one would have choices in activating p of them. This may be preferable since it gives flexibility in the control of the adaptive structure. It is especially useful when there is a likelihood that some of the actuators may fail or may need periodic maintenance during the continuous operations.

It is assumed here that the number of primary actuators q is decided by the designer of the adaptive structure to satisfy Eq.(5.54).

5.6.2 Placement and Control of Primary Actuators

This is the problem of deciding which q columns of the coefficient matrix $\mathbf{I}_\alpha^T\mathbf{C}'^T \in \mathrm{R}^{p\times m}$ in Eq.(5.50) are to be retained for the *controlled deflections versus controls* relationship of Eq.(5.53), namely, what the labels $\beta_1,\beta_2,\cdots,\beta_q$ that define the actuator placement matrix $\mathbf{I}_\beta \in(\text{Int})^{m\times q}$ are. The problem is a complicated one. The issues involved are (1) controllability, (2) control energy, and (3) control flexibility. These are discussed briefly below:

1. For controllability, one would like to select the locations of p of the q primary actuators as the ones which are the most instrumental for the control objective of creating desired response in the observed nodal deflection directions. This ensures the robustness of the control. For example, if one chooses elements ab and ef as the locations of the two primary actuators for controlling the displacements of node g of the truss shown in Fig.5.1a, clearly one would create an uncontrollable situation, since the actuators, as positioned, cannot create

displacements at node g. As discussed in the previous chapter, the measure of controllability is the magnitude of the determinant of the p selected columns of matrix $\mathbf{I}_\alpha^T\mathbf{C}'^T$. For the robustness of the control, one should select the columns of matrix $\mathbf{I}_\alpha^T\mathbf{C}'^T$ that make the magnitude of the determinant as large as possible.

2. It does not need much argument to convince one that the control energy must be as small as possible. As discussed in Section 3.8, the energy needed by the actuators consists of the energy to insert the control deformations against the preexisting element forces, and the energy to overcome the stiffness of the structure in the directions of actuator induced deformations. If there are preexisting element forces $\grave{\mathbf{s}}_o$, one may try to place the actuators such that the total positive energy need is as small as possible. However, if there is no bias in the variation of $\grave{\mathbf{s}}_o$ during the expected life time of the adaptive structure, one may have to ignore $\grave{\mathbf{s}}_o$ in the actuator placement problem. If the structure is statically indeterminate, one has to use secondary actuators to eliminate the stiffness of the structure against deformation insertion. As discussed in the previous section, there may be occasions when one may need no secondary actuators although the structure is statically indeterminate. For example, in Fig.5.1a, although the truss is statically indeterminate to degree 3, no secondary actuators are needed when the primary actuators are on elements ab and ef. The placement with the objective of minimizing the number of secondary actuators may not coincide with the objective of maximum controllability. This may be a problem in retrofitting adaptability to existing structures.

3. One may use more primary actuators than the minimum p, in order to achieve increased flexibility during the control of adaptive structure. When the number of primary actuators q is larger than the number of controlled deflection components p, this situation brings the additional problems of placing the extra $q - p$ primary actuators during the design process, and selecting p of them at each control step for the operations.

In the discussions of this section, the following are assumed: (a) There is no discernible bias to take advantage of, for actuator placement, in the distribution of the preexisting element forces $\grave{\mathbf{s}}_o$ and in their variation during the lifetime of the adaptive structure. (b) The aim is not to retrofit adaptability to an existing structure. Hence the controllability is the only criterion used for actuator placement. Once the first p of the primary actuators is placed solely with the criterion of maximum controllability, the secondary actuator placement problem can be tackled as discussed in the previous section.

Since the adaptive structure is statically indeterminate, per Eq.(3.26), one has

$$\mathbf{C}' \neq \mathbf{B}^{-1} \quad \text{since } r > 0 \tag{5.55}$$

One may obtain the definition of $\mathbf{C}' \in R^{m\times n}$ from Eq.(5.16) where matrix $\mathbf{B}_1 \in R^{n\times n}$ consists of those columns of matrix $\mathbf{B} \in R^{n\times m}$ which are associated with the primary structure. As discussed in Subsection 5.1.3, there may be as many as $\frac{m!}{n!(m-n)!}$ different primary structures associated with a given statically indeterminate structure. Which one should be used in determining $\mathbf{B}_1$ in order to generate $\mathbf{C}'$ per Eq.(5.16)? In Subsection 5.1.3, the *robust primary structure*, i.e., the one for which $|det(\mathbf{B}_1)|$ is not smaller than that of any other $\mathbf{B}_1$, is used. Is this the correct one to use in placing the primary actuators? The answer to this question is usually negative. If the statically indeterminate structure is designed to carry a dominant load, then the procedures of structural optimization direct one to a unique statically determinate structure which may be different than the robust primary structure. By incorporating new structural elements to this optimal primary structure, it may be made to meet the requirements of other loading conditions. In such cases, the statically determinate structure optimized for the dominant load is the primary structure that can be used for placing the primary actuators.

When there is no other guidance, the robust primary structure may be used to identify the locations of the primary actuators. In subsection 5.1.3, the procedure which identifies the robust primary structure is explained and an algorithm is provided in Table 5.1. The algorithm chooses the columns $\gamma_1, \gamma_2, \cdots, \gamma_n$ of matrix $\mathbf{B} \in R^{n\times m}$, in defining $\mathbf{B}_1 \in R^{n\times n}$, and also provides $\mathbf{B}_1^{-1}$ over the first n columns of $\mathbf{B}$ by destroying them. Using this $\mathbf{B}_1^{-1}$, from Eq.(5.16), one may write for $\mathbf{C}'^T$

$$\begin{bmatrix} \mathbf{C}'^T \\ \textit{column labels} \end{bmatrix} = \begin{bmatrix} \mathbf{B}_1^{-T} & \mathbf{0} \\ \gamma_1, \gamma_2, \cdots, \gamma_n & \gamma_{n+1}, \cdots, \gamma_m \end{bmatrix} \tag{5.56}$$

where the row labels $1, 2, \cdots, n$ are those of $\Delta\boldsymbol{\xi}$ (since there is no row interchange in the algorithm of Table 5.1), and labels $\gamma_1, \gamma_2, \cdots, \gamma_n, \gamma_{n+1}, \cdots, \gamma_m$ is a permutation of $1, 2, \cdots, m$. If one uses $\mathbf{C}'^T$ from Eq.(5.56) in Eq.(5.53), the latter may be displayed as

$$\Delta\mathbf{y} = -\mathbf{P}\Delta\mathbf{u} \tag{5.57}$$

where $\mathbf{P} \in R^{p\times q}$ is

$$\mathbf{P} = \mathbf{I}_\alpha^T \mathbf{C}'^T \mathbf{I}'_\beta = \mathbf{I}_\alpha^T \mathbf{B}_1^{-T} \mathbf{I}'_\beta = \begin{bmatrix} \leftarrow & \mathbf{i}_{\alpha_1}^T & \rightarrow \\ & \vdots & \\ \leftarrow & \mathbf{i}_{\alpha_p}^T & \rightarrow \end{bmatrix}$$

$$\begin{bmatrix} \uparrow & & \uparrow & \uparrow & & \uparrow \\ (\mathbf{B}_1^{-T})_{\gamma_1} & \cdots & (\mathbf{B}_1^{-T})_{\gamma_n} & \mathbf{o}_{\gamma_{n+1}} & \cdots & \mathbf{o}_{\gamma_m} \\ \downarrow & & \downarrow & \downarrow & & \downarrow \end{bmatrix} \begin{bmatrix} \uparrow & & \uparrow \\ \mathbf{i}'_{\beta_1} & \cdots & \mathbf{i}'_{\beta_q} \\ \downarrow & & \downarrow \end{bmatrix} \tag{5.58}$$

where $\mathbf{i}_j$ is the jth column of nth order identity matrix, $\mathbf{i}'_j$ is the jth column of the mth order identity matrix, $\mathbf{o}$ is a zero column of order n, and $(\mathbf{B}_1^{-T})_j$ is the jth column of $\mathbf{B}_1^{-T}$. Since the primary actuators are being placed on the primary structure, it may be observed from Eq.(5.58) that

$$\beta_j \neq \gamma_{n+j} \quad \text{for } j = 1, \cdots, r \tag{5.59}$$

Therefore the triple product in the right hand side of Eq.(5.58) selects q columns from the first n columns and p rows from the n rows of $\mathbf{C}'^T$. If for some reason one wants to place the jth primary actuator such that $\beta_j = \gamma_{n+j}$, then one should include $\mathbf{b}_{\beta_j}$, i.e., the β_jth column of $\mathbf{B}$, in $\mathbf{B}_1$, that is, it should be part of a new primary structure.

In general, at this stage, one does not know $\beta_1, \beta_2, \cdots, \beta_q$, other than the fact that they are q of $\gamma_1, \gamma_2, \cdots, \gamma_n$ which are the column labels of $\mathbf{B}_1^{-T} \in \mathrm{R}^{n\times n}$, i.e., the matrix associated with the primary structure. Noting that the primary structure is statically determinate, one may observe that the actuator placement in statically indeterminate adaptive structures is reduced to the actuator placement problem in statically determinate adaptive structures. The latter is studied in Sections 4.5, 4.6, and 4.7 for $q = p < n$, $p < q = n$, and $p < q < n$ cases, respectively, with sufficient detail, using the controllability criterion. Also studied in these sections are the computation of the controls assuming that $\Delta\mathbf{y}$ represents disturbances. These studies are adopted for statically determinate mechanical manipulators in Subsections 4.9.1, 4.9.2, and 4.9.3, where $\Delta\mathbf{y}$ represents the desired nodal deflection increments. The algorithms presented in Sections 4.5-4.7 and Subsections 4.9.1-4.9.3 can be used here by paying attention to the following.

1. Matrix $\mathbf{B}_1^{-T} \in \mathrm{R}^{n\times n}$ should replace matrix $\mathbf{B}^{-T} \in \mathrm{R}^{n\times n}$.

2. The column labels of matrix $\mathbf{B}_1^{-T} \in \mathrm{R}^{n\times n}$ are $\gamma_1, \gamma_2, \cdots, \gamma_n$, a permutation of $1, 2, \cdots, n$, whereas the column labels of $\mathbf{B}^{-T} \in \mathrm{R}^{n\times n}$ are $1, 2, \cdots, n$.

3. Vector $\Delta\mathbf{y}$ defined in this section is the same as $\Delta\mathbf{y}$ defined in Subsections 4.9.1-4.9.3; therefore, the computations for controls $\Delta\mathbf{u}$ in Subsections 4.9.1-4.9.3 can be used here without alteration.

4. Vector $\Delta\mathbf{y}$ defined in this section is the negative of $\Delta\mathbf{y}$ defined in Sections 4.5-4.7; therefore, the controls $\Delta\mathbf{u}$ of Sections 4.5-4.7 are the negatives of $\Delta\mathbf{u}$ here.

5.6.3 Generation of Matrix $\mathbf{I}_\alpha^T\mathbf{B}_1^{-T}$

The placement and also controls of the primary actuators require the generation of matrix $\mathbf{I}_\alpha^T\mathbf{B}_1^{-T} \in R^{p\times n}$. This can be achieved as discussed in Section 4.10. The rows of matrix $\mathbf{I}_\alpha^T\mathbf{B}_1^{-T} \in R^{p\times n}$ may be obtained one row at a time, as the element forces in the primary structure corresponding to unit loads in the directions of controlled deflections $\alpha_1, \alpha_2, \cdots, \alpha_p$. For example the jth row of $\mathbf{I}_\alpha^T\mathbf{B}_1^{-T} \in R^{p\times n}$ is the element force components, in the primary structure, with labels $\gamma_1, \gamma_2, \cdots, \gamma_n$ corresponding to a unit load acting in the direction of controlled deflection component α_j.

5.7 Recapitulation of Actuator Placement

The actuator placement in statically indeterminate adaptive structures follows procedures similar to those of the actuator placement in statically determinate adaptive structures. The main difference is in the use of secondary actuators in order to prevent stress build-up as a result of controls induced by the primary actuators. To summarize the discussions in the previous two sections, the following may be stated.

1. Identify a primary structure associated with the statically indeterminate adaptive structure. If the degree of indeterminacy is r, one may introduce r number of cuts into the structure in order to obtain a primary structure. Usually the dominant loading condition determines the primary structure. In the absence of any practical help in choosing the primary structure, one may use the robust primary structure by applying the algorithm given in Table 5.1 on matrix $\mathbf{B} \in R^{n\times m}$.

2. Use the primary structure with the methods of Chapter 4 to identify the locations $\beta_1, \beta_2, \cdots, \beta_q$ of the primary actuators.

3. Use the algorithm in Table 5.6 with matrix $\mathbf{C} \in R^{m\times r}$ corresponding to the selected primary structure and primary actuator locations $\beta_1, \beta_2, \cdots, \beta_q$ to identify the number s and the locations of the secondary actuators. Note that $s \leq r$. In fact, s may be zero when r is not. See, for example, Fig.5.1a, where $r = 3, q = 2$, and yet $s = 0$.

4. The controls to be induced by the secondary actuators can be computed by Eq.(5.45) as a function of the controls induced by the primary actuators. One may discard a secondary actuator by removing

the associated redundant force by a structural *cut*. For example, in Fig.5.1b, the secondary actuator associated with the primary actuator on bar gh may be on bar kf. By removing bar kf, the need for the secondary actuator can be eliminated.

5.8 Recapitulation

In this chapter the problems associated with the static feedback control of p of the n deflection quantities by means of q deformation inducing actuators in statically indeterminate discrete parameter adaptive structures are studied. The problems include control energy, robustness of control, actuator placement, identification of number and location of secondary actuators, and computation of controls. These problems are also studied when the static control is of open loop type as in slow moving mechanical manipulators with prescribed trajectories.

6 Excitation-Response Relations, Dynamic Case

For the incremental excitation-response relations studied in Chapter 2 it is assumed that the loading of the structure is sufficiently slow in order to prevent the development of any appreciable inertial forces in the structure. In this chapter such a restriction does not exist. However, starting with this chapter and in the remainder of this book, it is assumed that the response-excitation relationships are linear and the structural system is time invariant. As in the static case, here too the relationships are the quantitative descriptions of the equilibrium of internal and external forces as required by *Newton's laws*, the geometric compatibility of nodal deflections and element deformations required by the rules of *Euclidean geometry*, and the stiffness (or flexibility) relations that exist between element forces and element deformations due to constitutive laws of the structure's material (in the present case *Hooke's laws*). These will be discussed in the following sections using the definitions and discussions given in Chapter 2.

6.1 Equilibrium Equations

In the dynamic case, it is more convenient to consider the implications of Newton's laws in the form of *d'Alembert's principle* which basically states that a material particle of a structure in motion is in dynamic equilibrium if the sum of all the forces, including the motion-induced *inertial* and *frictional forces*, acting on the particle is zero. Here the term "force" is meant to mean both forces and moments. According to Newton's law the inertial

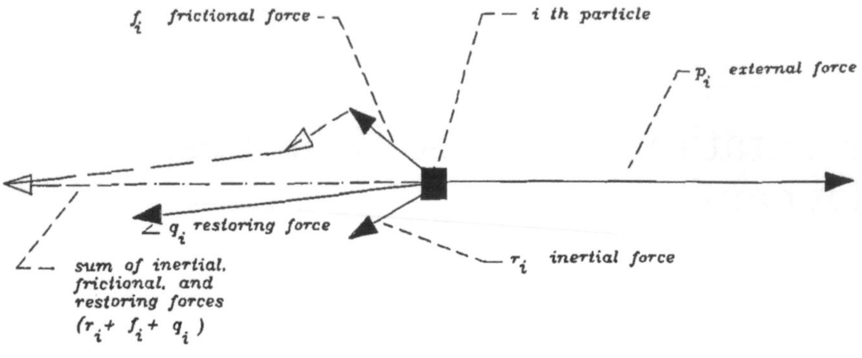

FIGURE 6.1. Free body diagram of a structural particle in dynamic equilibrium (the sum of the forces is zero according to d'Alembert's principle)

forces are proportional to the particle's acceleration. According to the experiments on structural materials the frictional forces are proportional to the particle's velocity. The forces acting on the particle that are caused by the movement of the particle relative to its neighbors are called *restoring forces.* All other forces that are acting on the particle, such as prescribed forces, loads, etc., are called *external forces.* In the free body diagram of the ith particle shown in Fig.(6.1), vectors $\mathbf{r}_i, \mathbf{f}_i, \mathbf{q}_i$, and $\mathbf{p}_i$ are shown and marked as the inertial, frictional, restoring, and external forces, respectively. The sum of inertial, frictional, and restoring forces is equal and opposite of the external force acting at the particle. Hence, by the d'Alembert's principle, the dynamic equilibrium equation of the ith particle may be stated as

$$\underset{inertial}{\mathbf{r}_i} + \underset{frictional}{\mathbf{f}_i} + \underset{restoring}{\mathbf{q}_i} = \underset{external}{\mathbf{p}_i} \tag{6.1}$$

Considering all material points of the structure, these equations may be rewritten as

$$\mathbf{q} + \mathbf{r} + \mathbf{f} = \mathbf{p} \tag{6.2}$$

where the vectors appear to be $R^{\infty \times 1}$! In the subsections below, it is shown these vectors are all of order n, since it may be sufficient to consider only the motion of the nodal particles in order to describe the motion of the whole structure.

6.1.1 Restoring Forces

In the absence of motion there are no inertial and frictional forces; hence, the dynamic force equilibrium equation reduces to

$$\underset{restoring}{\mathbf{q}_i} = \underset{external}{\mathbf{p}_i} \tag{6.3}$$

which is the static force equilibrium equation of the particle. In Chapter 2, it is shown that the static force equilibrium equations of all particles of a discrete parameter structure can be represented by only the nodal force equilibrium equations as

$$\mathbf{q} = \mathbf{p} \tag{6.4}$$

where $\mathbf{q} \in R^{n\times 1}$ are restoring forces and $\mathbf{p} \in R^{n\times 1}$ are external forces, both in the directions associated with the components of independent nodal deflections $\boldsymbol{\xi} \in R^{n\times 1}$. Chapter 2 discussions also lead to the conclusion that the restoring forces $\mathbf{q}$ can be expressed in terms of independent element forces $\mathbf{s} \in R^{m\times 1}$ and independent nodal deflections $\boldsymbol{\xi} \in R^{n\times 1}$as

$$\mathbf{q} = \grave{\mathbf{K}}\boldsymbol{\xi} + \mathbf{B}\mathbf{s} \tag{6.5}$$

where $\grave{\mathbf{K}} \in R^{n\times n}$ is the global stiffness matrix due to preexisting element forces and rank-n matrix $\mathbf{B} \in R^{n\times m}$ transforms $\mathbf{s}$ into global description of associated nodal internal forces. (Note that in Chapter 2, the total restoring forces were denoted by $\mathbf{q} + \grave{\mathbf{q}}$, a notation which is used also in Chapters 3, 4, and 5.) From now on the total restoring forces are shown by $\mathbf{q}$ as defined in Eq.(6.5).

6.1.2 Inertial Forces

Inertial forces are present at every mass point in motion. Although the mass points attached to the nodes of a discrete parameter structure do not create any particular problem since there are only a finite number of nodal points, the mass points of the structural elements pose a difficult problem since they are infinitely many. The motion of a structural element causes the creation of *force-type element loads.* These should be somehow lumped at the nodes. Strictly speaking, the procedure explained in Section 2.1.2 for lumping force-type element loads cannot be used here until appropriate time-varying interpolation rules describing the *state* of the particles in terms of the *state* of the nodes can be established. In the displacement method finite element analysis for the steady state vibrations of structures, it is customary to use the same interpolation rule for approximating both the position and the velocity of element particles in terms of those of the nodes. The *consistent mass matrices* are a result of such endeavors. Since the finite element method is devised for the *elliptic boundary value problems* where there are no *real characteristic directions* and therefore no discontinuity in the dependent variables, it is permissible in such problems to use fixed interpolation rules for dependent variables in order to express the values at non-nodal points in terms of those at the nodal points. Although the use of fixed interpolation rules causes no special difficulty in *parabolic problems* (since the single characteristic direction is parallel to the domain containing the elliptic part), in *hyperbolic problems* there are real characteristic directions (not parallel to the domain of elliptic part) which allow

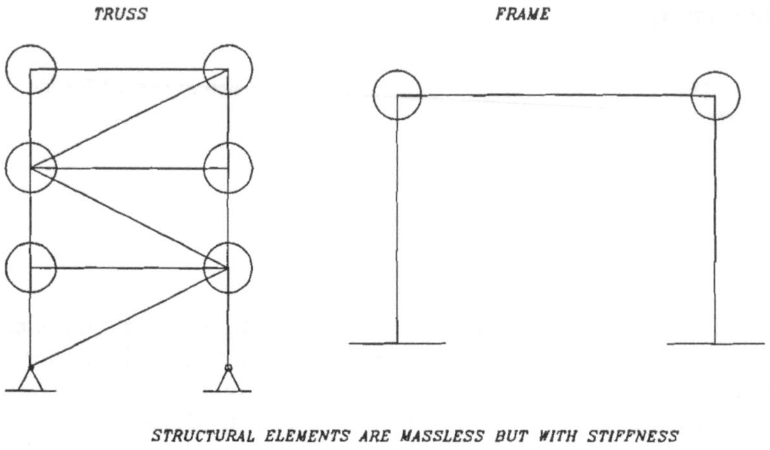

FIGURE 6.2. Idealizations implied by discrete parameter analysis: mass-spring sytems (dampers not shown)

discontinuity in the dependent variables, and thus make the use of fixed interpolation rules not permissible.[1] This may be seen from the sensitivity of the predicted high frequency structural response in steady state vibrations to the changes in the fixed interpolation rules. For transient problems, this is even more so.

The undesirable effects of the way the inertial forces are lumped at the nodes decrease as the ratio of mass of nodes over mass of elements increases. In fact, no such undesirable effects would be present if the structural members were all massless, i.e., if the structures were idealized as spring-mass systems as shown in Fig.(6.2).

In this work, it is assumed that most of the mass associated with a node belongs to the non-structural entities attached to the node, and the remaining small portion consists of part of the mass of structural elements incident to that node. Then the undesirable effect of using any convenient rule for lumping the mass of a structural element at its vertices becomes negligible. A very simple rule is to lump the total mass of a structural element to its vertices in proportion to the distance of the element's mass center to the opposing vertex. This rule is sketched in Fig. (6.3). Once the mass of elements are lumped at the nodes by using this simple rule, the dynamic force equilibrium equations of all nodes, in the directions of the

[1] See, for example, Chapter 6, pp. 337-405, *Engineering Analysis*, Stephen H. Crandall, McGraw-Hill Book Company, Inc., New York, 1956.

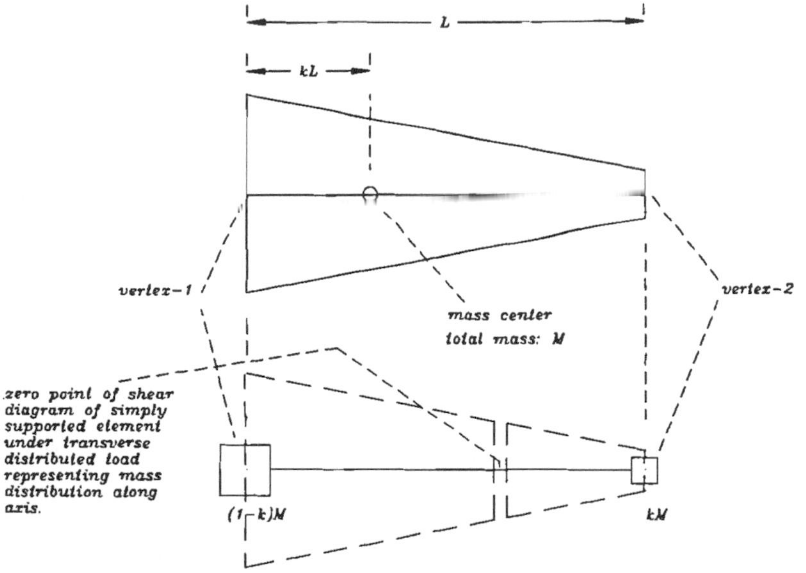

FIGURE 6.3. A simple rule for lumping mass of a structural element at its vertices

components of independent nodal deflections $\boldsymbol{\xi} \in R^{n\times 1}$, may be written as in Eq.(6.2), provided that the frictional forces $\mathbf{f}$ and the external forces $\mathbf{p}$ exist only at the nodes. The kth component of $\mathbf{r} \in R^{n\times 1}$, r_k , is the inertial force in the direction of the kth component of $\boldsymbol{\xi} \in R^{n\times 1}$, i.e., ξ_k . Using Newton's second law, this may be expressed as

$$r_k = m_k\ddot{\xi}_k \tag{6.6}$$

where m_k is the mass (or moment of inertia) of related nodal mass and $\ddot{\xi}_k$ is the acceleration ($\ddot{\xi}_k$ is used instead of $\frac{\partial^2 \xi_k}{\partial t^2}$). Considering all possible motion directions, Eq.(6.6) may rewritten compactly as

$$\mathbf{r} = \mathbf{M}\ddot{\boldsymbol{\xi}} \tag{6.7}$$

where diagonal matrix $\mathbf{M} \in R^{n\times n}$ is the *mass matrix of structure.* Since the kinetic energy of structure, i.e., $\frac{1}{2}\dot{\boldsymbol{\xi}}^T\mathbf{M}\dot{\boldsymbol{\xi}}$, is always a positive scalar which vanishes only if $\dot{\boldsymbol{\xi}} = \mathbf{o}$, then it follows that $\mathbf{M} \in R^{n\times n}$ is a real symmetric and positive definite matrix.

6.1.3 Frictional Forces

The quantification of the frictional forces $\mathbf{f} \in R^{n\times 1}$ is based on experimental evidence. In structures undergoing deformations at speeds far smaller than the speed of sound in the structure's material, it is reasonable to assume that the frictional forces $\mathbf{f}$ are proportional to nodal velocities $\dot{\boldsymbol{\xi}} \in R^{n\times 1}$, i.e.:

$$\mathbf{f} = \grave{\mathbf{C}}\dot{\boldsymbol{\xi}} \tag{6.8}$$

where $\grave{\mathbf{C}} \in R^{n\times n}$ is called the *damping matrix of structure*. Since the power leaking out of the system due to frictional forces, i.e., $-\frac{1}{2}\dot{\boldsymbol{\xi}}^T\mathbf{f} = -\frac{1}{2}\dot{\boldsymbol{\xi}}^T\grave{\mathbf{C}}\dot{\boldsymbol{\xi}}$, is a negative scalar which vanishes only if $\dot{\boldsymbol{\xi}} = \mathbf{o}$, then it follows that matrix $\grave{\mathbf{C}} \in R^{n\times n}$ is always a real symmetric and positive definite matrix. The entries of $\grave{\mathbf{C}}$ can be determined by $\frac{n(n+1)}{2}$ independent experiments. However, this is impractical. Instead, it is customary to express $\grave{\mathbf{C}}$ as a linear combination of known nth order positive definite matrices, such as $\mathbf{K}'$ and $\mathbf{M}$:

$$\grave{\mathbf{C}} = \alpha\mathbf{K}' + \beta\mathbf{M} \tag{6.9}$$

Such a definition requires only two experiments to determine constants α and β. Later in the next chapter, a variant of this, in the form of *modal damping*, will be explained (see Section 3.1.1 for the definition of matrix $\mathbf{K}' \in R^{n\times n}$, i.e., the *global stiffness matrix of structure*).

6.1.4 External Forces

It is assumed that the external forces, i.e., the loads of the structure, are acting at the nodes only. Any force-type loads acting on the elements should be suitably lumped at the nodes. For this the method described in Subsection 2.1.3 may be used, but with caution. The nodal load acting in the direction of ξ_k is denoted by p_k, and it is the kth component of $\mathbf{p} \in R^{n\times 1}$ which is the *load vector of the structure.*

6.1.5 Dynamic Equilibrium Equations of Nodes

These equations are given in Eq.(6.2) in terms of $\mathbf{q}, \mathbf{r}$, and $\mathbf{f}$ representing restoring, inertial, and frictional forces that are discussed in the preceding subsections. By using $\mathbf{q}$ from Eq.(6.5), $\mathbf{r}$ from Eq.(6.7), and $\mathbf{f}$ from Eq.(6.8), one may rewrite the dynamic equilibrium equations of the nodes in Eq.(6.2) as

$$\mathbf{M}\ddot{\boldsymbol{\xi}} + \grave{\mathbf{C}}\dot{\boldsymbol{\xi}} + \grave{\mathbf{K}}\boldsymbol{\xi} + \mathbf{B}\mathbf{s} = \mathbf{p} \tag{6.10}$$

In these equations the nth order vectors $\boldsymbol{\xi}, \dot{\boldsymbol{\xi}}$, and $\ddot{\boldsymbol{\xi}}$ are nodal deflections, velocities, and accelerations, respectively, and mth order vector $\mathbf{s}$ is the complete list of independent element forces.

6.2 Geometric Relations

These are the linear relations between nodal deflections and element deformations. They are discussed in length in Section 2.3. It was shown that

$$\mathbf{v} = \mathbf{B}^T \boldsymbol{\xi} + \mathbf{v}_o \tag{6.11}$$

where $\mathbf{v} \in \mathrm{R}^{m\times 1}$represents the element deformations and $\mathbf{v}_o \in \mathrm{R}^{m\times 1}$stands for the prescribed element deformations, if any. For the definitions of these quantities one may refer to Sections 1.6, 2.1, and 2.3. In the dynamic case $\boldsymbol{\xi}, \mathbf{v}$, and $\boldsymbol{\xi}$ are all functions of time t.

The prescribed element deformations $\mathbf{v}_o$ may consist of uncontrolled and controlled parts, i.e., $\mathbf{\grave{v}}_o$ and $\mathbf{v}_{oc} = \mathbf{I}_\beta \mathbf{u}$, respectively, such that

$$\mathbf{v}_o = \mathbf{\grave{v}}_o + \mathbf{I}_\beta \mathbf{u} \tag{6.12}$$

The part related to the actuators is $\mathbf{I}_\beta \mathbf{u}$, where $\mathbf{I}_\beta$ is the *actuator placement matrix* discussed in earlier chapters and also below. The uncontrolled and controlled parts of prescribed element deformations are discussed in the following subsections.

6.2.1 Uncontrolled Part of Prescribed Element Deformations

The uncontrolled part of prescribed element deformations $\mathbf{\grave{v}}_o \in \mathrm{R}^{m\times 1}$ may be caused by various effects, such as fabrication errors, thermal loads, support settlements, and actuators that are built as part of the element.

When they represent the errors during the manufacture of the structural element, they are called fabrication errors and are shown by $\mathbf{v}_{of} \in \mathrm{R}^{m\times 1}$ which assumes that all elements have fabrication errors; perhaps in many the error has a value of zero. The quantification of $\mathbf{v}^k_{of} \in \mathrm{R}^{a\times 1}$ is made by fixing the kth element at its first vertex and measuring the off-sets due to fabrication error at the second vertex, in the local coordinate system of the kth element.

When the prescribed element deformations represent the effects of thermal loads in free-standing elements, they are shown by $\mathbf{v}_{ot} \in \mathrm{R}^{m\times 1}$. Thermal loads are the difference between the temperatures and/or thermal gradients of the elements after the structure is constructed, i.e., the *working* temperatures and/or gradients, and those before the structure is assembled, i.e., the *ambient* temperatures and/or gradients. The quantification of $\mathbf{v}^k_{ot} \in \mathrm{R}^{a\times 1}$ is made by fixing the kth element at its first vertex in ambient thermal environment and measuring the off-sets at the second vertex, in the local coordinate system of the kth element, when the ambient thermal environment of the element is changed into its working thermal environment.

When the prescribed element deformations represent the effects of support settlements, they are shown by $\mathbf{v}_{os} \in \mathrm{R}^{m\times 1}$. This case is important in earthquake engineering where the excitations to the structure are a result of the movements of the supports attached to the foundations. Let $\boldsymbol{\xi}^s \in R^{b\times 1}$ denote the movements of the supports. From the kinematically determinate base system of the structure (i.e., the state of the structure when all nodes are locked), by applying unit settlements one at a time, one may obtain element elongations corresponding to the support settlements, and thus write

$$\mathbf{v}_{os} = \mathbf{B}_s^T \boldsymbol{\xi}^s \tag{6.13}$$

where $\boldsymbol{\xi}^s \in R^{b\times 1}$are the support movements, and $\mathbf{B}_s^T \in R^{m\times b}$is the matrix where the kth column corresponds to the element elongations due to unit movement in the kth support direction in the kinematically determinate base system. Sparse matrix $\mathbf{B}_s^T \in R^{m\times b}$ contains the direction cosines of the unit vectors of the element coordinate system (and element length if element forces contain moments) of those elements that are incident to the moving supports. For more information, the reader may refer to Section 2.3. Since matrix $\mathbf{B}_s^T$ is assumed time independent, the time dependency $\mathbf{v}_{os}$ and $\boldsymbol{\xi}^s$ are interrelated as

$$\mathbf{v}_{os}(t) = \mathbf{B}_s^T \boldsymbol{\xi}^s(t) \tag{6.14}$$

With the definitions above and superposition one may write

$$\grave{\mathbf{v}}_o = \mathbf{v}_{of} + \mathbf{v}_{ot} + \mathbf{v}_{os} \tag{6.15}$$

Defining

$$\grave{\mathbf{v}}_e = \mathbf{v}_{of} + \mathbf{v}_{ot} \tag{6.16}$$

and using $\mathbf{v}_{os}$ from Eq.(6.13), one may rewrite this equation as

$$\grave{\mathbf{v}}_o = \grave{\mathbf{v}}_e + \mathbf{B}_s^T \boldsymbol{\xi}^s \tag{6.17}$$

where $\grave{\mathbf{v}}_o \in R^{m\times 1}$ represents the uncontrolled part of prescribed element deformations.

6.2.2 Controlled Part of Prescribed Element Deformations

When the prescribed element deformations are caused by the imbedded actuators, they are shown by $\mathbf{v}_{oc} \in R^{m\times 1}$which represents the controlled part of the prescribed element deformations. The quantification of $\mathbf{v}_{oc}^k \in R^{a\times 1}$ is obtained by fixing the kth element at its first vertex and then measuring at the second vertex and in the local coordinate system of the kth element the induced deflections due to the imbedded actuator movements. If the actuators are placed along components $\beta_1, \beta_2, \cdots, \beta_q$ of $\mathbf{v} \in R^{m\times 1}$, and the actuations in these actuators are $u_1, u_2, \cdots, u_q$, then

$$\mathbf{v}_{oc} = \mathbf{I}_\beta \mathbf{u} \tag{6.18}$$

Matrix $\mathbf{I}_\beta \in (\text{Int})^{m\times q}$ is the *actuator placement matrix*, introduced in Chapters 3, 4, and 5, and defined as

$$\mathbf{I}_\beta = [\mathbf{i}_{\beta_1}, \mathbf{i}_{\beta_2}, \cdots, \mathbf{i}_{\beta_q}] \tag{6.19}$$

where $\mathbf{i}_k$ is the kth column of mth order identity matrix $\mathbf{I} \in (\text{Int})^{m\times m}$. Matrix column $\mathbf{u} \in \text{R}^{q\times 1}$ is called the control vector and defined as

$$\mathbf{u} = \begin{bmatrix} u_1 & u_2 & \cdots & u_q \end{bmatrix}^T \tag{6.20}$$

In discrete parameter adaptive structures discussed in this text $\mathbf{u} \in \text{R}^{q\times 1}$ is the only control entity.

6.3 Stiffness Relations of Elements

These are the linear relations between element forces $\mathbf{s} \in \text{R}^{m\times 1}$ and element deformations $\mathbf{v} \in \text{R}^{m\times 1}$that exist because of the mechanical properties of the structure's material. They are studied with sufficient detail in Section 2.4. The stiffness relation of the kth element is

$$\mathbf{s}^k = \mathbf{K}^k \mathbf{v}^k \tag{6.21}$$

where $\mathbf{s}^k$ and $\mathbf{v}^k \in \text{R}^{a\times 1}$are kth element forces and kth element deformations (defined for various types of discrete parameter structures in Tables 1.2 and 1.3), and $\mathbf{K}^k \in \text{R}^{a\times a}$is the kth *element stiffness matrix* which is always real, symmetric and positive definite. In Table 2.7 the definitions of element stiffness matrices are given for straight axis structural elements of various types (without the contributions of shear forces on deformations, and without the imposition of possible zero-element force constraints). Since the kth partitions of $\mathbf{s}$ and $\mathbf{v}$ are $\mathbf{s}^k$ and $\mathbf{v}^k$, respectively, Eq.(6.21) may be written for all elements compactly as $\mathbf{s} = diag(\mathbf{K}^i)\mathbf{v}$, or defining

$$\vec{\mathbf{K}} = diag(\mathbf{K}^i) \tag{6.22}$$

as

$$\mathbf{s} = \vec{\mathbf{K}}\mathbf{v} \tag{6.23}$$

where $\vec{\mathbf{K}} \in \text{R}^{m\times m}$ is real, symmetric, positive definite and block diagonal. [Note that $\vec{\mathbf{K}}$ is a new symbol for $diag(\mathbf{K}^i)$ which is a block diagonal matrix with diagonal blocks $\mathbf{K}^1, \mathbf{K}^2, \cdots, \mathbf{K}^M$.] It is called the *matrix of element stiffnesses.*

6.4 Excitation-Response Relations

By combining the dynamic equilibrium equations of Eq.(6.10), the geometric relations of Eq.(6.11), and the stiffness relations of Eq.(6.23), one may obtain the linear relation between the response quantities $\boldsymbol{\xi}, \mathbf{s}, \mathbf{v}$ and the excitation quantities $\mathbf{p}, \mathbf{v}_o$ as:

$$\begin{bmatrix} \grave{\mathbf{K}} & \mathbf{B} & \cdot \\ \mathbf{B}^T & \cdot & -\mathbf{I} \\ \cdot & -\mathbf{I} & \vec{\mathbf{K}} \end{bmatrix} \begin{Bmatrix} \boldsymbol{\xi} \\ \mathbf{s} \\ \mathbf{v} \end{Bmatrix} = \begin{Bmatrix} \mathbf{p} - \mathbf{M}\ddot{\boldsymbol{\xi}} - \grave{\mathbf{C}}\dot{\boldsymbol{\xi}} \\ -\mathbf{v}_o \\ \cdot \end{Bmatrix} \tag{6.24}$$

where $\grave{\mathbf{K}} \in \mathrm{R}^{n\times n}$ is the global stiffness matrix due to preexisting element forces, and $\vec{\mathbf{K}} \in \mathrm{R}^{m\times m}$ is the block diagonal matrix of element stiffnesses. This form of the excitation-response relationship is useful only if the accelerations $\ddot{\boldsymbol{\xi}}$ and velocities $\dot{\boldsymbol{\xi}}$ on the right-hand side are prescribed independent of $\boldsymbol{\xi}$, as in the so-called *semi-static dynamic analysis.* The nominal form of the excitation-response relations requires that only the excitation quantities remain on the right-hand side. An acceptable form would be:

$$\begin{bmatrix} \mathbf{M}\frac{d^2}{dt^2} + \grave{\mathbf{C}}\frac{d}{dt} + \grave{\mathbf{K}} & \mathbf{B} & \cdot \\ \mathbf{B}^T & \cdot & -\mathbf{I} \\ \cdot & -\mathbf{I} & \vec{\mathbf{K}} \end{bmatrix} \begin{Bmatrix} \boldsymbol{\xi} \\ \mathbf{s} \\ \mathbf{v} \end{Bmatrix} = \begin{Bmatrix} \mathbf{p} \\ -(\dot{\mathbf{v}}_e + \mathbf{B}_s^T\boldsymbol{\xi}^s + \mathbf{I}_\beta \mathbf{u}) \\ \cdot \end{Bmatrix} \tag{6.25}$$

where the definition of $\mathbf{v}_o$ in Eq.(6.12) is used after substitution from Eq.(6.17).

6.4.1 Dynamic Equilibrium Equations in terms of $\boldsymbol{\xi}$

The dynamic equilibrium equations of the nodes can be expressed only in terms of independent nodal deflections $\boldsymbol{\xi}$ and its time derivatives by eliminating $\mathbf{s}$ from the first partition of equations as is commonly done in the displacement method of analysis (see Subsection 3.1.1). From Eq.(6.25), by adding the $\mathbf{B}$ multiple of the third row partition and the $\mathbf{B}\vec{\mathbf{K}}$ multiple of the second row partition to the first row partition of equations, one may obtain:

$$\begin{bmatrix} \mathbf{M}\frac{d^2}{dt^2} + \grave{\mathbf{C}}\frac{d}{dt} + \mathbf{K}' & \cdot & \cdot \\ \mathbf{B}^T & \cdot & -\mathbf{I} \\ \cdot & -\mathbf{I} & \vec{\mathbf{K}} \end{bmatrix} \begin{Bmatrix} \boldsymbol{\xi} \\ \mathbf{s} \\ \mathbf{v} \end{Bmatrix} = \begin{Bmatrix} \mathbf{p} - \mathbf{B}\vec{\mathbf{K}}\mathbf{v}_o \\ -\mathbf{v}_o \\ \cdot \end{Bmatrix} \tag{6.26}$$

where

$$\mathbf{K}' = \grave{\mathbf{K}} + \mathbf{B}\vec{\mathbf{K}}\mathbf{B}^T \tag{6.27}$$

is the global stiffness matrix of the structure and

$$\mathbf{v}_o = \dot{\mathbf{v}}_e + \mathbf{B}_s^T\boldsymbol{\xi}^s + \mathbf{I}_\beta \mathbf{u} \tag{6.28}$$

In Eq.(6.26), the n equations in the first partition, i.e., the *dynamic equilibrium equations of the nodes* in the directions of the components of $\boldsymbol{\xi} \in \mathrm{R}^{n\times 1}$, are in terms of $\ddot{\boldsymbol{\xi}}, \dot{\boldsymbol{\xi}}, \boldsymbol{\xi}$ only. As in the displacement method of analysis, one may solve $\boldsymbol{\xi}$ from these equations. Then one may use $\boldsymbol{\xi}$ in the equations of the second partition, i.e., in the *geometric relations*, to obtain $\mathbf{v}$. Finally one may use $\mathbf{v}$ in the equations of the third partition, i.e., in the *stiffness relations*, to obtain $\mathbf{s}$.

The dynamic equilibrium equations of the nodes in terms of nodal deflections may be obtained from the first row partition of Eq.(6.26) and Eq.(6.28) as:

$$\mathbf{M}\ddot{\boldsymbol{\xi}} + \grave{\mathbf{C}}\dot{\boldsymbol{\xi}} + \mathbf{K}'\boldsymbol{\xi} = \mathbf{p} - \mathbf{B}\vec{\mathbf{K}}(\grave{\mathbf{v}}_e + \mathbf{B}_s^T\boldsymbol{\xi}^s + \mathbf{I}_\beta\mathbf{u}) \tag{6.29}$$

They are n coupled second order linear ordinary differential equations with constant coefficients. Mass matrix $\mathbf{M}$, damping matrix $\grave{\mathbf{C}}$, and stiffness matrix $\mathbf{K}'$ are all real symmetric, positive definite and of order n.

6.4.2 Equations of Motion of Nodes

Since $\mathbf{M}$ is positive definite, $\mathbf{M}^{-1}$ exists. By premultiplying both sides with $\mathbf{M}^{-1}$, one may obtain

$$\ddot{\boldsymbol{\xi}} = -\mathbf{M}^{-1}\grave{\mathbf{C}}\dot{\boldsymbol{\xi}} - \mathbf{M}^{-1}\mathbf{K}'\boldsymbol{\xi} + \mathbf{M}^{-1}\mathbf{p} - \mathbf{M}^{-1}\mathbf{B}\vec{\mathbf{K}}(\grave{\mathbf{v}}_e + \mathbf{B}_s^T\boldsymbol{\xi}^s + \mathbf{I}_\beta\mathbf{u}) \tag{6.30}$$

These equations are called the *second order equations of motion of the nodes.*

The second order equations of motion of the nodes can be reduced to a set of $2n$ first order equations by considering also the identity of $\dot{\boldsymbol{\xi}} = \dot{\boldsymbol{\xi}}$ and defining a new dependent variable $\boldsymbol{\zeta} \in R^{2n\times 1}$ as

$$\boldsymbol{\zeta} = \begin{Bmatrix} \dot{\boldsymbol{\xi}} \\ \boldsymbol{\xi} \end{Bmatrix} \tag{6.31}$$

which defines the state of the structure. This leads to the *first order equations of motion in the state space* as

$$\dot{\boldsymbol{\zeta}} = \mathbf{A}\boldsymbol{\zeta} + \grave{\mathbf{B}}\mathbf{u} + \mathbf{g} \tag{6.32}$$

where $\boldsymbol{\zeta} \in R^{2n\times 1}$is the *state*, and $\mathbf{u} \in R^{q\times 1}$is the *controls*.

$$\mathbf{A} = \begin{bmatrix} -\mathbf{M}^{-1}\grave{\mathbf{C}} & -\mathbf{M}^{-1}\mathbf{K}' \\ \mathbf{I} & \mathbf{0} \end{bmatrix} \tag{6.33}$$

$$\grave{\mathbf{B}} = \begin{bmatrix} -\mathbf{M}^{-1}\mathbf{B}\vec{\mathbf{K}}\mathbf{I}_\beta \\ \mathbf{0} \end{bmatrix} \tag{6.34}$$

and

$$\mathbf{g} = \begin{Bmatrix} \mathbf{M}^{-1}\mathbf{p} - \mathbf{M}^{-1}\mathbf{B}\vec{\mathbf{K}}(\grave{\mathbf{v}}_e + \mathbf{B}_s^T\boldsymbol{\xi}^s) \\ \mathbf{o} \end{Bmatrix} \tag{6.35}$$

In modern control theories the first order equations of motion in the form of Eq.(6.32) are preferred. The first order system is called *autonomous* if $\mathbf{g} = \mathbf{o}$; otherwise, it is called *non-autonomous.*

6.4.3 Initial Conditions

In the next chapter the inverse of the excitation-response relations of this chapter is studied. Since the dynamic equilibrium equations of nodes are the second order ordinary differential equations involving $\boldsymbol{\xi}, \dot{\boldsymbol{\xi}}, \ddot{\boldsymbol{\xi}}$, for the inverse relations one needs the initial values of $\boldsymbol{\xi}$, and $\dot{\boldsymbol{\xi}}$, i.e., the *initial position of nodes*

$$\boldsymbol{\xi}_o = \boldsymbol{\xi}_{t=0} = \boldsymbol{\xi}(t=0) = \boldsymbol{\xi}(0) \in \mathrm{R}^{n\times 1} \tag{6.36}$$

and the *initial velocities of nodes*

$$\boldsymbol{\xi}'_o = \dot{\boldsymbol{\xi}}_{t=0} = \dot{\boldsymbol{\xi}}(t=0) = \dot{\boldsymbol{\xi}}(0) \in \mathrm{R}^{n\times 1} \tag{6.37}$$

must be known.

Alternately, if the state space form of the equations of motions is considered, then the *initial state of nodes*, consisting of nodal velocities and nodal positions, i.e.:

$$\boldsymbol{\zeta}_o = \boldsymbol{\zeta}_{t=0} = \boldsymbol{\zeta}(t=0) = \boldsymbol{\zeta}(0) = \left\{ \begin{array}{c} \boldsymbol{\xi}'_o \\ \boldsymbol{\xi}_o \end{array} \right\} \in \mathrm{R}^{2n\times 1} \tag{6.38}$$

must be known. Then, mathematically speaking, the *initial value problem* related to Eq.(6.32):

$$\begin{array}{lll} \dot{\boldsymbol{\zeta}} = \mathbf{A}\boldsymbol{\zeta} + \grave{\mathbf{B}}\mathbf{u} + \mathbf{g} & \text{for } t > 0 & \text{and} \\ \boldsymbol{\zeta}(0) = \boldsymbol{\zeta}_o \quad \text{at} & t = 0 & \end{array} \tag{6.39}$$

is a well-posed problem.

If the initial state of the nodes is the *equilibrium state*, then $\boldsymbol{\zeta}_o = \mathbf{o}$. However, if the initial state is a *non-equilibrium state*, then $\boldsymbol{\zeta}_o \neq \mathbf{o}$.

As discussed in the next chapter, the dynamic response of a structure consists of the superposition of the contributions of loads $\mathbf{g} \in \mathrm{R}^{n\times 1}$, controls $\mathbf{u} \in \mathrm{R}^{q\times 1}$ (if any), and initial state $\boldsymbol{\zeta}_o \in \mathrm{R}^{2n\times 1}$.

Sometimes one is interested in the response under a repetitious load acting on the structure indefinitely. In such cases the contributions of the initial state will damp out and it will not be present in the long time response. The long time response under repetitious loads is called *steady-state response*. In steady state response problems, one does not need to know the initial state.

When the contribution of initial state is of the same order of magnitude as those of loads and controls, the response is called *transient response*. The determination of transient response requires the knowledge of the initial state.

7

Inverse Relations, Dynamic Case

In this chapter, the inverse relations are obtained for the linear excitation-response relations of time-invariant discrete parameter structures. In the inverse relations the response is expressed in terms of excitation quantities. They are needed for investigating the possibilities of controlling the response, assessing the effectiveness of any given control scheme, and for the placement actuators and sensors.

7.1 Method for Obtaining the Inverse Relations

The linear excitation-response relations of time-invariant discrete parameter structures are obtained in Section 6.4 as

$$\begin{bmatrix} \mathbf{M}\frac{d^2}{dt^2} + \dot{\mathbf{C}}\frac{d}{dt} + \dot{\mathbf{K}} & \mathbf{B} & \cdot \\ \mathbf{B}^T & \cdot & -\mathbf{I} \\ \cdot & -\mathbf{I} & \vec{\mathbf{K}} \end{bmatrix} \begin{Bmatrix} \boldsymbol{\xi} \\ \mathbf{s} \\ \mathbf{v} \end{Bmatrix} = \begin{Bmatrix} \mathbf{p} \\ -(\dot{\mathbf{v}}_e + \mathbf{B}_s^T \boldsymbol{\xi}^s + \mathbf{I}_\beta \mathbf{u}) \\ \cdot \end{Bmatrix} \tag{7.1}$$

for $t > 0$, and the necessary initial conditions at $t = 0$ are stated in Subsection 6.4.3 as

$$\left.\begin{aligned} \boldsymbol{\xi}(0) &= \boldsymbol{\xi}_o \\ \dot{\boldsymbol{\xi}}(0) &= \boldsymbol{\xi}'_o \end{aligned}\right\} \quad \text{given at } t = 0 \tag{7.2}$$

In these equations the response quantities are

$\boldsymbol{\xi}$	$\in \mathrm{R}^{n \times 1}$	nodal deflections
$\mathbf{s}$	$\in \mathrm{R}^{m \times 1}$	element forces
$\mathbf{v}$	$\in \mathrm{R}^{m \times 1}$	element deformations

and the system related parameters are

$\mathbf{B}$	$\in R^{n\times m}$	matrix defining internal nodal forces as $\mathbf{Bs}$
$\mathbf{B}_s^T$	$\in R^{m\times b}$	matrix relating $\boldsymbol{\xi}^s$ to element deformations
$\grave{\mathbf{C}}$	$\in R^{n\times n}$	damping matrix
$\grave{\mathbf{K}}$	$\in R^{n\times n}$	stiffness matrix due to preexisting stresses
$\vec{\mathbf{K}}$	$\in R^{m\times m}$	block diagonal matrix of element stiffnesses
$\mathbf{I}_\beta$	$\in (\mathrm{Int})^{m\times q}$	actuator placement matrix

The quantities

$\boldsymbol{\xi}_o$	$\in R^{n\times 1}$	initial deflections of the nodes
$\boldsymbol{\xi}'_o$	$\in R^{n\times 1}$	initial velocities of the nodes
$\mathbf{p}$	$\in R^{n\times 1}$	nodal loads
$\grave{\mathbf{v}}_e$	$\in R^{m\times 1}$	load type prescribed element deformations
$\boldsymbol{\xi}^s$	$\in R^{b\times 1}$	support movements
$\mathbf{u}$	$\in R^{q\times 1}$	actuator induced element deformations

are the excitation quantities. Although the initial conditions $\boldsymbol{\xi}_o$ and $\boldsymbol{\xi}'_o$ are constant vectors, the nodal loads $\mathbf{p}$, the load type prescribed element deformations $\grave{\mathbf{v}}_e$, support movements $\boldsymbol{\xi}^s$, and the actuator induced element deformations $\mathbf{u}$ are all functions of time. Sometimes it may be more useful to separately designate the time dependency of vectors $\mathbf{p}$, $\grave{\mathbf{v}}_e$, $\boldsymbol{\xi}^s$, and $\mathbf{u}$. The notation for this is explained next.

Let c_k denote any kth component of any of the vectors $\mathbf{p}$, $\grave{\mathbf{v}}_e$, $\boldsymbol{\xi}^s$, and $\mathbf{u}$. Let c'_k denote the largest magnitude ordinate in the graph of c_k as a function of time t. If the largest magnitude ordinate is not known *a priori*, then c'_k may be an estimate of it. The time dependency of c_k is identified by a physically dimensionless quantity $\hat{c}_k$ which is defined as $\hat{c}_k = c_k/c'_k$ when $c'_k \neq 0$, and as $\hat{c}_k = 0$ when $c'_k = 0$. It follows that $c_k(t) = \hat{c}_k(t)c'_k$. Applying this notation, one may write

$$\left.\begin{aligned} \mathbf{p} &= diag(\hat{p}_k)\ \mathbf{p}' \\ \grave{\mathbf{v}}_e &= diag(\hat{v}_e^k)\grave{\mathbf{v}}'_e \\ \boldsymbol{\xi}^s &= diag(\hat{\xi}_k^s)\boldsymbol{\xi}^{s\prime} \\ \mathbf{u} &= diag\ (\hat{u}_k)\ \mathbf{u}' \end{aligned}\right\} \tag{7.3}$$

where the time dependent matrices $diag\ (\hat{p}_k) \in R^{n\times n}$, $diag\ (\hat{v}_e^k) \in R^{m\times m}$, $diag\ (\hat{\xi}_k^s) \in R^{b\times b}$, and $diag(\hat{u}_k) \in R^{q\times q}$ are with no physical dimension, and constant vectors $\mathbf{p}' \in R^{n\times 1}$, $\grave{\mathbf{v}}'_e \in R^{m\times 1}$, $\boldsymbol{\xi}^{s\prime} \in R^{b\times 1}$ and $\mathbf{u}' \in R^{q\times 1}$ are with physical dimension.

As it is done in Subsection 6.4.1, by a displacement approach, in Eq.(7.1) one may substitute $\mathbf{v}$ from the equations of the second row partition, i.e., geometric relations, in the equations of the third row partition, i.e., the stiffness relations, and then use $\mathbf{s}$ from the latter in the equations of the first row partition, i.e., in the dynamic equilibrium equations, and thus express them in terms of $\boldsymbol{\xi}$ and its time derivatives only. This procedure transforms Eq.(7.1) into

$$\begin{bmatrix} \mathbf{M}\frac{d^2}{dt^2} + \grave{\mathbf{C}}\frac{d}{dt} + \mathbf{K}' & \cdot & \cdot \\ \mathbf{B}^T & \cdot & -\mathbf{I} \\ \cdot & \mathbf{I} & \vec{\mathbf{K}} \end{bmatrix} \begin{Bmatrix} \boldsymbol{\xi} \\ \mathbf{s} \\ \mathbf{v} \end{Bmatrix} = \begin{Bmatrix} \mathbf{p} - \mathbf{B}\vec{\mathbf{K}}\mathbf{v}_o \\ -\mathbf{v}_o \\ \cdot \end{Bmatrix} \tag{7.4}$$

where

$$\mathbf{K}' = \grave{\mathbf{K}} + \mathbf{B}\vec{\mathbf{K}}\mathbf{B}^T \tag{7.5}$$

is the *global stiffness matrix* of the structure, and

$$\mathbf{v}_o = \grave{\mathbf{v}}_e + \mathbf{B}_s^T\boldsymbol{\xi}^s + \mathbf{I}_\beta\mathbf{u} \tag{7.6}$$

is the *prescribed element deformations.* From the first row partition, one may obtain the dynamic equilibrium equations in terms of $\boldsymbol{\xi}$ and its time derivatives as:

$$\begin{aligned} &\mathbf{M}\ddot{\boldsymbol{\xi}} + \grave{\mathbf{C}}\dot{\boldsymbol{\xi}} + \mathbf{K}'\boldsymbol{\xi} = \mathbf{p} - \mathbf{B}\vec{\mathbf{K}}\grave{\mathbf{v}}_e - \mathbf{B}\vec{\mathbf{K}}\mathbf{B}_s^T\boldsymbol{\xi}^s - \mathbf{B}\vec{\mathbf{K}}\mathbf{I}_\beta\mathbf{u} \quad \text{for } t > 0\,, \\ &\boldsymbol{\xi}(0) = \boldsymbol{\xi}_o \text{ and } \dot{\boldsymbol{\xi}}(0) = \boldsymbol{\xi}_o' \quad \text{at } t = 0 \end{aligned} \tag{7.7}$$

From the second row partition one may write the geometric relations as

$$\mathbf{v} = \mathbf{B}^T\boldsymbol{\xi} + (\grave{\mathbf{v}}_e + \mathbf{B}_s^T\boldsymbol{\xi}^s + \mathbf{I}_\beta\mathbf{u}) \tag{7.8}$$

and from the third partition the stiffness relations of the elements may be written as

$$\mathbf{s} = \vec{\mathbf{K}}\mathbf{v} \tag{7.9}$$

Using superposition, the solution of the linear second order coupled ordinary differential equations with constant coefficients appearing in Eq.(7.7) may be expressed as the sum of the individual contributions of

$\boldsymbol{\xi}_o$	$\in \mathrm{R}^{n\times 1}$	initial deflections of the nodes
$\boldsymbol{\xi}_o'$	$\in \mathrm{R}^{n\times 1}$	initial velocities of the nodes
$\mathbf{p} = diag\ (\hat{p}_k)\ \mathbf{p}'$	$\in \mathrm{R}^{n\times 1}$	nodal loads
$\grave{\mathbf{v}}_e = diag\ (\hat{v}_e^k)\ \grave{\mathbf{v}}_e'$	$\in \mathrm{R}^{m\times 1}$	load type prescribed element deformations
$\boldsymbol{\xi}^s = diag\ (\hat{\xi}_k^s)\boldsymbol{\xi}^{s\prime}$	$\in \mathrm{R}^{b\times 1}$	support movements
$\mathbf{u} = diag\ (\hat{u}_k)\ \mathbf{u}'$	$\in \mathrm{R}^{q\times 1}$	actuator induced element deformations

in the form of

$$\boldsymbol{\xi}(t) = \begin{bmatrix} \mathbf{A}_{\xi_o} & \mathbf{A}_{\xi_o'} & \mathbf{A}_{p'} & \mathbf{A}_{\grave{v}_e'} & \mathbf{A}_{\xi^{s\prime}} & \mathbf{A}_{u'} \end{bmatrix} \begin{Bmatrix} \boldsymbol{\xi}_o \\ \boldsymbol{\xi}_o' \\ \mathbf{p}' \\ \grave{\mathbf{v}}_e' \\ \boldsymbol{\xi}^{s\prime} \\ \mathbf{u}' \end{Bmatrix} \tag{7.10}$$

where the time effects are included in matrices $\mathbf{A}_{\xi_o} \in \mathrm{R}^{n\times n}$, $\mathbf{A}_{\xi'_o} \in \mathrm{R}^{n\times n}$, $\mathbf{A}_{p'} \in \mathrm{R}^{n\times n}$, $\mathbf{A}_{\hat{v}'_e} \in \mathrm{R}^{n\times m}$, $\mathbf{A}_{\xi^{s\prime}} \in \mathrm{R}^{n\times b}$, and $\mathbf{A}_{u'} \in \mathrm{R}^{n\times q}$. Note that the multipliers $\boldsymbol{\xi}_o$, $\boldsymbol{\xi}'_o$, $\mathbf{p}'$, $\hat{\mathbf{v}}'_e$, $\boldsymbol{\xi}^{s\prime}$, and $\mathbf{u}'$ in the superposition equation are all constant vectors.

Using Eqs.(7.8, 7.9, 7.6), one may write

$$\left\{\begin{matrix}\boldsymbol{\xi}\\ \mathbf{s}\\ \mathbf{v}\end{matrix}\right\} = \begin{bmatrix}\cdot\\ \vec{\mathbf{K}}\\ \mathbf{I}\end{bmatrix}\mathbf{v}_o + \begin{bmatrix}\mathbf{I}\\ \vec{\mathbf{K}}\mathbf{B}^T\\ \mathbf{B}^T\end{bmatrix}\boldsymbol{\xi}(t) \tag{7.11}$$

The substitution of $\boldsymbol{\xi}(t)$ from Eq.(7.10) into this equation leads to the inverse relations as

$$\left\{\begin{matrix}\boldsymbol{\xi}\\ \mathbf{s}\\ \mathbf{v}\end{matrix}\right\} = \hat{\mathbf{D}}(t)\left\{\begin{matrix}\boldsymbol{\xi}_o\\ \boldsymbol{\xi}'_o\\ \mathbf{p}'\\ \hat{\mathbf{v}}'_e\\ \boldsymbol{\xi}^{s\prime}\\ \mathbf{u}'\end{matrix}\right\} \tag{7.12}$$

where

$$\hat{\mathbf{D}}(t) = \left[\begin{matrix}\mathbf{A}_{\xi_o} & \mathbf{A}_{\xi'_o} & \mathbf{A}_{p'} & \mathbf{A}_{\hat{v}'_e}\\ \vec{\mathbf{K}}\mathbf{B}^T\mathbf{A}_{\xi_o} & \vec{\mathbf{K}}\mathbf{B}^T\mathbf{A}_{\xi'_o} & \vec{\mathbf{K}}\mathbf{B}^T\mathbf{A}_{p'} & \vec{\mathbf{K}}[diag(\hat{v}^k_e)+\mathbf{B}^T\mathbf{A}_{\hat{v}'_e}]\\ \mathbf{B}^T\mathbf{A}_{\xi_o} & \mathbf{B}^T\mathbf{A}_{\xi'_o} & \mathbf{B}^T\mathbf{A}_{p'} & diag(\hat{v}^k_e)+\mathbf{B}^T\mathbf{A}_{\hat{v}'_e}\end{matrix}\right.$$
$$\left.\begin{matrix}\mathbf{A}_{\xi^{s\prime}} & \mathbf{A}_{u'}\\ \vec{\mathbf{K}}[\mathbf{B}_s^T diag(\hat{\xi}^s_k)+\mathbf{B}^T\mathbf{A}_{\xi^{s\prime}}] & \vec{\mathbf{K}}[\mathbf{I}_\beta diag(\hat{u}_k)+\mathbf{B}^T\mathbf{A}_{u'}]\\ \mathbf{B}_s^T diag(\hat{\xi}^s_k)+\mathbf{B}^T\mathbf{A}_{\xi^{s\prime}} & \mathbf{I}_\beta diag(\hat{u}_k)+\mathbf{B}^T\mathbf{A}_{u'}\end{matrix}\right] \tag{7.13}$$

Matrix $\hat{\mathbf{D}}(t)$ is called the *dynamic load factor matrix*, since some norm of $\hat{\mathbf{D}}(t)$ partitions may be used as a multiplier of $\boldsymbol{\xi}_o$, $\boldsymbol{\xi}'_o$, $\mathbf{p}'$, $\hat{\mathbf{v}}'_e$, $\boldsymbol{\xi}^{s\prime}$, and $\mathbf{u}'$ in approximating upper bound values for response magnitudes.

In the following sections the explicit solution of the well-posed initial value problem of Eq.(7.7) is discussed. The explicit solution in the form of Eq.(7.10) furnishes the definitions of matrices $\mathbf{A}_{\xi_o} \in \mathrm{R}^{n\times n}$, $\mathbf{A}_{\xi'_o} \in \mathrm{R}^{n\times n}$, $\mathbf{A}_{p'} \in \mathrm{R}^{n\times n}$, $\mathbf{A}_{\hat{v}'_e} \in \mathrm{R}^{n\times m}$, $\mathbf{A}_{\xi^{s\prime}} \in \mathrm{R}^{n\times b}$, and $\mathbf{A}_{u'} \in \mathrm{R}^{n\times q}$. The knowledge of these matrices, in turn, uniquely defines the inverse relations given in Eqs.(7.12 and 7.13).

7.2 Undamped Free Vibrations of Nodes

Consider a discrete parameter structure with zero damping. If the nodes of this structure are pulled slowly away from their equilibrium state by an amount $\boldsymbol{\xi}_o$ by an external agent who lets them go at time $t = 0$ with initial velocities $\boldsymbol{\xi}'_o$, the resulting motion is called the *undamped free vibrations of the structure.* It may be studied by using Eq.(7.7) with zero values of $\dot{\mathbf{C}}$, $\mathbf{p}$, $\dot{\mathbf{v}}_o$, and $\mathbf{u}$, i.e., by using

$$\begin{aligned} &\mathbf{M}\ddot{\boldsymbol{\xi}} + \mathbf{K}'\boldsymbol{\xi} = \mathbf{o} \quad \text{for } t > 0 \text{ and} \\ &\boldsymbol{\xi}(0) = \boldsymbol{\xi}_o \text{ and } \dot{\boldsymbol{\xi}}(0) = \boldsymbol{\xi}'_o \quad \text{at } t = 0 \end{aligned} \tag{7.14}$$

The periodic motions of the nodes may be expressed by

$$\boldsymbol{\xi}(t) \longrightarrow \mathbf{x}e^{j(\omega t+\phi)} \tag{7.15}$$

where ω is the frequency, ϕ is the phase angle, $\mathbf{x} \in \mathrm{R}^{n\times 1}$is the vector listing the amplitudes of the motion, and $j = \sqrt{-1}$. The use of $\boldsymbol{\xi}(t)$ from Eq.(7.15) in the differential equation of Eq.(7.14) leads to a *general algebraic eigenvalue* problem:

$$\mathbf{K}'\mathbf{x} = \lambda\mathbf{M}\mathbf{x} \tag{7.16}$$

where

$$\lambda = \omega^2 \tag{7.17}$$

Since this can be rewritten as $(\mathbf{K}'-\lambda\mathbf{M})\mathbf{x} = \mathbf{o}$, i.e., as a set of homogeneous algebraic equations, one may observe that $\mathbf{x} = \mathbf{o}$ is the trivial solution. Nontrivial solutions for $\mathbf{x}$ are possible only if matrix $(\mathbf{K}' - \lambda\mathbf{M})$ is singular, i.e., if the square of vibration frequencies, i.e., λ_i, satisfy the *characteristic equation*:

$$\det(\mathbf{K}' - \lambda\mathbf{M}) = p_n(\lambda) = 0 \tag{7.18}$$

By the *fundamental theorem of algebra*, the nth degree polynomial equation $p_n(\lambda) = 0$ has precisely n roots, $\lambda_1, \lambda_2, \cdots, \lambda_n$, which are the eigenvalues of the problem in Eq.(7.16). Since $\mathbf{K}' \in \mathrm{R}^{n\times n}$ and $\mathbf{M} \in \mathrm{R}^{n\times n}$ are real and symmetric matrices, it can be shown that all eigenvalues are real numbers. Since these matrices are also positive definite, it can be shown that the eigenvalues are all positive, hence

$$0 < \lambda_1 \leq \lambda_2 \leq \cdots \leq \lambda_n \tag{7.19}$$

Note that the smallest eigenvalue carries the label 1, and the largest one carries the label n. Using the definition in Eq.(7.17), one may write

$$\omega_i = \sqrt{\lambda_i} \qquad \textit{for} \quad i = 1, 2, \cdots, n \tag{7.20}$$

namely, the discrete structure has precisely n discrete vibration frequencies. These are the undamped natural vibration frequencies of structure. The smallest one, i.e., ω_1, is called the *fundamental frequency.*

Using λ_i in Eq.(7.16) one may compute the corresponding amplitudes $\mathbf{x}'_i$ from $\mathbf{K}'\mathbf{x}'_i = \lambda_i\mathbf{M}\mathbf{x}'_i$, albeit with at least one arbitrary scale factor. Vectors $\mathbf{x}'_i \in \mathrm{R}^{n\times 1}$, $i = 1, 2, \cdots, n$, are the eigenvectors associated with the eigenvalues $\lambda_1, \lambda_2, \cdots, \lambda_n$, respectively. Since the mass matrix $\mathbf{M}$ is positive definite, the quantity $\mathbf{x}_i'^T \mathbf{M} \mathbf{x}_i'$ is always larger than zero and therefore it may be used to obtain the *normalized eigenvector* $\mathbf{x}_i$ corresponding to $\mathbf{x}_i'$:

$$\mathbf{x}_i = \frac{1}{\sqrt{\mathbf{x}_i'^T \mathbf{M} \mathbf{x}_i'}} \mathbf{x}_i' \qquad \text{for} \quad i = 1, 2, \cdots, n \tag{7.21}$$

Note that for normalized eigenvectors, the expression

$$\mathbf{x}_i^T \mathbf{M} \mathbf{x}_i = 1 \qquad \text{for} \quad i = 1, 2, \cdots, n \tag{7.22}$$

holds. Matrix $\mathbf{X} \in \mathrm{R}^{n \times n}$, whose ith column $\mathbf{x}_i$ is the ith normalized eigenvector, may be displayed as

$$\mathbf{X} = [\mathbf{x}_1, \mathbf{x}_2, \cdots, \mathbf{x}_n] \tag{7.23}$$

It may be used to consider all of the eigenvectors. In the parlance of structural dynamics, the normalized eigenvectors are referred to as the *normalized mode shapes* of structure. Therefore the n columns of matrix $\mathbf{X} \in \mathrm{R}^{n \times n}$ represent also the n normalized mode shapes. The mode shape corresponding to the first eigenvalue, i.e., $\mathbf{x}_1$, is called the *fundamental mode shape.*

If an eigenvalue is of multiplicity s, then any vector in the subspace generated by the s eigenvectors of that eigenvalue is also an eigenvector. This may help one to choose any desired linearly independent s directions of the subspace as the s eigenvectors.

Since both $\mathbf{K}'$ and $\mathbf{M}$ are real and symmetric, it can be shown that (a) the eigenvectors are linearly independent, (b) they span the n-space, and (c) they are mutually orthogonal with respect to $\mathbf{K}'$ and $\mathbf{M}$ matrices. With Eq.(7.22) and the orthogonality of eigenvectors with respect to $\mathbf{M}$ one may write

$$\mathbf{x}_i^T \mathbf{M} \mathbf{x}_j = \delta_{ij} \qquad \text{for} \quad i = 1, 2, \cdots, n; \quad j = 1, 2, \cdots, n \tag{7.24}$$

where δ_{ij} is the Kronecker delta (which has a value of 1 if $i = j$, and it has a value of 0 if $i \neq j$). Note that the n^2 scalar equations of Eq.(7.24) may be rewritten compactly as

$$\mathbf{X}^T \mathbf{M} \mathbf{X} = \mathbf{I} \tag{7.25}$$

where $\mathbf{I} \in (\mathrm{Int})^{n \times n}$ is the identity matrix.

Since the ith eigenpair $(\lambda_i, \mathbf{x}_i)$ satisfies Eq.(7.16), one may write

$$\mathbf{K}' \mathbf{x}_i = \lambda_i \mathbf{M} \mathbf{x}_i \qquad \text{for} \quad i = 1, 2, \cdots, n \tag{7.26}$$

These n equations may be rewritten as

$$\mathbf{K}' \mathbf{X} = \mathbf{M} \mathbf{X} \, diag(\lambda_i) \tag{7.27}$$

Premultiplication of both sides of Eq.(7.27) by $\mathbf{X}^T$ and the use of Eq.(7.25) lead to

$$\mathbf{X}^T \mathbf{K}' \mathbf{X} = diag(\boldsymbol{\lambda}_i) \tag{7.28}$$

In view of this and Eq.(7.25) one may state that the *mode shapes are orthogonal with respect to global stiffness and mass matrices* $\mathbf{K}'$ *and* $\mathbf{M}$, respectively.

The orthogonality of mode shapes is a very important property in uncoupling of the coupled second order differential equations of the motion in Eq.(7.14). Since the n normalized mode shapes shown in Eq.(7.23) are linearly independent and span the n-space, any vector in the n-space, such as $\boldsymbol{\xi} = \boldsymbol{\xi}(t)$, may be expressed in terms of them. In fact in structural dynamics, it is customary to use the columns of $\mathbf{X}$ as the basis vectors of the n-space. One may express $\boldsymbol{\xi}(t)$ in terms of the columns of $\mathbf{X}$ as

$$\boldsymbol{\xi}(t) = \sum_{j=1}^{n} \eta_j \mathbf{x}_j = [\mathbf{x}_1, \mathbf{x}_2, \cdots, \mathbf{x}_n] \begin{Bmatrix} \eta_1 \\ \eta_2 \\ \vdots \\ \eta_n \end{Bmatrix} \tag{7.29}$$

or

$$\boldsymbol{\xi} = \mathbf{X}\boldsymbol{\eta} \tag{7.30}$$

where the components of $\boldsymbol{\eta} \in \mathrm{R}^{n\times 1}$, i.e., $\eta_1, \eta_2, \cdots, \eta_n$, are called modal participation factors. They are functions of time.

One may use $\boldsymbol{\xi}(t)$ from Eq.(7.29) in Eq.(7.14) to write the latter equations in the new variable $\boldsymbol{\eta}(t)$ as

$$\begin{aligned} &\mathbf{M}\mathbf{X}\ddot{\boldsymbol{\eta}} + \mathbf{K}'\mathbf{X}\boldsymbol{\eta} = \mathbf{o} \quad \text{for } t > 0 \text{ and} \\ &\boldsymbol{\eta}(0) = \boldsymbol{\eta}_o \text{ and } \dot{\boldsymbol{\eta}}(0) = \boldsymbol{\eta}'_o \quad \text{at } t = 0 \end{aligned} \tag{7.31}$$

Premultiplying both sides of the differential equation by $\mathbf{X}^T$ and using the orthogonality relations of Eqs.(7.25 and 7.28), one may obtain

$$\begin{aligned} &\ddot{\boldsymbol{\eta}} + diag(\lambda_k)\boldsymbol{\eta} = \mathbf{o} \quad \text{for } t > 0 \text{ and} \\ &\boldsymbol{\eta}(0) = \boldsymbol{\eta}_o \text{ and } \dot{\boldsymbol{\eta}}(0) = \boldsymbol{\eta}'_o \quad \text{at } t = 0 \end{aligned} \tag{7.32}$$

Since the eigenvectors span the n-space the columns of $\mathbf{X}$ are linearly independent; hence, $\mathbf{X}^{-1}$ exists. Then one may obtain from Eq.(7.25)

$$\mathbf{X}^{-1} = \mathbf{X}^T\mathbf{M} \tag{7.33}$$

This may be used to advantage whenever the computation $\mathbf{X}^{-1}$is required. Using Eq.(7.30), the initial values of the participation factors, i.e., $\boldsymbol{\eta}_o$ and $\boldsymbol{\eta}'_o$, may be expressed in terms of $\boldsymbol{\xi}_o$ and $\boldsymbol{\xi}'_o$ as

$$\boldsymbol{\eta}_o = \mathbf{X}^{-1}\boldsymbol{\xi}_o \tag{7.34}$$

and

$$\boldsymbol{\eta}'_o = \mathbf{X}^{-1}\boldsymbol{\xi}'_o \tag{7.35}$$

The inverse of Eq.(7.30) may be stated as

$$\boldsymbol{\eta} = \mathbf{X}^{-1}\boldsymbol{\xi} \tag{7.36}$$

One may rewrite Eq.(7.32) in scalar form as:

$$\left.\begin{array}{ll} \ddot{\eta}_k + \lambda_k \eta_k = 0 & \text{for } t > 0 \text{ and} \\ \eta_k(0) = \eta_{ok} \text{ and } \dot{\eta}_k(0) = \eta'_{ok} & \text{at } t = 0 \end{array}\right\} \quad \text{for } k = 1, 2, \cdots, n \tag{7.37}$$

which are uncoupled n second order ordinary differential equations with constant coefficients. The solutions of these equations are

$$\eta_k(t) = \eta_{ok} \cos \omega_k t + \eta'_{ok} \frac{\sin \omega_k t}{\omega_k} \quad \text{for } k = 1, 2, \cdots, n \tag{7.38}$$

which can be rewritten as

$$\boldsymbol{\eta}(t) = diag\ (\cos \omega_k t)\ \boldsymbol{\eta}_o + diag\ (\frac{\sin \omega_k t}{\omega_k})\ \boldsymbol{\eta}'_o \tag{7.39}$$

Using Eqs.(7.34 and 7.35) for $\boldsymbol{\eta}_o$ and $\boldsymbol{\eta}'_o$, respectively, and then substituting $\boldsymbol{\eta}$ from this equation into Eq.(7.30), one may obtain the solution of Eq.(7.14) as

$$\boldsymbol{\xi}(t) = \mathbf{X} diag(\cos \omega_k t) \mathbf{X}^{-1} \boldsymbol{\xi}_o + \mathbf{X} diag(\frac{\sin \omega_k t}{\omega_k}) \mathbf{X}^{-1} \boldsymbol{\xi}'_o \tag{7.40}$$

7.3 Damped Free Vibrations of Nodes

Consider a discrete parameter structure with damping. If the nodes of this structure are pulled away from their equilibrium state slowly an amount of $\boldsymbol{\xi}_o$ by an external agent who lets them go at time $t = 0$ with initial velocities $\boldsymbol{\xi}'_o$, the resulting motion is called the *damped free vibrations of the structure.* It may be studied by using Eq.(7.7) with zero values of $\mathbf{p}$, $\dot{\mathbf{v}}_o$, $\boldsymbol{\xi}^s$, and $\mathbf{u}$, i.e., by using

$$\begin{array}{ll} \mathbf{M}\ddot{\boldsymbol{\xi}} + \grave{\mathbf{C}}\dot{\boldsymbol{\xi}} + \mathbf{K}'\boldsymbol{\xi} = \mathbf{o} & \text{for } t > 0 \text{ and} \\ \boldsymbol{\xi}(0) = \boldsymbol{\xi}_o \text{ and } \dot{\boldsymbol{\xi}}(0) = \boldsymbol{\xi}'_o & \text{at } t = 0 \end{array} \tag{7.41}$$

This problem is very similar to the one displayed by Eq.(7.14), except for the presence of the $\grave{\mathbf{C}}\dot{\boldsymbol{\xi}}$ term in the differential equation.

In order to uncouple the differential equations, as in the previous section, one expresses $\boldsymbol{\xi}$ in terms of basis vectors $[\mathbf{x}_1, \mathbf{x}_2, \cdots, \mathbf{x}_1]$ which are the normalized mode shapes of the undamped free vibration problem of the structure. As studied in the previous section, these are the n columns of the matrix of eigenvectors $\mathbf{X} \in R^{n \times n}$ of the general eigenvalue problem $\mathbf{K}'\mathbf{x} = \lambda \mathbf{M}\mathbf{x}$. The description of the motion in the new basis is $\boldsymbol{\eta} \in R^{n \times 1}$ such that

$$\boldsymbol{\xi} = \mathbf{X}\boldsymbol{\eta} \tag{7.42}$$

The n components of $\boldsymbol{\eta}$ are the participation factors of the n normalized mode shapes of the freely vibrating structure. Using $\boldsymbol{\xi}$ from this equation in Eq.(7.41) leads to

$$\begin{aligned} &\mathbf{M}\mathbf{X}\ddot{\boldsymbol{\eta}} + \grave{\mathbf{C}}\mathbf{X}\dot{\boldsymbol{\eta}} + \mathbf{K}'\mathbf{X}\boldsymbol{\eta} = \mathbf{o} \quad \text{for } t > 0 \text{ and} \\ &\boldsymbol{\eta}(0) = \boldsymbol{\eta}_o \text{ and } \dot{\boldsymbol{\eta}}(0) = \boldsymbol{\eta}'_o \quad \text{at } t = 0 \end{aligned} \tag{7.43}$$

As discussed in Subsection 6.1.3, it is reasonable to assume that the damping matrix $\grave{\mathbf{C}} \in \mathrm{R}^{n\times n}$ may be taken as a linear function of mass and stiffness matrices, i.e.,

$$\grave{\mathbf{C}} = \alpha\mathbf{M} + \beta\mathbf{K}' \tag{7.44}$$

Matrix $\grave{\mathbf{C}}$ is called the *proportional damping matrix.* Using $\grave{\mathbf{C}}$ from this equation in Eq.(7.43), then premultiplying both sides of the latter by $\mathbf{X}^T$ and using the orthogonality of the modes to replace $\mathbf{X}^T\mathbf{M}\mathbf{X}$ with $\mathbf{I}$ and $\mathbf{X}^T\mathbf{K}'\mathbf{X}$ with $diag(\omega_k^2)$ one may obtain

$$\begin{aligned} &\ddot{\boldsymbol{\eta}} + diag(\alpha + \beta\omega_k^2)\dot{\boldsymbol{\eta}} + diag(\omega_k^2)\boldsymbol{\eta} = \mathbf{o} \quad \text{for } t > 0 \text{ and} \\ &\boldsymbol{\eta}(0) = \boldsymbol{\eta}_o \text{ and } \dot{\boldsymbol{\eta}}(0) = \boldsymbol{\eta}'_o \quad \text{at } t = 0 \end{aligned} \tag{7.45}$$

Note that Eq.(7.42) may be used to express the initial conditions $\boldsymbol{\eta}_o$ and $\boldsymbol{\eta}'_o$ in terms of $\boldsymbol{\xi}_o$ and $\boldsymbol{\xi}'_o$ as

$$\boldsymbol{\eta}_o = \mathbf{X}^{-1}\boldsymbol{\xi}_o \tag{7.46}$$

and

$$\boldsymbol{\eta}'_o = \mathbf{X}^{-1}\boldsymbol{\xi}'_o \tag{7.47}$$

The equations in Eq.(7.45) may be rewritten in component form as

$$\left.\begin{aligned} &\ddot{\eta}_k + (\alpha + \beta\omega_k^2)\dot{\eta}_k + \omega_k^2\eta_k = 0, \quad t > 0 \\ &\eta_k(0) = \eta_{ok} \text{ and } \dot{\eta}_k(0) = \eta'_{ok} \quad \text{at } t = 0 \end{aligned}\right\} \text{for } k = 1, 2, \cdots, n \tag{7.48}$$

Because of the ease in experimental evaluations with more options in data fitting, the proportional damping coefficient $(\alpha + \beta\omega_k^2)$ may be replaced by $(2\gamma_k\omega_k)$ where the quantity γ_k is called the *modal damping ratio* of the kth mode.[1] It can be shown that this is equivalent to taking the damping matrix as

$$\grave{\mathbf{C}} = \mathbf{X}^{-T} diag(2\gamma_k\omega_k)\mathbf{X}^{-1} \tag{7.49}$$

which is called the *modal damping matrix.*

The use of modal damping alters Eqs.(7.48) into:

$$\left.\begin{aligned} &\ddot{\eta}_k + (2\gamma_k\omega_k)\dot{\eta}_k + \omega_k^2\eta_k = 0, \quad t > 0 \\ &\eta_k(0) = \eta_{ok} \text{ and } \dot{\eta}_k(0) = \eta'_{ok} \quad \text{at } t = 0 \end{aligned}\right\} \text{for } k = 1, 2, \cdots, n \tag{7.50}$$

---

[1] See "Classical Normal Modes in Damped Linear Dynamic Systems," T. K. Caughey, *Journal of Applied Mechanics*, Vol.27, June 1960.

These are uncoupled n second order ordinary differential equations with constant coefficients. The solution of these equations are:

$$\eta_k(t) = e^{-\gamma_k\omega_k t}(\cos\grave{\omega}_k t + \tfrac{\gamma_k\omega_k}{\grave{\omega}_k}\sin\grave{\omega}_k t)\eta_{ok} + (\tfrac{\sin\grave{\omega}_k t}{\grave{\omega}_k}e^{-\gamma_k\omega_k t})\eta'_{ok} \quad \text{for } k = 1, 2, \cdots, n \tag{7.51}$$

where

$$\grave{\omega}_k = \omega_k\sqrt{1-\gamma_k^2} \tag{7.52}$$

The frequencies $\grave{\omega}_k$, $k = 1, 2, \cdots, n$ are the *damped natural vibration frequencies.* Note that when $\gamma_k = 0$, i.e., when the system is undamped, $\grave{\omega}_k = \omega_k$. On the other hand, when $\gamma_k = 1$ the motion becomes non-oscillatory, and the kth mode becomes *critically damped.* For example γ_1 in earthbound structures is usually less than 10%. It is much less for space structures. In general, the damping ratio is higher for higher frequency modes. If one assumes an average of 10% value for the damping ratios, it may be seen from Eq.(7.52) that the damped vibration frequencies are about $\frac{1}{2}$% smaller than their undamped counterparts.

The scalar equations in Eq.(7.51) may be rewritten as

$$\boldsymbol{\eta}(t) = diag\ [e^{-\gamma_k\omega_k t}(\cos\grave{\omega}_k t + \frac{\gamma_k\omega_k}{\grave{\omega}_k}\sin\grave{\omega}_k t)]\boldsymbol{\eta}_o + diag\ (\frac{\sin\grave{\omega}_k t}{\grave{\omega}_k}e^{-\gamma_k\omega_k t})\boldsymbol{\eta}'_o \tag{7.53}$$

Substituting $\boldsymbol{\eta}(t)$ from this equation into Eq.(7.42) and using Eqs.(7.46 and 7.47), one may obtain the solution of Eq.(7.41) as:

$$\boldsymbol{\xi}(t) = \mathbf{A}_{\xi_o}\boldsymbol{\xi}_o + \mathbf{A}_{\xi'_o}\boldsymbol{\xi}'_o \tag{7.54}$$

where

$$\mathbf{A}_{\xi_o} = \mathbf{X}diag\ [e^{-\gamma_k\omega_k t}(\cos\grave{\omega}_k t + \frac{\gamma_k\omega_k}{\grave{\omega}_k}\sin\grave{\omega}_k t)]\ \mathbf{X}^{-1} \tag{7.55}$$

and

$$\mathbf{A}_{\xi'_o} = \mathbf{X}diag\ [e^{-\gamma_k\omega_k t}\frac{\sin\grave{\omega}_k t}{\grave{\omega}_k}]\ \mathbf{X}^{-1} \tag{7.56}$$

Note that Eq.(7.54) is the special case of Eq.(7.10) when $\mathbf{p}' = \grave{\mathbf{v}}'_o = \mathbf{u}' = \mathbf{o}$. Therefore the definitions given for $\mathbf{A}_{\xi_o}$ and $\mathbf{A}_{\xi'_o}$ above are also for those appearing in Eq.(7.10).

7.4 Nodal Motion due to Nodal Forces

Consider a discrete parameter structure with damping in its at-rest state. The nodes of this structure are subjected to a force type nodal loading of

$$\mathbf{p} = \mathbf{p}(t) = diag\ [\hat{p}_k(t)]\ \mathbf{p}' \quad \text{for } t \geq 0 \tag{7.57}$$

where $\mathbf{p}(t) \in R^{n\times 1}$ nodal loads, *diag* $[\hat{p}_k(t)] \in R^{n\times n}$ non-dimensional time functions showing the time variations of nodal load components, and vector $\mathbf{p}' \in R^{n\times 1}$ maximum magnitudes (or estimates of maximum magnitudes) of nodal load components, as discussed for Eq.(7.3). The resulting motion is called the *forced motion of damped structure due to force type nodal loads.* It may be studied by using Eq.(7.7) with zero values of $\boldsymbol{\xi}_{\circ}$, $\boldsymbol{\xi}'_{\circ}$, $\dot{\mathbf{v}}_{\circ}$, $\boldsymbol{\xi}^{s}$, and $\mathbf{u}$, i.e., by using

$$\begin{aligned} &\mathbf{M}\ddot{\boldsymbol{\xi}} + \grave{\mathbf{C}}\dot{\boldsymbol{\xi}} + \mathbf{K}'\boldsymbol{\xi} = \mathbf{p} \quad \text{for } t > 0 \text{ and} \\ &\boldsymbol{\xi}(0) = \mathbf{o} \text{ and } \dot{\boldsymbol{\xi}}(0) = \mathbf{o} \quad \text{at } t = 0 \end{aligned} \tag{7.58}$$

These are a set of coupled non-homogeneous second order ordinary differential equations with constant coefficients and with homogeneous initial conditions. In order to uncouple the differential equations, as in the previous section, one expresses $\boldsymbol{\xi}$ in terms of basis vectors $[\mathbf{x}_1, \mathbf{x}_2, \cdots, \mathbf{x}_1]$ which are the normalized mode shapes of the undamped free vibration problem of structure. As studied in Section 7.2, these are the n columns of the matrix of eigenvectors $\mathbf{X} \in R^{n\times n}$ of the associated general eigenvalue problem of $\mathbf{K}'\mathbf{x} = \lambda\mathbf{M}\mathbf{x}$. Assuming modal damping, the damping matrix $\grave{\mathbf{C}}$ in the above equations may be taken as in Eq.(7.49):

$$\grave{\mathbf{C}} = \mathbf{X}^{-T} diag(2\gamma_k\omega_k)\mathbf{X}^{-1} \tag{7.59}$$

where γ_k is the modal damping ratio of the kth mode, and ω_k is the kth mode's free undamped vibration frequency. If the description of motion in basis $[\mathbf{x}_1, \mathbf{x}_2, \cdots, \mathbf{x}_n]$ is $\boldsymbol{\eta} \in R^{n\times 1}$, then

$$\boldsymbol{\xi} = \mathbf{X}\boldsymbol{\eta} \tag{7.60}$$

The n components of $\boldsymbol{\eta}$ are the participation factors of the n normalized mode shapes of the freely vibrating undamped structure. Substituting $\boldsymbol{\xi}$ from this equation into Eq.(7.58) and then premultiplying both sides of the latter by $\mathbf{X}^T$ leads to

$$\begin{aligned} &\ddot{\boldsymbol{\eta}} + diag(2\gamma_k\omega_k)\,\dot{\boldsymbol{\eta}} + diag(\omega_k^2)\boldsymbol{\eta} = \mathbf{X}^T diag[\hat{p}_k(t)]\;\mathbf{p}' \quad \text{for } t > 0 \text{ and} \\ &\boldsymbol{\eta}(0) = \mathbf{o} \text{ and } \dot{\boldsymbol{\eta}}(0) = \mathbf{o} \quad \text{at } t = 0 \end{aligned} \tag{7.61}$$

where the use of Eqs.(7.25, 7.28, and 7.57) is made. These equations may be rewritten in component form as

$$\left.\begin{aligned} &\ddot{\eta}_k + (2\gamma_k\omega_k)\dot{\eta}_k + \omega_k^2\eta_k = \mathbf{x}_k^T diag[\hat{p}_i(t)]\;\mathbf{p}', \quad t > 0 \\ &\eta_k(0) = 0 \text{ and } \dot{\eta}_k(0) = 0 \quad \text{at } t = 0 \end{aligned}\right\} \text{ for } k = 1, 2, \cdots, n \tag{7.62}$$

These are the uncoupled n second order differential equations with constant coefficients. The solution of these equations is

$$\eta_k(t) = \mathbf{x}_k^T diag[\frac{1}{\grave{\omega}_i}\int_0^t e^{-\gamma_k\omega_k\tau}\;\hat{p}_i(\tau)\sin\grave{\omega}_i\tau\;d\tau]\;\mathbf{p}' \text{ for } k = 1, 2, \cdots, n \tag{7.63}$$

which can be rewritten as

$$\boldsymbol{\eta}(t) = \mathbf{X}^T diag\ [\frac{1}{\grave{\omega}_k} \int_0^t e^{-\gamma_k \omega_k \tau}\ \hat{p}_k(\tau) \sin \grave{\omega}_k \tau\ d\tau]\ \mathbf{p}' \tag{7.64}$$

The use of this in Eq.(7.60) provides the solution of Eq.(7.58) as

$$\boldsymbol{\xi}(t) = \mathbf{A}_{p'}\mathbf{p}' \tag{7.65}$$

where

$$\mathbf{A}_{p'} = \mathbf{X}\mathbf{X}^T diag\ [\frac{1}{\grave{\omega}_k} \int_0^t e^{-\gamma_k \omega_k \tau}\ \hat{p}_k(\tau) \sin \grave{\omega}_k \tau\ d\tau] \tag{7.66}$$

Observing that the orthogonality condition of modes, i.e., $\mathbf{X}^T\mathbf{M}\mathbf{X} = \mathbf{I}$, can be expressed also as

$$\mathbf{X}\mathbf{X}^T = \mathbf{M}^{-1} \tag{7.67}$$

Eq.(7.66) can be also written as

$$\mathbf{A}_{p'} = \mathbf{M}^{-1} diag\ [\frac{1}{\grave{\omega}_k} \int_0^t e^{-\gamma_k \omega_k \tau}\ \hat{p}_k(\tau) \sin \grave{\omega}_k \tau\ d\tau] \tag{7.68}$$

Note that Eq.(7.65) is the special case of Eq.(7.10) when $\boldsymbol{\xi}_o = \boldsymbol{\xi}'_o = \mathbf{v}'_o = \mathbf{u}' = \mathbf{o}$. Therefore the definition given for $\mathbf{A}_{p'}$ above is also for the one appearing in Eq.(7.10).

7.5 Nodal Motion due to Prescribed Element Deformations

Consider a discrete parameter structure with damping in its at-rest state. The elements of this structure are subjected to prescribed element deformations type loading of

$$\grave{\mathbf{v}}_e = \grave{\mathbf{v}}_e(t) = diag\ [\hat{v}_e^k(t)]\ \grave{\mathbf{v}}'_e \quad \text{for } t \geq 0 \tag{7.69}$$

where $\grave{\mathbf{v}}_e(t)\ \in R^{m\times 1}$prescribed element deformations, $diag\ [\hat{v}_e^k(t)]\ \in R^{m\times m}$ physically non-dimensional time functions showing the time variations of the prescribed element deformations, and $\grave{\mathbf{v}}'_e\ \in R^{m\times 1}$ maximum magnitudes (or estimates of maximum magnitudes) of prescribed element deformation components, as discussed for Eq.(7.3). The resulting motion is called the *forced motion of damped structure due to prescribed element deformations.* It may be studied by using Eq.(7.7) with zero values of $\boldsymbol{\xi}_o$, $\boldsymbol{\xi}'_o$, $\mathbf{p}$, $\boldsymbol{\xi}^s$, and $\mathbf{u}$, i.e., by using

$$\begin{aligned} &\mathbf{M}\ddot{\boldsymbol{\xi}} + \grave{\mathbf{C}}\dot{\boldsymbol{\xi}} + \mathbf{K}'\boldsymbol{\xi} = -\mathbf{B}\vec{\mathbf{K}}\grave{\mathbf{v}}_e \quad \text{for } t > 0 \text{ and} \\ &\boldsymbol{\xi}(0) = \mathbf{o} \text{ and } \dot{\boldsymbol{\xi}}(0) = \mathbf{o} \quad \text{at } t = 0 \end{aligned} \tag{7.70}$$

Assuming modal damping, the solution of these equations may be written from the solution of Eq.(7.58) given in Eqs.(7.65 and 7.68) by noting that one needs to substitute $-\mathbf{B}\vec{\mathbf{K}}\dot{\mathbf{v}}_e$ for $\mathbf{p}$, and recalling that $\dot{\mathbf{v}}_e = diag[\hat{v}_e^k(t)]\dot{\mathbf{v}}_e'$. This leads to the solution of Eq.(7.70) as

$$\boldsymbol{\xi}(t) = \mathbf{A}_{\dot{v}_e'}\dot{\mathbf{v}}_e' \tag{7.71}$$

where

$$\mathbf{A}_{\dot{v}_e'} = -\mathbf{M}^{-1}\mathbf{B}\vec{\mathbf{K}}diag\,[\frac{1}{\check{\omega}_k}\int_0^t e^{-\gamma_k\omega_k\tau}\,\hat{v}_e^k(\tau)\sin\check{\omega}_k\tau\,d\tau] \tag{7.72}$$

Note that Eq.(7.71) is the special case of Eq.(7.10) when $\boldsymbol{\xi}_o = \boldsymbol{\xi}_o' = \mathbf{p}' = \boldsymbol{\xi}^{s\prime} = \mathbf{u}' = \mathbf{o}$. Therefore the definition given for $\mathbf{A}_{\dot{v}_e'}$ above is also for the one appearing in Eq.(7.10).

7.6 Nodal Vibrations due to Support Movements

Consider a discrete parameter structure with damping in its at-rest state. The elements of this structure are subjected to support movements type loading of

$$\boldsymbol{\xi}^s = \boldsymbol{\xi}^s(t) = diag\,[\hat{\xi}_k^s(t)]\boldsymbol{\xi}^{s\prime} \quad \text{for } t \geq 0 \tag{7.73}$$

where $\boldsymbol{\xi}^s(t)\ \in R^{b\times 1}$support movements, $diag\ [\hat{\xi}_k^s(t)]\ \in R^{b\times b}$ physically nondimensional time functions showing the time variations of the support movements, and $\boldsymbol{\xi}^{s\prime}\ \in R^{b\times 1}$ maximum magnitudes (or estimates of maximum magnitudes) of support movements, as discussed for Eq.(7.3). This case is important in earthquake engineering where the excitations to the structure are a result of the movements of the supports attached to the foundations. The resulting motion is called the *forced motion of damped structure due to support movements.* It may be studied by using Eq.(7.7) with zero values of $\boldsymbol{\xi}_o$, $\boldsymbol{\xi}_o'$, $\mathbf{p}$, $\mathbf{v}_e$, and $\mathbf{u}$, i.e., by using

$$\begin{array}{l} \mathbf{M}\ddot{\boldsymbol{\xi}} + \check{\mathbf{C}}\dot{\boldsymbol{\xi}} + \mathbf{K}'\boldsymbol{\xi} = -\mathbf{B}\vec{\mathbf{K}}\mathbf{B}_s^T\boldsymbol{\xi}^s \quad \text{for } t > 0 \text{ and} \\ \boldsymbol{\xi}(0) = \mathbf{o} \text{ and } \dot{\boldsymbol{\xi}}(0) = \mathbf{o} \quad \text{at } t = 0 \end{array} \tag{7.74}$$

Assuming modal damping, the solution of these equations may be written from the solution of Eq.(7.58) given in Eqs.(7.65 and 7.68) noting that $-\mathbf{B}\vec{\mathbf{K}}\mathbf{B}_s^T\boldsymbol{\xi}^s$ is the excitation instead of $\mathbf{p}$. This leads to the solution of Eq.(7.74) as

$$\boldsymbol{\xi}(t) = \mathbf{A}_{\xi^{s\prime}}\boldsymbol{\xi}^{s\prime} \tag{7.75}$$

where

$$\mathbf{A}_{\xi^{s\prime}} = -\mathbf{M}^{-1}\mathbf{B}\vec{\mathbf{K}}\mathbf{B}_s^T diag\,[\frac{1}{\grave{\omega}_k}\int_0^t e^{-\gamma_k\omega_k\tau}\;\hat{\xi}_k^s(\tau)\sin\grave{\omega}_k\tau\;d\tau] \qquad (7.76)$$

Note that Eq.(7.75) is the special case of Eq.(7.10) when $\boldsymbol{\xi}_o = \boldsymbol{\xi}_o' = \mathbf{p}' = \mathbf{u}' = \mathbf{v}_e' = \mathbf{o}$. Therefore the definition given for $\mathbf{A}_{\xi^{s\prime}}$ above is also for the one appearing in Eq.(7.10).

7.7 Nodal Motion due to Actuator Induced Deformations

Consider a discrete parameter structure with damping in its at-rest state. The elements of this structure are subjected to actuator induced element deformations of

$$\mathbf{u} = \mathbf{u}(t) = diag\,[\hat{u}_k(t)]\;\mathbf{u}' \quad \text{for } t \geq 0 \qquad (7.77)$$

where $\mathbf{u}(t) \in R^{q\times 1}$actuator induced deformations, $diag\,[\hat{u}_k(t)] \in R^{q\times q}$ physically non-dimensional time functions showing the time variations of the actuator induced deformations, and $\mathbf{u}' \in R^{q\times 1}$ maximum magnitudes (or estimates of maximum magnitudes) of actuator induced deformation components, as discussed for Eq.(7.3). The resulting motion is called the *forced vibrations of damped structure due to actuator induced deformations.* It may be studied by using Eq.(7.7) with zero values of $\boldsymbol{\xi}_o$, $\boldsymbol{\xi}_o'$, $\mathbf{p}$, $\grave{\mathbf{v}}_e$, and $\boldsymbol{\xi}^s$, i.e., by using

$$\begin{array}{l} \mathbf{M}\ddot{\boldsymbol{\xi}} + \grave{\mathbf{C}}\dot{\boldsymbol{\xi}} + \mathbf{K}'\boldsymbol{\xi} = -\mathbf{B}\vec{\mathbf{K}}\mathbf{I}_\beta\mathbf{u} \quad \text{for } t > 0 \text{ and} \\ \boldsymbol{\xi}(0) = \mathbf{o} \text{ and } \dot{\boldsymbol{\xi}}(0) = \mathbf{o} \quad \text{at } t = 0 \end{array} \qquad (7.78)$$

Assuming modal damping, the solution of these equations may be written from the solution of Eq.(7.58) given in Eqs.(7.65 and 7.68) noting that $-\mathbf{B}\vec{\mathbf{K}}\mathbf{I}_\beta\mathbf{u}$ is the excitation instead of $\mathbf{p}$. This leads to the solution of Eq.(7.78) as

$$\boldsymbol{\xi}(t) = \mathbf{A}_{u'}\mathbf{u}' \qquad (7.79)$$

where

$$\mathbf{A}_{u'} = -\mathbf{M}^{-1}\mathbf{B}\vec{\mathbf{K}}\mathbf{I}_\beta diag\,[\frac{1}{\grave{\omega}_k}\int_0^t e^{-\gamma_k\omega_k\tau}\;\hat{u}_k(\tau)\sin\grave{\omega}_k\tau\;d\tau] \qquad (7.80)$$

Note that Eq.(7.79) is the special case of Eq.(7.10) when $\boldsymbol{\xi}_o = \boldsymbol{\xi}_o' = \mathbf{p}' = \mathbf{v}_o' = \boldsymbol{\xi}^{s\prime} = \mathbf{o}$. Therefore the definition given for $\mathbf{A}_{u'}$ above is also for the one appearing in Eq.(7.10).

7.8 Trajectory of Nodes in State Space

As it may be observed from Eq.(7.11), the motion of the nodes, $\boldsymbol{\xi}(t) \in \mathrm{R}^{n\times 1}$, plays a dominant role in obtaining the inverse relations. The motion of the nodes is expressed in Eqs.(7.12 and 7.13) in terms of the excitation quantities. The submatrices $\mathbf{A}_{\xi_o} \in \mathrm{R}^{n\times n}$, $\mathbf{A}_{\xi'_o} \in \mathrm{R}^{n\times n}$, $\mathbf{A}_{\eta'} \in \mathrm{R}^{n\times n}$, $\mathbf{A}_{\eta'_b} \in \mathrm{R}^{n\times m}$, $\mathbf{A}_{\xi''} \in \mathrm{R}^{n\times 1}$, and $\mathbf{A}_{u'} \in \mathrm{R}^{n\times q}$ appearing in Eq.(7.13) are expressed in Sections 7.3 - 7.7, in terms of the time variation of excitation quantities. Having explicit expressions for $\boldsymbol{\xi}(t) \in \mathrm{R}^{n\times 1}$, one may plot it as a curve in the n space, having time t as the curve parameter. Such a curve defines the position of the nodes as successive time points. The metric of this space, i.e., the rule to measure distance between any two points of the space without violating our common perception of the one we use in the space we are born into, can be used as a measure of the strain energy of structure. Since the nodal velocities are not directly available, the kinetic energy is not in this representation of the nodal motion.

In a more modern treatment one may plot the state, i.e., $\boldsymbol{\zeta} \in \mathrm{R}^{2n\times 1}$

$$\boldsymbol{\zeta} = \left\{ \begin{matrix} \dot{\boldsymbol{\xi}} \\ \boldsymbol{\xi} \end{matrix} \right\} \tag{7.81}$$

of the nodes in $2n$ space. Since the state in a given instant t provides both the position $\boldsymbol{\xi}$ and the velocities of $\dot{\boldsymbol{\xi}}$ of the nodes, the *total energy* E of the structure at that instant is available as the sum of its *strain energy* U and *kinetic energy* K:

$$E = K + U \tag{7.82}$$

Since the total strain energy of structure is the work done by restoring forces $\mathbf{K}'\boldsymbol{\xi}$ under $\boldsymbol{\xi}$, as the nodes move from their at rest position to the position $\boldsymbol{\xi}$, one may write

$$U = \frac{1}{2}\boldsymbol{\xi}^T\mathbf{K}'\boldsymbol{\xi} \tag{7.83}$$

Since the total kinetic energy K gained by the structure during the time interval $(0-t)$ is the work done by inertial forces $\mathbf{M}\ddot{\boldsymbol{\xi}}$ under velocities $\dot{\boldsymbol{\xi}}$, one may write

$$K = \frac{1}{2}\dot{\boldsymbol{\xi}}^T\mathbf{M}\dot{\boldsymbol{\xi}} \tag{7.84}$$

because $K = \int_0^t \dot{\boldsymbol{\xi}}^T\mathbf{M}\ddot{\boldsymbol{\xi}}d\tau = \int_0^t d(\frac{1}{2}\dot{\boldsymbol{\xi}}^T\mathbf{M}\dot{\boldsymbol{\xi}})$ and $\mathbf{M} = \mathbf{M}^T$. Using Eqs.(7.81-7.84), one may express

$$E = \frac{1}{2}\boldsymbol{\zeta}^T diag\ (\mathbf{M}, \mathbf{K}')\boldsymbol{\zeta} \tag{7.85}$$

Since $diag\ (\mathbf{M}, \mathbf{K}') \in \mathrm{R}^{2n\times 2n}$ is a positive definite matrix, one may define the *norm* of state vector $\boldsymbol{\zeta}$ as the square root of its energy, i.e.,

$$\|\boldsymbol{\zeta}\| = \|\boldsymbol{\zeta}\|_{energy} = \sqrt{E} \tag{7.86}$$

This definition meets *homogeneity*, *positivity*, and *triangle inequality* rules.[2] [Note that the *energy norm* of $\boldsymbol{\zeta}$ is the same as the Euclid norm of $\boldsymbol{\zeta}' = diag(\mathbf{M}^{1/2}, \mathbf{K}'^{1/2})\, \boldsymbol{\zeta}$ and $(\boldsymbol{\zeta}, \boldsymbol{\zeta}')$ mapping is one-to-one.]

By using the energy norm defined in Eq.(7.86) as the *metric* of the state space,[3] the distance of trajectory point $\boldsymbol{\zeta}$ from the origin is the measure of energy associated with that state, relative to one at the origin. Use of nodal trajectories in state space is very instructional for describing nodal motion. The state space approach is extensively used in modern dynamics and modern control fields.

By having explicit expressions for $\boldsymbol{\xi}$ as a function of time, one may easily obtain expressions for the state vector by using its definition in Eq.(7.81). However, it is more convenient to obtain the state vector $\boldsymbol{\zeta}$ from the first order equations of nodal motion given in Subsection 6.4.2, i.e., from

$$\begin{aligned} \dot{\boldsymbol{\zeta}} &= \mathbf{A}\boldsymbol{\zeta} + \grave{\mathbf{B}}\mathbf{u} + \mathbf{g} \qquad \text{for } t > 0 \\ \boldsymbol{\zeta}(0) &= \boldsymbol{\zeta}_o \qquad \text{at } t = 0 \end{aligned} \tag{7.87}$$

where $\mathbf{A} \in R^{2n\times 2n}$ and $\grave{\mathbf{B}} \in R^{2n\times q}$ represent system parameters, $\boldsymbol{\zeta}_o \in R^{2n\times 1}$ is the initial state, $\mathbf{g} \in R^{2n\times 1}$ is the loading type excitations, and $\mathbf{u} \in R^{q\times 1}$ is the control excitations. The definitions of matrices $\mathbf{A}, \grave{\mathbf{B}}$, and $\mathbf{g}$ are given in Subsection 6.4.2 and repeated here:

$$\mathbf{A} = \begin{bmatrix} -\mathbf{M}^{-1}\grave{\mathbf{C}} & -\mathbf{M}^{-1}\mathbf{K}' \\ \mathbf{I} & \cdot \end{bmatrix} \tag{7.88}$$

$$\grave{\mathbf{B}} = \begin{bmatrix} -\mathbf{M}^{-1}\mathbf{B}\vec{\mathbf{K}}\mathbf{I}_\beta \\ \cdot \end{bmatrix} \tag{7.89}$$

$$\mathbf{g} = \begin{Bmatrix} \mathbf{M}^{-1}\mathbf{p} - \mathbf{M}^{-1}\mathbf{B}\vec{\mathbf{K}}(\dot{\mathbf{v}}_e + \mathbf{B}_s^T\boldsymbol{\xi}_s) \\ \cdot \end{Bmatrix} \tag{7.90}$$

In the following subsections, the solutions of state vector $\boldsymbol{\zeta}$ from Eqs.(7.87) for various excitations are given.

7.8.1 Uncontrolled Autonomous System

Since the structure is not controlled, one has $\mathbf{u} = \mathbf{o}$; and since the system is autonomous, one has $\mathbf{g} = \mathbf{o}$. With $\mathbf{u} = \mathbf{g} = \mathbf{o}$, the state equations may be obtained from Eq.(7.87) as

$$\begin{aligned} \dot{\boldsymbol{\zeta}} &= \mathbf{A}\boldsymbol{\zeta} \qquad \text{for } t > 0 \\ \boldsymbol{\zeta}(0) &= \boldsymbol{\zeta}_o \qquad \text{at } t = 0 \end{aligned} \tag{7.91}$$

---

[2] See, for example, p.55, *The Algebraic Eigenvalue Problem*, J. H. Wilkinson, Oxford University Press, 1965.

[3] See, for example, Chapter 1, *Introduction to Approximation Theory*, E. W. Cheney, McGraw-Hill Book Co., 1966.

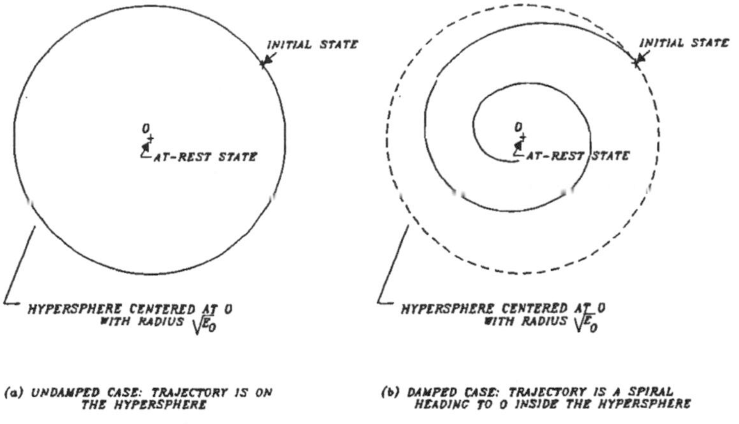

FIGURE 7.1. Trajectory of nodes in state space R^{2n} (uncontrolled autonomous case)

In the state space, the trajectory defined by the tip of state vector $\boldsymbol{\zeta}$ starts at initial point $\boldsymbol{\zeta}_o$ and evolves as dictated by the equations of nodal motion $\dot{\boldsymbol{\zeta}} = \mathbf{A}\boldsymbol{\zeta}$. The initial energy of the system E_o is

$$E_o = E(0) = \frac{1}{2}\boldsymbol{\zeta}_o^T diag(\mathbf{M}, \mathbf{K}')\boldsymbol{\zeta}_o \tag{7.92}$$

which represents the square of the distance of the initial state point $\boldsymbol{\zeta}_o$ from point $\mathbf{o}$ of the state space, i.e., from the point corresponding to the at-rest state of the structure.

If there is no damping in the structure, i.e., if $\dot{\mathbf{C}} = 0$ in Eq.(7.88), then the system energy would remain always the same, i.e.,

$$E_o = E(t) \quad \text{for } t > 0 \tag{7.93}$$

implying that the trajectory starting from non-equilibrium initial state $\boldsymbol{\zeta}_o$ is on a hypersphere with radius $\sqrt{E_o}$ and center at $\mathbf{o}$. This is sketched in Fig.(7.1 a).

If the system is with damping, i.e., if $\dot{\mathbf{C}} \neq 0$, the energy will decrease with increasing time, and the trajectory will be a spiral, starting at $\boldsymbol{\zeta}_o$ and asymptotically approaching to $\mathbf{o}$ as sketched in Fig.(7.1 b). Since $\dot{\mathbf{C}}$ is positive definite the power leaking from the system due to frictional forces is negative, i.e., $-\frac{1}{2}\dot{\boldsymbol{\xi}}^T\dot{\mathbf{C}}\dot{\boldsymbol{\xi}}$ <0, indicating that the distance of trajectory points from $\mathbf{o}$ will uniformly decrease.

A better proof of the fact that the trajectory of a damped uncontrolled autonomous system is a helical hyper/curve, that approaches to the at rest state at point $\mathbf{o}$ with increasing time, can be given by showing that

$$\frac{dE}{dt} < 0 \quad \text{for } t > 0 \tag{7.94}$$

This is so, since from Eq.(7.85) by differentiation, one may write

$$\frac{dE}{dt} = \frac{d}{dt}(\frac{1}{2}\dot{\xi}^T \mathbf{M}\dot{\xi} + \frac{1}{2}\xi^T \mathbf{K}'\xi) = \dot{\xi}^T(\mathbf{M}\ddot{\xi} + \mathbf{K}'\xi) \tag{7.95}$$

observing that $\mathbf{M} = \mathbf{M}^T$ and $\mathbf{K}' = \mathbf{K}'^T$. From the equilibrium equations of the nodes in terms of ξ given in Eq.(7.41) one observes that

$$(\mathbf{M}\ddot{\xi} + \mathbf{K}'\xi) = -\grave{\mathbf{C}}\dot{\xi} \quad \text{for } t > 0 \tag{7.96}$$

The use of this in Eq.(7.95) gives

$$\frac{dE}{dt} = -\frac{1}{2}\dot{\xi}^T \grave{\mathbf{C}}\dot{\xi} \quad \text{for } t > 0 \tag{7.97}$$

Due to positive definiteness of $\grave{\mathbf{C}}$, the quadratic form $\dot{\xi}^T \grave{\mathbf{C}}\dot{\xi}$ is always larger than zero for nonzero $\dot{\xi}$. Hence the assertion in Eq.(7.94) is correct, implying that, indeed, the trajectory is a spiral, starting at point ζ_o and monotonically approaching to point $\mathbf{o}$.

The solution of Eq.(7.91) can be expressed in terms of the eigenvectors of the *special eigenvalue problem* $\mathbf{A}\mathbf{z} = \mu\mathbf{z}$, i.e., in terms of:

$$\mathbf{Z} = [\mathbf{z}_1, \mathbf{z}_2, \cdots, \mathbf{z}_{2n}] \tag{7.98}$$

where $\mathbf{Z} \in C^{2n \times 2n}$. The eigenvalues corresponding to $\mathbf{Z}$ are $\mu_1, \mu_2, \cdots, \mu_{2n}$. Then one may write

$$\mathbf{AZ} = \mathbf{Z} diag\ (\mu_k) \tag{7.99}$$

or

$$\mathbf{Z}^{-1}\ \mathbf{AZ} = diag\ (\mu_k) \tag{7.100}$$

We assume that the labels of eigenvalues increase as their distances from the origin of complex plane increase. The eigenpairs $(\mu_k\ , \mathbf{z}_k)$, $k = 1, 2, \cdots, 2n$, are closely related to the undamped free vibration frequencies and normalized mode shapes $(\omega_j\ , \mathbf{x}_j)$, $j = 1, 2, \cdots, n$ of the structure. This is shown next.

The eigenvalues μ_k , $k = 1, 2, \cdots, 2n$, are the roots of the characteristic equation

$$\det\ (\mathbf{A} - \mu\mathbf{I}) = 0 \tag{7.101}$$

Using the definition of $\mathbf{A}$ from Eq.(7.88) and assuming $\mu \neq 0$, one may write

$$\det\ (\begin{bmatrix} -\mathbf{M}^{-1}\grave{\mathbf{C}} - \mu\mathbf{I} & -\mathbf{M}^{-1}\mathbf{K}' \\ \mathbf{I} & -\mu\mathbf{I} \end{bmatrix}) = \det [\begin{bmatrix} -\mathbf{M}^{-1}\mathbf{X}^{-T} & \cdot \\ \cdot & \mathbf{X} \end{bmatrix}$$

$$\begin{bmatrix} diag\ (\mu^2 + 2\gamma_k\omega_k\mu + \omega_k^2) & diag\ (\omega_k^2) \\ \cdot & -\mu\mathbf{I} \end{bmatrix} \begin{bmatrix} \mathbf{X}^{-1} & \cdot \\ -\frac{1}{\mu}\mathbf{X}^{-1} & \mathbf{X}^{-1} \end{bmatrix}] =$$

$$c[\Pi_{k=1}^n(\mu^2 + 2\gamma_k\omega_k\mu + \omega_k^2)][\Pi_{k=1}^n(-\mu)] = 0 \tag{7.102}$$

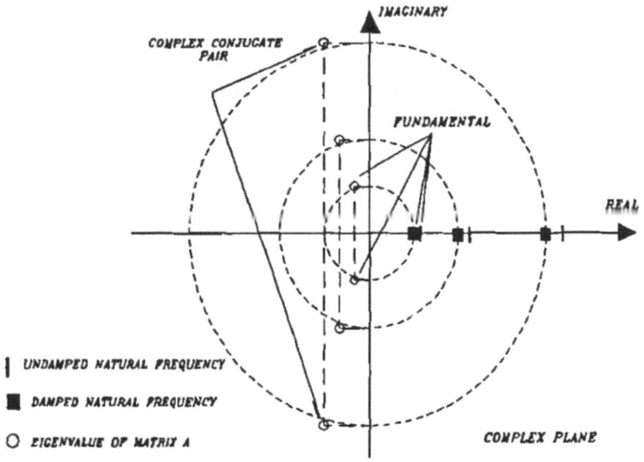

FIGURE 7.2. Relations among vibration frequencies for undamped (ω_k) and damped ($\grave{\omega}_k$) structure, and eigenvalues of $\mathbf{A} \in R^{2n\times 2n}$

and obtain the eigenvalues as

$$\mu = \grave{\mu}_k = -\gamma_k\omega_k \pm j\grave{\omega}_k \text{ , for } k = 1,2,\cdots,n \tag{7.103}$$

where the damped free vibration frequency $\grave{\omega}_k$ is as defined in Eq.(7.52). Note that the $2n$ eigenvalues are n complex conjugate numbers with negative real parts. Typical locations in the complex plane are sketched in Fig.(7.2) for n undamped free vibration frequencies ω_k, damped free vibration frequencies $\grave{\omega}_k$, and $2n$ eigenvalues μ_k of $\mathbf{A}$. By inspection one may observe that

$$\begin{aligned} \mathbf{z}_{2k-1} &= \left\{ \begin{array}{c} \mu_{2k-1}\mathbf{x}_k \\ \mathbf{x}_k \end{array} \right\} \\ \mathbf{z}_{2k} &= \left\{ \begin{array}{c} \bar{\mu}_{2k}\mathbf{x}_k \\ \mathbf{x}_k \end{array} \right\} \end{aligned} \text{ , for } k = 1,2,\cdots,n \tag{7.104}$$

which may be written in matrix form as

$$\mathbf{Z} = diag(\mathbf{X},\mathbf{X}) \left[\begin{array}{c} diag([\mu_{2k-1},\mu_{2k}]) \\ diag([1,1]) \end{array} \right] \tag{7.105}$$

where

$$\begin{aligned} \mu_{2k-1} &= -\gamma_k\omega_k + j\grave{\omega}_k \\ \mu_{2k} &= -\gamma_k\omega_k - j\grave{\omega}_k \end{aligned} \text{ , for } k = 1,2,\cdots,n \tag{7.106}$$

These expressions, relating the eigenpairs of $\mathbf{Kx} = \lambda\mathbf{Mx}$ to those of $\mathbf{Az} = \mu\mathbf{z}$, are important for the economical identification of $\mathbf{Z}$ and $diag(\mu_k)$.

In obtaining the solution of Eq.(7.91), the eigenvectors matrix $\mathbf{Z} \in R^{2n\times 2n}$ can be used in the transformation

$$\zeta = \mathbf{Z}\grave{\eta} \tag{7.107}$$

which expresses the state in terms of the eigenvectors $\mathbf{z}_1, \mathbf{z}_2, \cdots, \mathbf{z}_{2n}$. The new state vector $\check{\eta} \in C^{2n\times 1}$describes the motion relative to the new basis vectors $\mathbf{z}_1, \mathbf{z}_2, \cdots, \mathbf{z}_{2n}$. Using $\boldsymbol{\zeta}$ from Eq.(7.107) in Eq.(7.91) and premultiplying both sides with $\mathbf{Z}^{-1}$, one may obtain

$$\frac{d}{dt}\check{\eta} = \mathbf{Z}^{-1}\mathbf{A}\mathbf{Z}\check{\eta} \tag{7.108}$$

Using Eq.(7.100) in this equation and the initial state

$$\check{\eta}_o = \check{\eta}(0) = \mathbf{Z}^{-1}\boldsymbol{\zeta}_o \tag{7.109}$$

one may write

$$\begin{aligned} &\tfrac{d}{dt}\check{\eta} = diag\ (\mu_k)\check{\eta} \quad \text{for } t > 0 \\ &\check{\eta}(0) = \check{\eta}_o \quad \text{at } t = 0 \end{aligned} \tag{7.110}$$

which, in scalar form, becomes

$$\begin{aligned} &\tfrac{d}{dt}\check{\eta}_k = \mu_k\check{\eta}_k \quad \text{for } t > 0 \\ &\check{\eta}_k(0) = \check{\eta}_{ok} \quad \text{at } t = 0 \end{aligned} \qquad k = 1, 2, \cdots, 2n \tag{7.111}$$

The solution of Eq.(7.111) is

$$\check{\eta}_k(t) = \check{\eta}_{ok}\, e^{\mu_k t}\ , \qquad k = 1, 2, \cdots, 2n \tag{7.112}$$

or in matrix form

$$\check{\eta} = diag\ (e^{\mu_k t})\check{\eta}_o \quad \text{for } t \geq 0 \tag{7.113}$$

Using Eqs.(7.107 and 7.109) in this equation, one may obtain the solution of Eq.(7.91) as

$$\boldsymbol{\zeta} = \mathbf{Z} diag\ (e^{\mu_k t})\ \mathbf{Z}^{-1}\boldsymbol{\zeta}_o \quad \text{for } t \geq 0 \tag{7.114}$$

Since the eigenvalues $\mu_1, \mu_2, \cdots, \mu_{2n}$ are all with negative real parts, every component of $\boldsymbol{\zeta}$ uniformly decreases in magnitude with increasing time, and the trajectory of an uncontrolled damped system is a spiral that asymptotically approaches to zero point, as sketched in Fig. (7.1).

7.8.2 Uncontrolled Non-Autonomous System

Since the structure is not controlled, $\mathbf{u} = \mathbf{o}$; and since it is non-autonomous, $\mathbf{g} \neq \mathbf{o}$. With these from Eq.(7.87), one may write:

$$\begin{aligned} &\dot{\boldsymbol{\zeta}} = \mathbf{A}\boldsymbol{\zeta} + \mathbf{g} \qquad \text{for } t > 0 \\ &\boldsymbol{\zeta}(0) = \boldsymbol{\zeta}_o \qquad \text{at } t = 0 \end{aligned} \tag{7.115}$$

As shown in Eq.(7.90), the excitations $\mathbf{g}$ may be due to nodal forces, temperature type element deformations, and support motions. Again using the transformation in Eq.(7.107), the state at time t and the initial state at time 0 may be described by $\grave{\eta}(t) \in C^{2n\times 1}$ and $\grave{\eta}_o \in C^{2n\times 1}$, respectively, by using as basis vectors the eigenvectors $\mathbf{z}_1, \mathbf{z}_2, \cdots, \mathbf{z}_{2n}$ of $\mathbf{A}$, i.e., the columns of matrix $\mathbf{Z} \in R^{2n\times 2n}$. The reader may notice that the basis vectors used in describing $\boldsymbol{\zeta}, \mathbf{g}$, and $\mathbf{z}_k, k = 1, \cdots, 2n$ are $\mathbf{i}_1, \mathbf{i}_2, \cdots, \mathbf{i}_{2n}$, i.e., the columns of identity matrix $\mathbf{I} \in (\mathrm{Int})^{2n\times 2n}$. As before, using Eq.(7.100), the equations of state, in terms of new state variable $\grave{\eta}$, may be restated as:

$$\begin{aligned} &\tfrac{d}{dt}\grave{\eta} = diag\ (\mu_k)\ \grave{\eta} + \rho \quad \text{for } t > 0 \\ &\grave{\eta}(0) = \grave{\eta}_o \quad \text{at } t = 0 \end{aligned} \tag{7.116}$$

where

$$\rho = \mathbf{Z}^{-1}\mathbf{g} \tag{7.117}$$

is the description of the excitations $\mathbf{g}$ in the basis established by the eigenvectors. In scalar form:

$$\begin{aligned} &\tfrac{d}{dt}\grave{\eta}_k = \mu_k\grave{\eta}_k + \rho_k \quad \text{for } t > 0 \\ &\grave{\eta}_k(0) = \grave{\eta}_{ok} \quad \text{at } t = 0 \end{aligned} \qquad k = 1, 2, \cdots, 2n \tag{7.118}$$

and their solution:

$$\grave{\eta}_k(t) = \grave{\eta}_{ok}\, e^{\mu_k t} + \int_0^t \rho_k(\tau)e^{\mu_k\tau}d\tau\ , \qquad k = 1, 2, \cdots, 2n \tag{7.119}$$

which may be rewritten in matrix form:

$$\grave{\eta} = diag\ (e^{\mu_k t})\ \grave{\eta}_o + \int_0^t diag\ (e^{\mu_k\tau})\rho(\tau)d\tau \quad \text{for } t \geq 0 \tag{7.120}$$

or using Eqs.(7.107 and 7.117):

$$\boldsymbol{\zeta} = \mathbf{Z}diag\ (e^{\mu_k t})\ \mathbf{Z}^{-1}\boldsymbol{\zeta}_o + \int_0^t \mathbf{Z}diag\ (e^{\mu_k\tau})\mathbf{Z}^{-1}\mathbf{g}(\tau)d\tau \quad \text{for } t \geq 0 \tag{7.121}$$

It is possible to rewrite this solution more compactly. From the inverse of Eq.(7.100) one may obtain:

$$\mathbf{A} = \mathbf{Z}diag\ (\mu_k)\ \mathbf{Z}^{-1} \tag{7.122}$$

which may help to define the *transition matrix* $e^{\mathbf{A}t} \in R^{2n\times 2n}$ as

$$e^{\mathbf{A}t} = \mathbf{Z}diag\ (e^{\mu_k t})\ \mathbf{Z}^{-1} \tag{7.123}$$

since this definition concurs well with the formal Maclaurin series expansion of $e^{\mathbf{A}t}$:

$$e^{\mathbf{A}t} = \sum_{k=0}^{\infty} \mathbf{A}^k \frac{t^k}{k!} \tag{7.124}$$

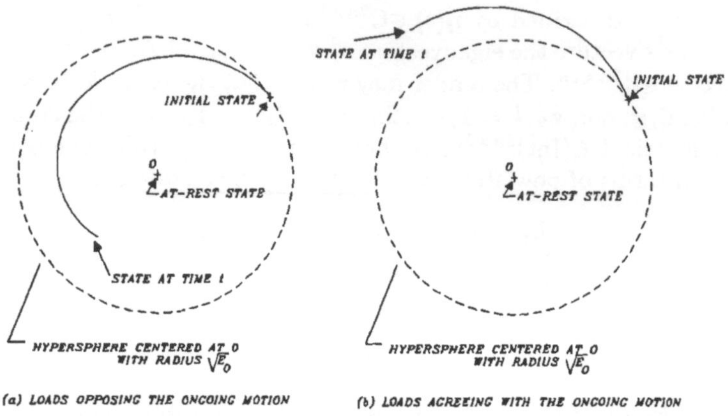

FIGURE 7.3. Nodal trajectory in state space R^{2n} (uncontrolled non-autonomous system)

In terms of the transition matrix $e^{\mathbf{A}t} \in R^{2n\times 2n}$, one may rewrite Eq.(7.121) as

$$\boldsymbol{\zeta}(t)=e^{\mathbf{A}t}\boldsymbol{\zeta}_o + \int_0^t e^{\mathbf{A}\tau}\mathbf{g}(\tau)d\tau \quad \text{for } 0 \le t \tag{7.125}$$

This formalism of expressing the response in terms of the transition matrix is very convenient. For example, suppose excitation $\mathbf{g}$ starts acting on structure at time $t' > 0$, rather than at time $t = 0$ of the initial state $\boldsymbol{\zeta}_o$. Then one may express the response as

$$\begin{aligned} \boldsymbol{\zeta}(t)&=e^{\mathbf{A}(t-0)}\boldsymbol{\zeta}_o \quad \text{for } 0 \le t \le t' \\ \boldsymbol{\zeta}(t)&=e^{\mathbf{A}(t-t')}\boldsymbol{\zeta}(t') + \textstyle\int_{t'}^t e^{\mathbf{A}(\tau-t')}\mathbf{g}(\tau - t')d\tau \quad \text{for } t' \le t \end{aligned} \tag{7.126}$$

The reader may observe that the transition matrix $e^{\mathbf{A}(t-t')} \in R^{2n\times 2n}$ carries over to time t an event that took place at an earlier time $t' \le t$. This generalized definition of the transition matrix may be obtained from the one in Eq.(7.123) by replacing $t-0$ by $t-t'$:

$$e^{\mathbf{A}(t-t')} = \mathbf{Z}diag\left(e^{\mu_k(t-t')}\right)\mathbf{Z}^{-1} \tag{7.127}$$

It is a powerful instrument in expressing the response of time invariant linear discrete parameter systems.

The trajectory of nodes for an uncontrolled non-autonomous discrete parameter damped structure is sketched in Fig.(7.3). From the known initial condition $\boldsymbol{\zeta}_o$ one can compute the initial energy of the system $E(0) = E_o$ by using Eq.(7.92). For times $t > 0$, the system energy $E(t)$ changes as the state $\boldsymbol{\zeta}(t)$ changes:

$$E(t) = \frac{1}{2}\boldsymbol{\zeta}^T(t)diag\left(\mathbf{M}, \mathbf{K}'\right)\boldsymbol{\zeta}(t) \tag{7.128}$$

Since the system energy at time t is the initial energy plus the work done by the loads, one may also write

$$E(t) = E(0) + \Delta E \tag{7.129}$$

where

$$\Delta E = \int_0^t \boldsymbol{\zeta}^T(\tau)\mathbf{g}(\tau)d\tau \tag{7.130}$$

If $\Delta E < 0$ and $E_o > 0$, then the point of trajectory at time t will be closer to the origin than it was at time 0 [see Fig. (7.3a)]. This is the case when loads are opposing to the ongoing motion. On the other hand, if $\Delta E > 0$ then the point of trajectory at time t will be farther away from the origin than it was at time 0 [see Fig. (7.3b). This is the case when loads are applied to agree with the ongoing motion.

7.8.3 Controlled Autonomous System

Since the structure is controlled, $\mathbf{u} \neq \mathbf{o}$; and since it is autonomous, $\mathbf{g} = \mathbf{o}$. With these from Eq.(7.87), one may write:

$$\begin{aligned} \dot{\boldsymbol{\zeta}} &= \mathbf{A}\boldsymbol{\zeta} + \grave{\mathbf{B}}\mathbf{u} && \text{for } t > 0 \\ \boldsymbol{\zeta}(0) &= \boldsymbol{\zeta}_o && \text{at } t = 0 \end{aligned} \tag{7.131}$$

Suppose controls $\mathbf{u}$ started at $t' > 0$ and continued until $t'' > t'$ and then stopped. The quantity T defined as

$$T = t'' - t' \tag{7.132}$$

is called the *control time*. Due to the presence of the excitation $\grave{\mathbf{B}}\mathbf{u}$ term on the right-hand side, the term "controlled autonomous system" appears to be a misnomer in general. However in feedback control systems, as discussed in the next chapter, through an appropriately selected *control law*, $\mathbf{u}$ is expressed as a function of $\boldsymbol{\zeta}$, and thus makes the system autonomous.

By using the transition matrix $e^{\mathbf{A}(t-t')} \in R^{2n\times 2n}$, studied in the previous subsection, the solution of $\boldsymbol{\zeta}(t)$ for this case can be readily expressed as

$$\begin{aligned} &\boldsymbol{\zeta}(t) = e^{\mathbf{A}(t-0)}\boldsymbol{\zeta}_o \quad \text{for } 0 \leq t \leq t' \\ &\boldsymbol{\zeta}(t) = e^{\mathbf{A}(t-t')}\boldsymbol{\zeta}(t') + \textstyle\int_{t'}^{t} e^{\mathbf{A}(\tau-t')}\grave{\mathbf{B}}\mathbf{u}(\tau - t')d\tau \quad \text{for } t' \leq t \leq t'' \\ &\boldsymbol{\zeta}(t) = e^{\mathbf{A}(t-t'')}\boldsymbol{\zeta}(t'') \quad \text{for } t'' \leq t \end{aligned} \tag{7.133}$$

The trajectory of nodes for this case is sketched in Fig.(7.4) assuming that the controls are opposing the ongoing motion. It is assumed that the initial energy of the system is a positive quantity, i.e., $E(0) > 0$. During the time interval $(0 - t')$ the system loses some energy due to damping; hence the system's energy at time t' is less than the initial energy, i.e.,

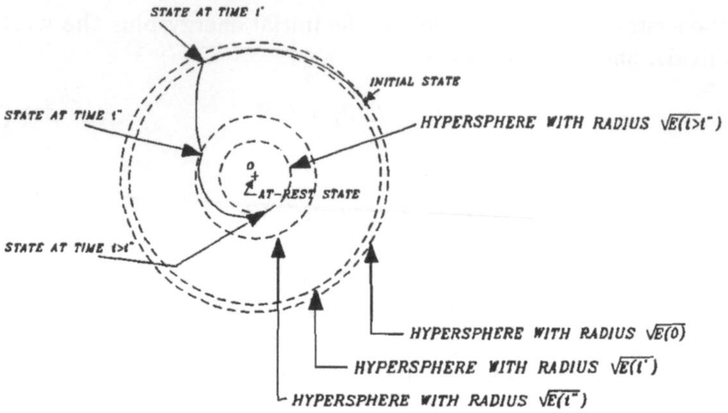

FIGURE 7.4. Trajectory of nodes in state space R^{2n} (controlled autonomous case)

$E(t') < E(0)$. During the control time T, the system's energy decreased considerably to $E(t'')$, i.e., $E(t'') << E(t') < E(0)$. After the controls are turned off, i.e., at times $t > t$", the system's energy continues to decrease due to damping.

7.8.4 Controlled Non-Autonomous System

Since the structure is controlled, $\mathbf{u} \neq \mathbf{o}$; and since it is non-autonomous, $\mathbf{g} \neq \mathbf{o}$. This is the general case given by Eq.(7.87). It is reproduced below:

$$\begin{aligned} \dot{\boldsymbol{\zeta}} &= \mathbf{A}\boldsymbol{\zeta} + \grave{\mathbf{B}}\mathbf{u} + \mathbf{g} \qquad \text{for } t > 0 \\ \boldsymbol{\zeta}(0) &= \boldsymbol{\zeta}_o \qquad \text{at } t = 0 \end{aligned} \tag{7.134}$$

Suppose the controls are turned on at $t' > 0$ and turned off after T time, i.e., at $t'' = t' + T$. Assuming that we have *a priori* knowledge of $\mathbf{u} = \mathbf{u}(t)$, by using the transition matrix $e^{\mathbf{A}(t-t')} \in R^{2n\times 2n}$ of the system, the solution $\boldsymbol{\zeta}(t)$ for this general case may be expressed as

$$\begin{aligned} &\boldsymbol{\zeta}(t) = e^{\mathbf{A}(t-0)}\boldsymbol{\zeta}_o + \int_0^t e^{\mathbf{A}(\tau-0)}\mathbf{g}(\tau - 0)d\tau, \quad \text{for } 0 \leq t \leq t' \\ &\boldsymbol{\zeta}(t) = e^{\mathbf{A}(t-t')}\boldsymbol{\zeta}(t') + \int_{t'}^t e^{\mathbf{A}(\tau-t')}[\grave{\mathbf{B}}\mathbf{u}(\tau - t') + \mathbf{g}(\tau - t')]d\tau, \text{ for } t' \leq t \leq t'' \\ &\boldsymbol{\zeta}(t) = e^{\mathbf{A}(t-t'')}\boldsymbol{\zeta}(t''), \quad \text{for } t\text{"} \leq t \end{aligned} \tag{7.135}$$

The trajectory of nodes for this case is sketched in Fig. (7.5). The scenario described in the figure assumes that the excitations $\mathbf{g}(t)$ agree with the ongoing motion, controls $\mathbf{u}(t)$ oppose the ongoing motion, and the initial energy of the system is a positive quantity, i.e., $E(0) > 0$. Accordingly, the system's energy increases from $E(0)$ to $E(t')$, then, due to controls, it decreases from $E(t')$ to $E(t'')$. After the controls are turned off, the

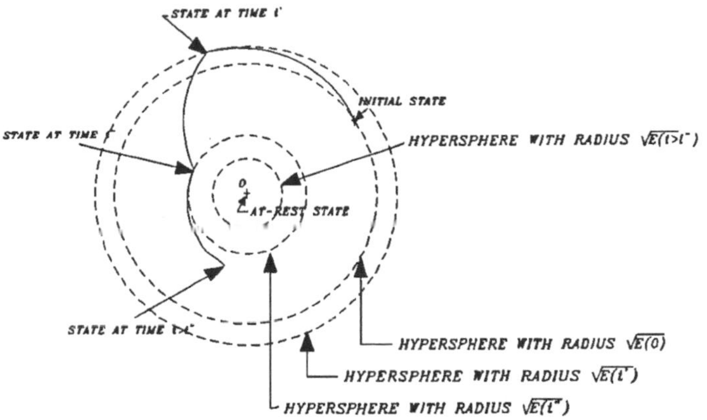

FIGURE 7.5. Trajectory of nodes in state space R^{2n} (controlled non-autonomous case)

system's energy starts to increase again. From an energy standpoint, if the object of the controls is to keep the system at the vicinity of its at-rest state, then the controls should eliminate not only the system's energy at the time when they are turned on but also the additional energy that may be caused by the excitations during the control time T.

7.9 Steady State of Nodal Deflections

In Sections 7.4, 7.5, 7.6 and 7.7, the motion of nodes $\boldsymbol{\xi}(t)$ is obtained as a function of nodal loads $\mathbf{p} = diag(\hat{p}_k)\ \mathbf{p}'$, load type prescribed element deformations $\mathring{\mathbf{v}}_e = diag(\hat{v}_e^k)\ \mathring{\mathbf{v}}_e'$, support movements $\boldsymbol{\xi}^s = diag(\hat{\xi}_k^s)\ \boldsymbol{\xi}^{s\prime}$, and actuator induced deformations $\mathbf{u} = diag(\hat{u}_k)\ \mathbf{u}'$, respectively. In the mathematical description of these loadings, as explained for Eq.(7.3), the diagonal matrices represent the time dependency, and their post multipliers represent the largest magnitudes (or an approximation of the largest magnitudes). In $\boldsymbol{\xi}(t)$ expressions for nodal loads, prescribed element deformations, support movements, and actuator induced deformations [given in Eqs.(7.65, 7.68), (7.71, 7.72), (7.75, 7.76), and (7.79, 7.80) respectively], the diagonal entries of time dependency matrices appear inside the integral operators with limits from 0 to t. Depending upon the character of the excitations, for sufficiently large values of time, say for t', the independent nodal deflections $\boldsymbol{\xi}(t')$ may be (a) a constant vector, (b) a vector with a fixed periodicity, or (c) a vector that never settles to a predictable pattern. Cases (a) and (b) constitute the steady state of the nodes. In the following subsections these two cases are discussed.

7.9.1 Time Function of Loads Becomes Constant

This is the case where the nodal motion ceases to exist at a sufficiently large time t' and beyond. For such times, the structure reaches to its final equilibrium state under its no longer changing loads. This is the scenario for most static loadings.

Suppose constant vectors $\mathbf{p}'$, $\mathbf{\mathring{v}}'_e$, $\boldsymbol{\xi}^{s'}$, $\mathbf{u}'$ consist of the largest magnitude components of loads $\mathbf{p}(t) = diag\ [\hat{p}_k(t)]\ \mathbf{p}'$, $\mathbf{\mathring{v}}_e(t) = diag\ [\hat{v}_e^k(t)]\ \mathbf{\mathring{v}}'_e$, $\boldsymbol{\xi}^s(t) = diag\ [\hat{\xi}_k(t)]\ \boldsymbol{\xi}^{s'}$, and $\mathbf{u}(t) = diag\ [\hat{p}_k(t)]\ \mathbf{u}'$, respectively, as explained for Eq.(7.3). By using identity matrices $\mathbf{I}$, one may write

$$\begin{aligned} &diag\ [\hat{p}_k(t)] = \mathbf{I} \in (\text{Int})^{n\times n}, \quad diag\ [\hat{v}_e^k(t)] = \mathbf{I} \in (\text{Int})^{m\times m} \\ &diag\ [\hat{\xi}_k(t)] = \mathbf{I} \in (\text{Int})^{b\times b}, \quad diag\ [\hat{u}_k(t)] = \mathbf{I} \in (\text{Int})^{q\times q} \end{aligned} \tag{7.136}$$

for $t \geq t'$. Then it follows that

$$\begin{aligned} &\mathbf{p}(t) = \mathbf{p}', \quad \mathbf{\mathring{v}}_e(t) = \mathbf{\mathring{v}}'_e, \\ &\boldsymbol{\xi}^s(t) = \boldsymbol{\xi}^{s'}, \quad \mathbf{u}(t) = \mathbf{u}', \end{aligned} \tag{7.137}$$

for $t \geq t'$. Because of damping the effects of initial conditions and transients all dissipate, and it may be observed that

$$\left.\begin{aligned} \boldsymbol{\xi}(t) &= \boldsymbol{\xi}_s \\ \dot{\boldsymbol{\xi}}(t) &= \mathbf{o} \\ \ddot{\boldsymbol{\xi}}(t) &= \mathbf{o} \end{aligned}\right\} \quad \text{for } t \geq t' \tag{7.138}$$

where $\boldsymbol{\xi}_s$ is the constant steady state value of the independent nodal deflections. With these, one may rewrite Eq.(7.7) as

$$\mathbf{K}'\boldsymbol{\xi}_s = \mathbf{p}' - \mathbf{B}\vec{\mathbf{K}}\mathbf{\mathring{v}}'_e - \mathbf{B}\vec{\mathbf{K}}\mathbf{B}_s^T\boldsymbol{\xi}^{s'} - \mathbf{B}\vec{\mathbf{K}}\mathbf{I}_\beta\mathbf{u}' \quad \text{for } t \geq t' \tag{7.139}$$

or

$$\boldsymbol{\xi}_s = \mathbf{K}'^{-1}(\mathbf{p}' - \mathbf{B}\vec{\mathbf{K}}\mathbf{\mathring{v}}'_e - \mathbf{B}\vec{\mathbf{K}}\mathbf{B}_s^T\boldsymbol{\xi}^{s'} - \mathbf{B}\vec{\mathbf{K}}\mathbf{I}_\beta\mathbf{u}') \quad \text{for } t \geq t' \tag{7.140}$$

which are the independent nodal deflections were the loads applied statically.

When Eqs.(7.136) are valid, the static analysis with final load values shown in Eqs.(7.137) provides the constant steady state of nodal deflections.

7.9.2 Time Function of Loads is Cyclic

The expressions given in Sections 7.4 - 7.7 for $\boldsymbol{\xi}(t)$ are very useful for studying transient motions of the nodes; however not so informative when these loadings are cyclic and applied on the structure indefinitely. The steady state expressions are very important for designs dominated by steady state conditions. It is also useful in earthquake and wind engineering, since it helps the engineer to understand the behavior of his structure when a given earthquake episode or a storm record were to repeat itself indefinitely.[4] The following is a short discussion of *steady state vibrations* caused by cyclic support movements.

Suppose the time function of support movements, i.e., $diag(\hat{\xi}^s_k)\in C^{b\times b}$, is given as

$$diag(\hat{\xi}^s_k) = \mathbf{I}e^{j\Omega t} \tag{7.141}$$

where frequency Ω is called the *driving frequency* and $\mathbf{I} \in (\mathrm{Int})^{b\times b}$ is the identity matrix. Using this, the dynamic equilibrium equations of the nodes may be written from Eq.(7.74) as

$$\mathbf{M}\ddot{\boldsymbol{\xi}} + \dot{\mathbf{C}}\dot{\boldsymbol{\xi}} + \mathbf{K}'\boldsymbol{\xi} = \mathbf{f}e^{j\Omega t} \quad \text{for } t > 0 \tag{7.142}$$

where

$$\mathbf{f} = -\mathbf{B}\vec{\mathbf{K}}\mathbf{B}_s^T\boldsymbol{\xi}^{s'} \tag{7.143}$$

is a constant vector in R^n. Since for large t, the effect of initial conditions is already dissipated [see Eqs.(7.54, 7.55, and 7.56)], one needs only a particular solution to this matrix differential equation. It may be easily verified that the time dependency of $\boldsymbol{\xi}(t)$ for large t is also cyclic with frequency Ω. Denoting modal amplitudes by $\mathbf{a}$, one may propose:

$$\boldsymbol{\xi} = \mathbf{X}\mathbf{a}e^{j\Omega t} \tag{7.144}$$

as a particular solution of Eq.(7.142). Using this in Eq.(7.142), and taking advantage of Eqs.(7.20, 7.25, 7.28, and 7.49), one may obtain for modal amplitudes

$$\mathbf{a} = diag\ [\frac{1}{\omega_k^2[1-(\frac{\Omega}{\omega_k})^2 + 2j\gamma_k(\frac{\Omega}{\omega_k})]}]\ \mathbf{X}^T\mathbf{f} \tag{7.145}$$

and the nodal motion as

$$\boldsymbol{\xi} = \mathbf{X}diag\ [\frac{1}{\omega_k^2[1-(\frac{\Omega}{\omega_k})^2 + 2j\gamma_k(\frac{\Omega}{\omega_k})]}]\ \mathbf{X}^T\mathbf{f}e^{j\Omega t} \tag{7.146}$$

or

$$\boldsymbol{\xi} = \mathbf{X}diag\ [c_k e^{j(\Omega t-\psi_k)}]\mathbf{X}^T\mathbf{f} \tag{7.147}$$

where

$$\psi_k = \tan^{-1}\frac{2\gamma_k(\frac{\Omega}{\omega_k})}{1-(\frac{\Omega}{\omega_k})^2} \tag{7.148}$$

and

$$c_k = \frac{1}{\omega_k^2\sqrt{[1-(\frac{\Omega}{\omega_k})^2]^2 + [2\gamma_k(\frac{\Omega}{\omega_k})]^2}} \tag{7.149}$$

---

[4]See Section 10.10, pp. 364-367, *Dynamics of Structures*, Walter C. Hurty and Moshe F. Rubinstein, Prentice-Hall, Inc., 1964.

According to Eq.(7.147)motion follows the cyclic excitation with the same frequency but with a phase angle of $-\psi_k$ in the kth modal component with $k = 1, 2, \cdots, n$. Note that if there is no damping for the kth mode, i.e., if $\gamma_k = 0$, then coefficient c_k, hence the motion itself, becomes unbounded when the driving frequency Ω coincides with the kth natural vibration frequency $\omega_k, k = 1, 2, \cdots, n$. This is the *resonance* phenomenon. It is evident from Eq.(7.149) that in mechanical systems if resonance cannot be avoided, it must be compensated by high damping.

7.10 Recapitulation of Inverse Relations

In this chapter, confined to time invariant linear systems, the inverse relations for the discrete parameter adaptive structures are obtained for the dynamic case. Using a displacement approach, and heavily relying on *superposition*, the structural response $(\boldsymbol{\xi}, \mathbf{s}, \mathbf{v})$ is expressed in terms of the motion of the nodes $\boldsymbol{\xi}$ [see Eq.(7.11)].

In Sections 7.4 - 7.7, the *transient motion of the nodes* is expressed in terms of *position of the nodes* $\boldsymbol{\xi}(t)$ for various loading conditions, by solving the *dynamic equilibrium equations* of Eq.(7.7) for $\boldsymbol{\xi}(t)$. In Section 7.8, the transient motion of the nodes is alternately expressed in terms of the *state of nodes* $\boldsymbol{\zeta}(t) = [\dot{\boldsymbol{\xi}}^T(t), \boldsymbol{\xi}^T(t)]^T$ for the same loading conditions, by solving the *equations of state* given by Eq.(7.87). The equivalency of these two groups of expressions may be easily shown by using the Euler's formula for complex conjugate exponentials, i.e., by considering the identity of

$$(e^{a+jb} + e^{a-jb})/2 = e^a Cos\, b$$

The *steady state of nodal deflections* is discussed briefly in Section 7.9 where asymptotic values of $\boldsymbol{\xi}(t)$ are obtained for loadings that become constant eventually or are cyclic.

8

Active Control of Response, Autonomous Case

In this chapter active control of response in discrete parameter adaptive structures is studied for the dynamic case. As in earlier chapters, the emphasis is on controlling nodal deflections $\boldsymbol{\xi}(t)$. It is assumed that at time $t = 0$, the structure is in a non-equilibrium state, and during control, excluding the actuator induced element deformations, there is no active loading on the structure. With the notation of Section 7.8, $\boldsymbol{\zeta}_o \neq \mathbf{o}$, $\mathbf{u} \neq \mathbf{o}$, and $\mathbf{g} = \mathbf{o}$, for $0 \leq t \leq T$. The equations of motion in state space are given in Subsection 7.8.3, and a sketch of the trajectory of nodes is given in Fig. (7.4). This chapter deals with the real time determination of $\mathbf{u}(t)$.

8.1 Response to Actuator Induced Element Deformations

Using a displacement approach, the inverse of excitation-response relations is obtained in Chapter 7. Specialization of the relations for the zero values of nodal forces $\mathbf{p} = diag[\hat{p}_k(t)]\mathbf{p}'$, load type prescribed element deformations $\mathring{\mathbf{v}}_e = diag[\hat{v}^k_e(t)]\mathring{\mathbf{v}}_e$, and support movements $\boldsymbol{\xi}^s = diag[\hat{\xi}^s_k(t)]\boldsymbol{\xi}^{s'}$ yields:

$$\left\{ \begin{array}{c} \boldsymbol{\xi}(t) \\ \mathbf{s}(t) \\ \mathbf{v}(t) \end{array} \right\} = \left[\begin{array}{c} \cdot \\ \vec{\mathbf{K}}\mathbf{I}_\beta \\ \mathbf{I}_\beta \end{array} \right] \mathbf{u}(t) + \left[\begin{array}{c} \mathbf{I} \\ \vec{\mathbf{K}}\mathbf{B}^T \\ \mathbf{B}^T \end{array} \right] \boldsymbol{\xi}(t) \qquad (8.1)$$

where

$$\boldsymbol{\xi}(t) = \left[\ \mathbf{A}_{\xi_o}(t) \quad \mathbf{A}_{\xi'_o}(t)\ \right] \left\{ \begin{array}{c} \boldsymbol{\xi}_o \\ \boldsymbol{\xi}'_o \end{array} \right\} + \mathbf{A}_{u'}(t)]\mathbf{u}' \tag{8.2}$$

and

$$\begin{aligned} \mathbf{A}_{\xi_o}(t) &= \mathbf{X} diag[e^{-\gamma_k\omega_k t}(\cos\grave{\omega}_k t + \tfrac{\gamma_k\omega_k}{\grave{\omega}_k}\sin\grave{\omega}_k t)]\ \mathbf{X}^{-1} \\ \mathbf{A}_{\xi'_o}(t) &= \mathbf{X} diag[e^{-\gamma_k\omega_k t}\tfrac{\sin\grave{\omega}_k t}{\grave{\omega}_k}]\mathbf{X}^{-1} \\ \mathbf{A}_{u'}(t) &= -\mathbf{M}^{-1}\mathbf{B}\vec{\mathbf{K}}\mathbf{I}_\beta diag[\tfrac{1}{\grave{\omega}_k}\int_0^t e^{-\gamma_k\omega_k\tau}\hat{u}_k(\tau)d\tau] \end{aligned} \tag{8.3}$$

In these expressions the response quantities are

$\boldsymbol{\xi}(t)$	$\in R^{n\times n}$	nodal deflections
$\mathbf{s}(t)$	$\in R^{m\times m}$	element forces
$\mathbf{v}(t)$	$\in R^{m\times m}$	element deformations

and the excitation quantities are

$\boldsymbol{\xi}_o$	$\in R^{n\times n}$	initial nodal deflections
$\boldsymbol{\xi}'_o$	$\in R^{n\times n}$	initial nodal velocities
$\mathbf{u}(t) = diag[\hat{u}_k(t)]\mathbf{u}'$	$\in R^{q\times 1}$	actuator induced deformations

where the components of constant vector $\mathbf{u}' \in R^{q\times 1}$ are the largest magnitudes or an estimate of the largest magnitudes of actuator induced element deformations, and the diagonal elements of $diag[\hat{u}_k(t)] \in R^{q\times q}$ are the time dependencies of these actuator induced deformations. The system related parameters are

$\mathbf{B}$	$\in R^{n\times m}$	matrix defining nodal internal forces as $\mathbf{Bs}$
$\mathbf{I}_\beta$	$\in (\mathrm{Int})^{m\times q}$	actuator placement matrix
$\vec{\mathbf{K}}$	$\in R^{m\times m}$	block diagonal matrix of element stiffnesses
$\mathbf{M}$	$\in R^{n\times n}$	mass matrix
$\mathbf{X}$	$\in R^{n\times n}$	matrix of M-normalized vibration mode shapes
γ_k	$\in R^{1\times 1}$	modal damping ratio of the kth mode
ω_k	$\in R^{1\times 1}$	natural vibration frequency of the kth mode
$\grave{\omega}_k$	$\in R^{1\times 1}$	damped vibration frequency of the kth mode

In addition to these, integers m, n, and q are also system parameters:

$m = Ma - f$	$\in (\mathrm{Int})^{1\times 1}$	number of independent element forces
$n = Ne - b$	$\in (\mathrm{Int})^{1\times 1}$	number of independent nodal deflections
q	$\in (\mathrm{Int})^{1\times 1}$	number of actuators

where e is the number of deflection degrees of freedom at a node, a is the number of element deformations per element, f is the number of element forces prescribed as zero, and b is the number of deflection constraints. Since the structure is time invariant, all system parameters are true constants, i.e., they do not change with time.

From Eqs.(8.1 and 8.2) one may observe that the response is due to initial nodal deflections $\boldsymbol{\xi}_o$, initial nodal velocities $\boldsymbol{\xi}'_o$, and actuator induced element deformations $\mathbf{u}(t)$. Because of the linear behavior, as shown in these equations, the individual contributions are superimposed. For example, if one is interested in the response due to $\mathbf{u}(t)$ only, by assigning $\boldsymbol{\xi}_o = \mathbf{o}$ and $\boldsymbol{\xi}'_o = \mathbf{o}$, from these equations one may obtain

$$\left\{ \begin{array}{c} \boldsymbol{\xi}(t) \\ \mathbf{s}(t) \\ \mathbf{v}(t) \end{array} \right\} = \left[\begin{array}{c} \mathbf{A}_{u'}(t) \\ \vec{\mathbf{K}} \left[\mathbf{I}_\beta diag[\hat{u}_k(t)] + \mathbf{B}^T \mathbf{A}_{u'}(t) \right] \\ \mathbf{I}_\beta diag[\hat{u}_k(t)] + \mathbf{B}^T \mathbf{A}_{u'}(t) \end{array} \right] \mathbf{u}' \qquad (8.4)$$

where $\mathbf{A}_{u'}(t)$ is, as defined in Eq.(8.3), a function of the time dependency $diag[\hat{u}_k(t)]$ of the actuator induced deformations $\mathbf{u}$, and the components of constant vector $\mathbf{u}'$ are the largest magnitudes or an estimate of the largest magnitudes of these induced deformations. The response given in Eq.(8.4) corresponds to the case where the actuators are activated when the structure is at rest.

As shown by Eq.(8.4), one can create a response as a function of actuator induced deformations. By controlling the actuator induced deformations one may control, at least partially, the response of the structure.

8.2 Energy Cost of Inducing Element Deformations

The human body is a very good example of adaptive structures. By contracting and extending our muscles we control our movements. To do this we need energy, since the muscles have to work against the forces trying to prevent their contractions and extensions. These forces may be due to preexisting stresses caused by gravitational and other kinds of acceleration, or they may be due to the presence of other muscles in the neighborhood. For the latter, nature solved the problem by synchronizing muscle contractions and extensions. A discussion of how this is done in discrete parameter adaptive structures is given in Section 3.8 for static loading. Below is the extension of that discussion for the dynamic case.

The resistance an actuator sees in inducing a required element deformation is the main cause of its energy demand. The resistance comes from the presence of preexisting element forces $\grave{\mathbf{s}}_o$, and the stiffness of the structure in directions associated with the induced element deformations. As discussed in Section 3.8 for static loading, very little can be said for minimizing the resistance due to preexisting stresses without knowing $\grave{\mathbf{s}}_o$. This matter is not discussed any further in this text. However, as discussed in Section 3.9 for static loading, by ensuring the compatibility of the induced strains, the resistance of the structure against the insertion of actuator induced element deformations may be overcome.

8.2.1 Element Forces by the Displacement Method

From Eq.(8.4), the element forces created as a result of actuator induced element deformations $\mathbf{u}(t) = diag[\hat{u}_k(t)]\mathbf{u}'$ may be obtained as

$$\mathbf{s}(t) = \vec{\mathbf{K}}\mathbf{I}_\beta\mathbf{u}(t) - \vec{\mathbf{K}}\mathbf{B}^T\mathbf{M}^{-1}\mathbf{B}\vec{\mathbf{K}}\mathbf{I}_\beta diag[\frac{1}{\hat{\omega}_k}\int_0^t e^{-\gamma_k\omega_k\tau}\hat{u}_k(\tau)d\tau](t)\mathbf{u}' \quad (8.5)$$

where the first term on the right-hand side shows the effect when the nodes are fixed, i.e., when $\boldsymbol{\xi}(t) = \mathbf{o}$, and the second term shows the effect of the motion of the nodes, i.e., the effect of $\boldsymbol{\xi}(t)$. These element forces $\mathbf{s}(t)$ are the resistance of the structure to inserting $\mathbf{u}(t)$. The presence of actuator placement matrix $\mathbf{I}_\beta$ in the expression shows clearly that the resistance is, among many other factors, a function of the location of the actuators. Since Eq.(8.5) is obtained by the displacement method, it provides no information about how to reduce the resistance of structure by the compatibility of induced element deformations, when the structure is statically indeterminate, i.e., when $r = m - n > 0$.

8.2.2 Element Forces by the Force Method

The resistance to insertion of actuator induced element deformations in a statically indeterminate structure may be much larger than the one in a statically determinate structure. To see this, one has to express the element forces by the force method. For this one starts with the response-excitation relations discussed in Chapter 6. These relations are reproduced below:

$$\begin{bmatrix} \cdot & \mathbf{B} & \cdot \\ \mathbf{B}^T & \cdot & -\mathbf{I} \\ \cdot & -\mathbf{I} & \vec{\mathbf{K}} \end{bmatrix} \begin{Bmatrix} \boldsymbol{\xi} \\ \mathbf{s} \\ \mathbf{v} \end{Bmatrix} = \begin{Bmatrix} \mathbf{p} - \mathbf{M}\ddot{\boldsymbol{\xi}} - \grave{\mathbf{C}}\dot{\boldsymbol{\xi}} - \grave{\mathbf{K}}\boldsymbol{\xi} \\ -\mathbf{v}_o \\ \cdot \end{Bmatrix} \quad (8.6)$$

where $\mathbf{p}$ is the nodal loads, $\grave{\mathbf{C}}$ is the modal damping matrix, and $\grave{\mathbf{K}}$ is the global stiffness matrix due to preexisting element forces. The equations in the first row partition are dynamic equilibrium equations of the nodes, the equations in the second row partition are the geometric relations, and the equations in the third partition are the stiffness relations. As discussed in Subsection 3.1.2, the expression for element forces $\mathbf{s}$ may be obtained from Eq.(8.6) by the force method as:

$$\mathbf{s} = (\mathbf{I} - \mathbf{K}_c\vec{\mathbf{F}})\mathbf{C}'(\mathbf{p} - \mathbf{M}\ddot{\boldsymbol{\xi}} - \grave{\mathbf{C}}\dot{\boldsymbol{\xi}} - \grave{\mathbf{K}}\boldsymbol{\xi}) + \mathbf{K}_c\mathbf{v}_o \quad (8.7)$$

where

$$\mathbf{K}_c = \mathbf{C}[\mathbf{C}^T\vec{\mathbf{F}}\mathbf{C}]^{-1}\mathbf{C}^T \qquad \vec{\mathbf{F}} = \vec{\mathbf{K}}^{-1} \quad (8.8)$$

and $\mathbf{C} \in \mathrm{R}^{m\times r}$ is a basis in the null space of $\mathbf{B} \in \mathrm{R}^{n\times m}$, and $\mathbf{C}' \in \mathrm{R}^{m\times n}$ is a generalized inverse of $\mathbf{B}$ (see Subsection 3.1.2). For the present case $\mathbf{p} = \mathbf{o}$;

hence

$$\mathbf{s} = -(\mathbf{I} - \mathbf{K}_c\vec{\mathbf{F}})\mathbf{C}'(\,\mathbf{M}\ddot{\boldsymbol{\xi}} + \grave{\mathbf{C}}\dot{\boldsymbol{\xi}} + \grave{\mathbf{K}}\boldsymbol{\xi}) + \mathbf{C}[\mathbf{C}^T\vec{\mathbf{F}}\mathbf{C}]^{-1}\mathbf{C}^T\mathbf{v}_o \tag{8.9}$$

where, from Eqs. (8.2 and 8.3),

$$\boldsymbol{\xi} = -\mathbf{M}^{-1}\mathbf{B}\vec{\mathbf{K}}\mathbf{I}_\beta diag[\frac{1}{\grave{\omega}_k}\int_0^t e^{-\gamma_k\omega_k\tau}\hat{u}_k(\tau)d\tau]\mathbf{u}' \tag{8.10}$$

and

$$\begin{aligned} \mathbf{v}_o &= \mathbf{I}_\beta\mathbf{u} \\ \mathbf{u} &= diag[\hat{u}_k(t)]\mathbf{u}' \end{aligned} \tag{8.11}$$

This form of the expression for element forces is very informative. It states that the element forces are created by the

- actuator induced element deformations, assuming that the inertial forces, dissipative forces, and restoring forces caused by preexisting element forces are all zero[the second term in Eq.(8.9)], and
- motion of the nodes due to the inertial forces, dissipative forces, and restoring forces caused by preexisting element forces [the first term in Eq.(8.9)].

As discussed in Section 3.9, one may see from Eq.(8.9) that there is no contribution from the second term if the actuator induced element deformations satisfy the compatibility requirement at each instant of time, i.e., if

$$\mathbf{C}^T\mathbf{v}_o = \mathbf{o} \tag{8.12}$$

Since $\mathbf{v}_o = \mathbf{I}_\beta\mathbf{u}$, this can be restated alternatively as

$$\mathbf{C}^T\mathbf{I}_\beta diag[\hat{u}_k(t)]\mathbf{u}' = \mathbf{o} \tag{8.13}$$

The reader may observe that if the structure is statically determinate, i.e., if $m = n$, the degree of statical indeterminacy r becomes zero and the null space of $\mathbf{B}$ and therefore matrix $\mathbf{C}$ become non-existing, implying that all inserted element deformations are compatible; hence, the second term in Eq.(8.9) is always zero.

If we assume that actuator placement matrix $\mathbf{I}_\beta$ is such that the compatibility requirement stated in Eq.(8.13) is satisfied, then the element forces can be computed from Eq.(8.9) without the second term on the right-hand side, i.e., from

$$\mathbf{s}(t) = -(\mathbf{I} - \mathbf{K}_c\vec{\mathbf{F}})\mathbf{C}'(\,\mathbf{M}\ddot{\boldsymbol{\xi}} + \grave{\mathbf{C}}\dot{\boldsymbol{\xi}} + \grave{\mathbf{K}}\boldsymbol{\xi}) \tag{8.14}$$

For computation purposes, one may prefer Eq.(8.5), since it may involve less operations.

In statically indeterminate structures, by using synchronized *secondary actuators*, actuator induced element deformations can be made compatible. In other words, the use of secondary actuators can eliminate the resistance of the structure to insertion of element deformations. Chapter 5 discusses this problem with sufficient detail.

8.3 Response Control

It is obvious from the excitation-response relations of discrete parameter structures that one may affect the response by altering some of the parameters of the structural system, such as material constants, element geometries, support conditions, etc., or by applying nodal forces or element deformations. Similar to controlling our moves by contracting or extending our muscles, in adaptive structures, the control of response is done by actuators that insert deformation into the structural elements.

In this chapter it is assumed that, at time $t = 0$ when the controls are turned on, the structure is in a non-equilibrium state with nodal positions $\boldsymbol{\xi}(0) = \boldsymbol{\xi}_o$ and nodal velocities $\dot{\boldsymbol{\xi}}(0) = \boldsymbol{\xi}'_o$, comprising the initial state $\boldsymbol{\zeta}(0) = \boldsymbol{\zeta}_o = [\boldsymbol{\xi}_o^T, \boldsymbol{\xi}_o'^T]^T$. At this initial state, the total energy $E(0)$ of the system is

$$E(0) = E_o = \frac{1}{2}\boldsymbol{\zeta}_o^T diag(\mathbf{M}, \mathbf{K}')\boldsymbol{\zeta}_o \tag{8.15}$$

according to Eq.(7.128). The trajectory of the nodes preceding $t = 0$ is not known, and there is no active loading of any kind on the structure. Thanks to modern sensors, the capability of measuring the initial state and the succeeding ones in *real time* is available, i.e., at the time the event is taking place, continuously. The task here is to bring the structure to its at-rest state, in control time T, by using the deformation inserting actuators of the structure. Let $\mathbf{u}(t) \in R^{q\times 1}$denote the level of cumulative deformation insertion at time t. Knowing the excitation-response relations, the problem is how to quantify $\mathbf{u}$ at every moment, in real time, in a manner that it will monotonically drive the structure from its non-equilibrium initial state $\boldsymbol{\zeta}_o$ to its final equilibrium state $\boldsymbol{\zeta}(T) = \mathbf{o}$, as depicted in Fig.(8.1), during the time interval of $0 \leq t \leq T$.

Control of other response quantities is not directly discussed. However, it is clear from Eq.(8.1) that once the nodal deflections $\boldsymbol{\xi}(t)$ are controlled with the knowledge of controls $\mathbf{u}(t), 0 \leq t \leq T$, so are element elongations $\mathbf{v}(t)$ and element forces $\mathbf{s}(t)$:

$$\mathbf{v}(t) = \mathbf{I}_\beta\ \mathbf{u}(t) + \mathbf{B}^T\boldsymbol{\xi}(t) \tag{8.16}$$

and

$$\mathbf{s}(t) = \vec{\mathbf{K}}\mathbf{v}(t) \tag{8.17}$$

according to Eq.(8.1).

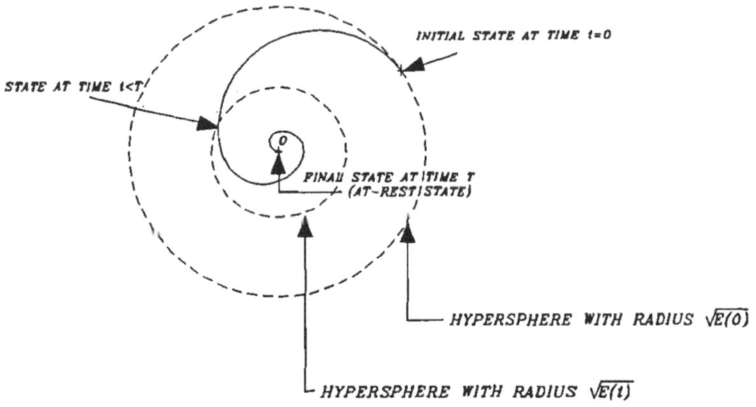

FIGURE 8.1. Trajectory of nodes, controlled autonomous system

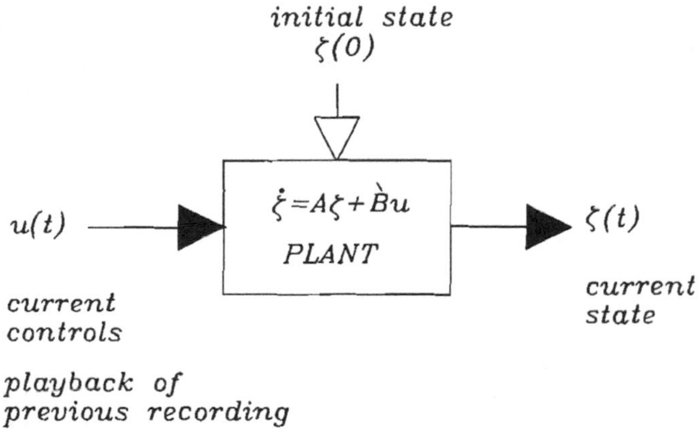

FIGURE 8.2. Open-loop state control

8.4 Open-Loop Control

Suppose the initial state $\boldsymbol{\zeta}_o \in \mathrm{R}^{2n \times 1}$ is known, and we have enough time to compute, a priori, controls $\mathbf{u}(t), 0 \leq t \leq T$. Then with this *a priori* knowledge of $\mathbf{u}(t) \in \mathrm{R}^{q \times 1}$, one may play it back in real time, when the need arises, to bring the nodes from their initial non-equilibrium state of $\boldsymbol{\zeta}_o$ to the final equilibrium state of $\boldsymbol{\zeta}(T) = \mathbf{o}$, as sketched in Fig. (8.1). This is called *open-loop control*. In Fig.(8.2), open-loop control is sketched in block diagram form. In the class of problems related to adaptive structures, this type of control is rarely of interest, since it heavily depends on *a priori* knowledge of the initial state. It is discussed here in order to gain insight for the state feedback control problem.

Once relieved from the requirement of real time determination, one may proceed to solve the following mathematical problem at a leisurely pace, rather than in real time. Given the equation of motion of nodes in state space

$$\dot{\boldsymbol{\zeta}} = \mathbf{A}\boldsymbol{\zeta} + \grave{\mathbf{B}}\mathbf{u} \text{ , for } 0 \leq t \leq T \tag{8.18}$$

where $\mathbf{A} \in R^{2n\times 2n}$, $\mathbf{B} \in R^{2n\times 2n}$ are as defined in Eqs.(6.33 and 6.34), and the initial and final states

$$\left.\begin{array}{l} \zeta(0) = \zeta_o \\ \zeta(T) = \mathbf{o} \end{array}\right\} \tag{8.19}$$

find controls $\mathbf{u}(t), 0 \leq t \leq T$ such that the system energy, i.e., the distance from the origin of the energy-normed state space [see Fig. (8.1)] ,

$$E(t) = \frac{1}{2}\zeta(t)^T diag(\mathbf{M}, \mathbf{K}')\zeta(t) \tag{8.20}$$

monotonically decreases with increasing time. The easiest way of solving this problem is by converting it to a two-point boundary value problem by expressing $\mathbf{u}$ as a known function of $\boldsymbol{\zeta}$ that satisfies the constraint of

$$\frac{d}{dt}E(t) < 0 \text{ for } \leq t \leq T \tag{8.21}$$

(see Subsection 7.8.1) and using it in Eq.(8.18). There are infinitely many choices for the function; hence, there are infinitely many solutions. To keep the problem linear, one may choose

$$\mathbf{u} = \mathbf{G}\boldsymbol{\zeta} \tag{8.22}$$

where $\mathbf{G} \in R^{q\times 2n}$. The equation is called a *linear control law.* One may use it in Eq.(8.18) to arrive at the following two-point linear boundary value problem:

$$\left.\begin{array}{r} \dot{\boldsymbol{\zeta}} = (\mathbf{A} + \grave{\mathbf{B}}\mathbf{G})\boldsymbol{\zeta} \text{ , for } 0 \leq t \leq T \\ \zeta(0) = \zeta_o \\ \zeta(T) = \mathbf{o} \end{array}\right\} \tag{8.23}$$

which possesses a unique solution. Although the requirement of *monotonic energy decreasing by increasing time*, as stated in Eq.(8.21), limits the admissible choices for $\mathbf{G} \in R^{q\times 2n}$, one still has infinitely many admissible choices for it. For each admissible choice, there is a unique trajectory that takes the system from point $\boldsymbol{\zeta}_o$ to point $\mathbf{o}$.

Which one is the *best* trajectory? The question, of course, implies the existence of some norm to order these trajectories. Energy is an important entity that can be used as a norm. Since the total work done in the actuators during the control time T dissipates the initial energy E_o of the system and thus brings it to the at-rest state, it can be considered as an index to trajectories. Unfortunately, so long as Eq.(8.21) is satisfied, all trajectories are associated with the same total energy E_o, i.e., they are all best in this sense. One needs some other norm, or using the jargon of the optimization field, *objective functional*, in order to select the best trajectory, or again using the optimization jargon, the *optimal trajectory*.

A widely used objective functional is

$$J = \frac{1}{2}\int_0^T [\zeta(t)^T \mathbf{Q}\zeta(t) + \mathbf{u}(t)^T \mathbf{R}\mathbf{u}(t)]dt \tag{8.24}$$

where symmetric matrices $\mathbf{Q} \in R^{2n \times 2n}$ and $\mathbf{R} \in R^{2n \times 2n}$ are positive and positive definite, respectively, otherwise completely arbitrary. This positive scalar J is called the *quadratic performance functional.* The first term in the integrand is a measure, at time t, of how far the nodes are from their equilibrium state, and the second term is a measure of how much energy is used by the actuators up to time t. By changing the norms of $\mathbf{Q}$ and $\mathbf{R}$, one may control the rate of dissipating the total system energy E_o as time increases from 0 to T. If J is minimum for the optimal trajectory, then, for example, the use of larger-norm $\mathbf{R}$ and smaller-norm $\mathbf{Q}$ biases the trajectory towards "smaller energy-dissipation rates initially," whereas the use of smaller-norm $\mathbf{R}$ and larger-norm $\mathbf{Q}$ biases the trajectory towards "slower state-dissipation rates initially." This provides an important option in control system design.

By trying all proper $\mathbf{G}$ matrices one by one in Eq.(8.23) and solving it for the trajectory, and then using the criterion of *performance* defined in Eq.(8.24) for indexing it, theoretically all trajectories may be found and indexed, and thus the optimal trajectory and the corresponding optimal $\mathbf{G} \in R^{q \times 2n}$ may be identified.

This hypothetical trial-and-error procedure may direct one to a more practical approach in identifying the optimal trajectory and its optimal $\mathbf{G}$ corresponding to given $\mathbf{Q}$ and $\mathbf{R}$. How this is done is discussed in the next section.

8.5 Optimal State Feedback Control

The discussions of the previous section lead to the idea that one may find the current value of control $\mathbf{u}(t)$ from the current value of state $\zeta(t)$, by using an appropriate *control law*, such as the one in Eq.(8.22). At a given instant t, we may

- physically measure $\zeta(t)$,
- use it in the control law to evaluate $\mathbf{u}(t)$, and
- physically implement it.

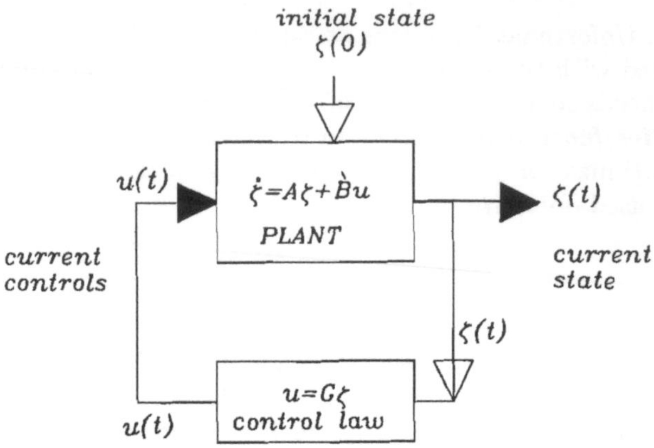

FIGURE 8.3. State feedback control

Thanks to the advances in sensor, actuator, and digital processing technologies, these can be done in many systems almost instantaneously. By repeating this three-step process, i.e., the *control cycle*, with high rates, the structure can be brought along an optimal trajectory from point $\boldsymbol{\zeta}_o$ to point $\mathbf{o}$ in time T. This process is called *state feedback control.* It is an important tool for modern technology. A sketch of closed-loop state control is shown in Fig. (8.3).

Among the important problems of state feedback control, the determination of control law of Eq.(8.22) comes first. As mentioned earlier, one may define easily many control laws that satisfy Eq.(8.21).[1] However, when one has the options, one should choose the one which produces a trajectory that is optimal with respect to a performance criteria such as the one given in Eq.(8.24).

8.5.1 Identification of Optimal Trajectory

For convenience and practical importance, one may first extend the performance functional in Eq.(8.24) to include the cost of a possible terminal error of $\boldsymbol{\zeta}(T) \neq 0$:

$$J = \frac{1}{2}\boldsymbol{\zeta}(T)^T\mathbf{F}\boldsymbol{\zeta}(T) + \frac{1}{2}\int_0^T [\boldsymbol{\zeta}(t)^T\mathbf{Q}\boldsymbol{\zeta}(t) + \mathbf{u}(t)^T\mathbf{R}\mathbf{u}(t)]dt \tag{8.25}$$

where symmetric matrix $\mathbf{F} \in \mathrm{R}^{2n\times 2n}$ is positive. Then, given $\mathbf{R}, \mathbf{Q}, \mathbf{F}$ matrices, initial state $\boldsymbol{\zeta}(0)$, and not-necessarily-zero final state $\boldsymbol{\zeta}(T)$, we try to find $\mathbf{u}(t)$ in the closed interval $0 \le t \le T$ that minimizes the performance

[1] See Section 8.8.

functional J and satisfies the constraints

$$\begin{aligned} \dot{\boldsymbol{\zeta}} &= \mathbf{f} \ , \ \text{for } 0 \leq t \leq T \\ \boldsymbol{\zeta}(0) &= \boldsymbol{\zeta}_o \text{ and } \boldsymbol{\zeta}(T) = \text{are given} \end{aligned} \tag{8.26}$$

where

$$\mathbf{f} = \mathbf{A}\boldsymbol{\zeta} + \grave{\mathbf{B}}\mathbf{u} \tag{8.27}$$

This is a well-posed constrained optimization problem. By using Lagrange multipliers $\boldsymbol{\lambda} \in \mathrm{R}^{2n\times 1}$, it may be converted to an unconstrained problem:

$$\text{find } \mathbf{u} \text{ to minimize } J = \frac{1}{2}\boldsymbol{\zeta}(T)^T\mathbf{F}\boldsymbol{\zeta}(T) + \int_0^T L' dt \tag{8.28}$$

where

$$\left.\begin{aligned} L' &= H - \boldsymbol{\lambda}^T\dot{\boldsymbol{\zeta}} \\ H &= L + \boldsymbol{\lambda}^T\mathbf{f} \\ L &= \tfrac{1}{2}[\boldsymbol{\zeta}(t)^T\mathbf{Q}\boldsymbol{\zeta}(t) + \mathbf{u}(t)^T\mathbf{R}\mathbf{u}(t)] \end{aligned}\right\} \tag{8.29}$$

H is the *state function of Pontryagin.* Since at the minimum point of J in $\mathrm{R}^{(4n+q)\times 1}$ space of $\boldsymbol{\eta}$:

$$\boldsymbol{\eta} = \left\{ \begin{matrix} \boldsymbol{\zeta} \\ \boldsymbol{\lambda} \\ \mathbf{u} \end{matrix} \right\} \tag{8.30}$$

the first variation of J with respect to the components of $\boldsymbol{\eta}$ vanishes, i.e., $\delta J = \delta_{\eta} J = 0$ or

$$\delta\boldsymbol{\eta}^T J_{,\eta} = 0 \tag{8.31}$$

where δ is the symbol for *first variation*, and the comma in the subscript is used to indicate the partial differentiation of J with respect to the quantities following the comma; hence $J_{,\eta} \in \mathrm{R}^{(4n+q)\times 1}$, and the kth entry $\frac{\partial J}{\partial \eta_k}$ is $J_{,\eta_k} \in \mathrm{R}^1$. Since $\boldsymbol{\eta}(0)$ and $\boldsymbol{\eta}(T)$ are both prescribed, Eq.(8.31) leads to the *Lagrange equation*

$$L'_{,\eta} - \frac{d}{dt}L'_{,\dot{\eta}} = \mathbf{o} \ , \ \text{for } 0 < t < T \tag{8.32}$$

which may be written in component form as

$$\left.\begin{aligned} \dot{\boldsymbol{\lambda}} &= -H_{,\zeta} \\ \dot{\boldsymbol{\zeta}} &= H_{,\lambda} \\ \mathbf{o} &= H_{,\mathbf{u}} \end{aligned}\right\} \tag{8.33}$$

or using Eqs.(8.29 and 8.27):

$$\left.\begin{aligned} \dot{\boldsymbol{\lambda}} &= -\mathbf{Q}\boldsymbol{\zeta} - \mathbf{A}^T\boldsymbol{\lambda} \\ \dot{\boldsymbol{\zeta}} &= \mathbf{A}\boldsymbol{\zeta} + \grave{\mathbf{B}}\mathbf{u} \end{aligned}\right\} \tag{8.34}$$

and

$$\mathbf{u} = -\mathbf{R}^{-1}\grave{\mathbf{B}}^T\boldsymbol{\lambda} \tag{8.35}$$

By substituting **u** from this equation, Eqs.(8.34) may be rewritten as

$$\begin{Bmatrix} \dot{\boldsymbol{\lambda}} \\ \dot{\boldsymbol{\zeta}} \end{Bmatrix} = \begin{bmatrix} -\mathbf{A}^T & -\mathbf{Q} \\ -\grave{\mathbf{B}}\mathbf{R}^{-1}\grave{\mathbf{B}}^T & \mathbf{A} \end{bmatrix} \begin{Bmatrix} \boldsymbol{\lambda} \\ \boldsymbol{\zeta} \end{Bmatrix}, \text{ for } 0 < t < T \tag{8.36}$$

and

$$\begin{Bmatrix} \boldsymbol{\lambda}(T) \\ \boldsymbol{\zeta}(0) \end{Bmatrix} = \begin{Bmatrix} \mathbf{F}\boldsymbol{\zeta}(T) \\ \boldsymbol{\zeta}_o \end{Bmatrix} \tag{8.37}$$

where the first of Eq.(8.37) is a result of the condition that J in Eq.(8.28) is stationary at $t = T$. These equations define a well-posed *two-point* boundary value problem for $(\boldsymbol{\zeta},\boldsymbol{\lambda})$. These equations identify optimal $(\boldsymbol{\zeta},\boldsymbol{\lambda})$, i.e., $(\boldsymbol{\zeta}^*,\boldsymbol{\lambda}^*)$, and, using $\boldsymbol{\lambda}^*$ in Eq.(8.35), the optimal controls $\mathbf{u}^*$. However there is no identification for the optimal control law. This is discussed next.

8.5.2 *Identification of the Optimal Control Law*

If one could express $\boldsymbol{\lambda}$ in Eq.(8.35) in terms of $\boldsymbol{\zeta}$, as $\boldsymbol{\lambda} = \mathbf{P}\boldsymbol{\zeta}$, one would obtain a control law similar to the one in Eq.(8.22). In fact, if one were to use the optimal values of $(\boldsymbol{\zeta},\boldsymbol{\lambda})$, i.e., $(\boldsymbol{\zeta}^*,\boldsymbol{\lambda}^*)$, one would obtain the optimal control law. By the Hamilton-Jacobi approach, it has been shown that, if the problem is time invariant and linear, then there is a unique time-dependent positive definite matrix $\mathbf{P}(t) \in \mathrm{R}^{2n\times 2n}$ such that the following is true:

$$\boldsymbol{\lambda}^*(t) = \mathbf{P}(t)\boldsymbol{\zeta}^*(t) \tag{8.38}$$

$\mathbf{P}(t)$ is called the *time-dependent Riccati matrix.* Since linearity and time-invariance are assumed, we can use Eq.(8.38) in Eqs.(8.36 and 8.37) to obtain a well-posed terminal value problem for $\mathbf{P}(t)$ as follows. Using Eq.(8.38), Eqs.(8.36) may be rewritten as:

$$\begin{Bmatrix} \dot{\mathbf{P}}\boldsymbol{\zeta}^* + \mathbf{P}\dot{\boldsymbol{\zeta}}^* \\ \dot{\boldsymbol{\zeta}}^* \end{Bmatrix} = \begin{bmatrix} -\mathbf{A}^T\mathbf{P} & -\mathbf{Q} \\ -\grave{\mathbf{B}}\mathbf{R}^{-1}\grave{\mathbf{B}}^T\mathbf{P} & \mathbf{A} \end{bmatrix} \begin{Bmatrix} \boldsymbol{\zeta}^* \\ \boldsymbol{\zeta}^* \end{Bmatrix} \tag{8.39}$$

Then subtracting **P** multiple of the second equation from the first, the latter equation becomes:

$$(\dot{\mathbf{P}} - \mathbf{P}\grave{\mathbf{B}}\mathbf{R}^{-1}\grave{\mathbf{B}}^T\mathbf{P} + \mathbf{P}\mathbf{A} + \mathbf{A}^T\mathbf{P} + \mathbf{Q})\boldsymbol{\zeta}^* = \mathbf{o} \tag{8.40}$$

In this identity, since $\boldsymbol{\zeta}^*$ is arbitrary (for it depends on arbitrary initial state $\boldsymbol{\zeta}_o$), its coefficient must vanish; hence

$$\dot{\mathbf{P}} = \mathbf{P}\grave{\mathbf{B}}\mathbf{R}^{-1}\grave{\mathbf{B}}^T\mathbf{P} - \mathbf{P}\mathbf{A} - \mathbf{A}^T\mathbf{P} - \mathbf{Q}, \text{ for } 0 < t < T \tag{8.41}$$

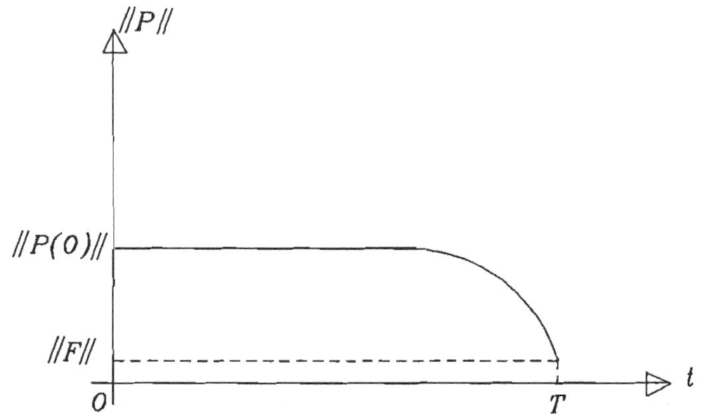

FIGURE 8.4. Qualitative behavior of time-dependent Riccati matrix

and from the first of Eq.(8.37) and Eq.(8.38) with a similar argument:

$$\mathbf{P}(T) = \mathbf{F} \tag{8.42}$$

Eq.(8.41) and Eq.(8.42) constitute a well-posed terminal value problem for finding the *time-dependent Riccati matrix* $\mathbf{P}(t)$. The numerical solution of these $\frac{(2n)[(2n)+1]}{2}$ coupled nonlinear first order ordinary differential equations from $t = T$ to $t = 0$ identifies $\mathbf{P}(t)$, $0 \leq t \leq T$. A typical plot of $\mathbf{P}(t)$ as a function of t is shown in Fig. (8.4).

By using the time-dependent Riccati matrix in Eq.(8.38) and then $\boldsymbol{\lambda}^*(t)$ from this equation in Eq.(8.35) the optimal control law for time-invariant autonomous linear systems is provided as:

$$\mathbf{u}^*(t) = \mathbf{G}^*(t)\boldsymbol{\zeta}^*(t) \tag{8.43}$$

where

$$\mathbf{G}^*(t) = -\mathbf{R}^{-1}\dot{\mathbf{B}}^T\mathbf{P}(t) \tag{8.44}$$

Matrix $\mathbf{G}^*(t) \in \mathrm{R}^{q\times 2n}$ is called the *optimal gain matrix.*

The reader may notice that the important issue of *controllability* is not brought in to the picture. A careful analysis[2] indicates that a time-invariant linear system is controllable if and only if the matrix

$$\dot{\mathbf{P}} = \left[\begin{array}{cccc} \mathbf{A}^0\dot{\mathbf{B}} & \mathbf{A}^1\dot{\mathbf{B}} & \cdots & \mathbf{A}^{2n-1}\dot{\mathbf{B}} \end{array}\right] \tag{8.45}$$

is not rank deficient, i.e., if

$$rank(\dot{\mathbf{P}}) = 2n \tag{8.46}$$

[2]See, for example, Michael Athans and Peter L. Falb, *Optimal Control, An Introduction to the Theory and Its Applications*, McGraw-Hill, New York, NY, 1966.

In a discrete parameter structure this is always so, since, as shown in Subsection 7.8.1, **A** is always *diagonalizable* and $\grave{\mathbf{B}}$ is never zero. A feedback system for which Eq.(8.46) is true is called *controllable*. Hence, one may state that the feedback control of the nodal vibrations problem of a discrete parameter adaptive structure is always controllable.

8.5.3 Practical Difficulties with Optimal State Control

The development of optimal state feedback control theory was in response to the need for implementations on discrete parameter systems where the order of state space $2n \leq 6$, the number of actuators $q \leq 3$, and analogue control devices were being used instead of digital ones. The main difficulty in that environment was the time dependency of the optimal gain matrix defined in Eq.(8.44). Already strained resources of analogue computers for the matrix multiplication in Eq.(8.43) at that time could not accommodate the variability of a gain matrix. It needed to be constant. One may observe from Fig. (8.4) that the Riccati matrix becomes almost constant if the control time T is sufficiently large. The constant value of the Riccati matrix may be obtained from Eq.(8.41) by assuming $\dot{\mathbf{P}} = \mathbf{0}$, i.e., from

$$\mathbf{P}\grave{\mathbf{B}}\mathbf{R}^{-1}\grave{\mathbf{B}}^T\mathbf{P} - \mathbf{P}\mathbf{A} - \mathbf{A}^T\mathbf{P} - \mathbf{Q} = \mathbf{0} \tag{8.47}$$

which is called the *algebraic Riccati equation*. The solution of this equation is the *time-invariant Riccati matrix* $\mathbf{P} \in \mathrm{C}^{2n \times 2n}$. With this, from Eq.(8.44), the gain matrix becomes constant

$$\mathbf{G} = -\mathbf{R}^{-1}\grave{\mathbf{B}}^T\mathbf{P} \tag{8.48}$$

and the control law of Eq.(8.43) transforms into

$$\mathbf{u}(t) = \mathbf{G}\boldsymbol{\zeta}(t) \tag{8.49}$$

which is, strictly speaking, no longer optimal.

Thank to the technological breakthroughs in digital technology, the modern trend is to use digital controls where digital signals replace the analogue currents, and the implementation of control law is performed in digital processors that contain appropriate software. This enables the industry to change the control system by changing the software, but not the hardware. There are no longer serious problems in handling time varying gain matrices in small discrete parameter systems.

For large n the computational complexity of solving $\mathbf{P}$ from Eq.(8.47) is $O[(2n)^3]$. Solving $\mathbf{P}(t)$ from Eq.(8.41) is still more costlier.[3] Real-time computation of these matrices within a control cycle can be only a hypothetical consideration.

---

[3] See, for example, "Computational Complexities and Storage Requirements of Some Riccati Equation Solvers," A. V. Ramesh, S. Utku, and J. Garba, *AIAA Journal of Guidance, Control and Dynamics*, Vol.12, No.4, pp.469-479, July-August 1989.

For large discrete parameter systems, however, there are other very serious problems with multi-input multi-output control electronics when the order of state space $2n$ exceeds a couple hundred, and the number of actuators q goes beyond a couple dozen. To overcome such difficulties, at least partially, one may observe only a subset of state variables. Under certain circumstances, this may be as good as observing the whole state. This is called *output feedback control*. The optimal output feedback control is discussed in next section.

8.6 Optimal Output Feedback Control

In the previous section, the optimal state feedback control is explained. One may see from Fig(8.3) that in order to evaluate the optimal controls $\mathbf{u}^*$, one needs to observe the whole state $\boldsymbol{\zeta} \in \mathrm{R}^{2n\times 1}$. This is a very expensive proposition if n is large. As it is done for the static case in Chapter 4, one may observe only the components $\alpha_1, \alpha_2, \cdots, \alpha_p$ of $\boldsymbol{\zeta} \in \mathrm{R}^{2n\times 1}$. Since the observations are done by various sensors, and sensor output $\mathbf{y} \in \mathrm{R}^{p\times 1}$ and the observed components are assumed correlated linearly and one-to-one, one may write

$$\mathbf{y} = \hat{\mathbf{C}}\boldsymbol{\zeta} \tag{8.50}$$

Here matrix $\hat{\mathbf{C}} \in \mathrm{R}^{p\times 2n}$ plays the role $\mathbf{I}_\alpha^T$ in static control as explained in Section 4.2. Here, if output $\mathbf{y}(t) \in \mathrm{R}^{p\times 1}$ is to represent state $\boldsymbol{\zeta}(t) \in \mathrm{R}^{2n\times 1}$, then one should be able to compute $\boldsymbol{\zeta}(t)$ from $\mathbf{y}$ data uniquely, perhaps not from the instantaneous $\mathbf{y}(t)$ but from $\mathbf{y}(\tau)$ for $t - \Delta t \leq \tau \leq t$ with sufficiently large Δt. It can be shown[4] that this is possible if the matrix

$$\hat{\mathbf{P}}^T = \left[\; (\hat{\mathbf{C}}\mathbf{A}^0)^T \;\; (\hat{\mathbf{C}}\mathbf{A}^1)^T \;\; \cdots \;\; (\hat{\mathbf{C}}\mathbf{A}^{2n-1})^T \; \right] \tag{8.51}$$

is not rank deficient, i.e., if

$$rank(\hat{\mathbf{P}}) = 2n \tag{8.52}$$

In the dynamics of discrete parameter structures this is always so, since, as shown in Subsection 7.8.1, $\mathbf{A}$ is always *diagonalizable* and $\hat{\mathbf{C}}$ is never zero. A feedback system for which Eq.(8.52) is true is called *observable*. Hence, one may state that the feedback control of a nodal vibrations problem of a discrete parameter adaptive structure is always observable.

The output feedback control is sketched in Fig.(8.5). The diagram is very similar to the one given in Fig.(8.3) for state feedback control. Here, instead of state $\boldsymbol{\zeta}(t)$, one feeds back the sensor output $\mathbf{y}(t)$. However, since

---

[4] See, for example, Donald W. Wiberg, *Theory and Problems of State Space and Linear Systems*, Schaum's Outline Series, McGraw-Hill Book Co., New York, NY, 1971.

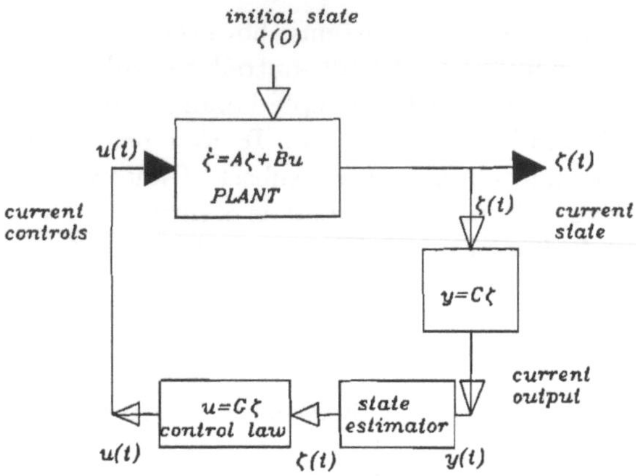

FIGURE 8.5. Output feedback control

in the gain block one still needs the state $\zeta(t)$ as shown below, a *state estimator* block is required. A state estimator block may be possible only if the system is observable.

In order to find the optimal gain matrix in output feedback control, one first expresses the performance functional J in Eq.(8.25) in terms of output $\mathbf{y}(t) \in \mathrm{R}^{p\times 1}$ as

$$J = \frac{1}{2}\mathbf{y}(T)^T\mathbf{F}\ \mathbf{y}(T) + \frac{1}{2}\int_0^T [\mathbf{y}(t)^T\mathbf{Q}\ \mathbf{y}(t) + \mathbf{u}(t)^T\mathbf{R}\ \mathbf{u}(t)]dt \qquad (8.53)$$

and continues on the same lines as in the previous section. The optimal control is obtained exactly the same as in Eqs.(8.43 and 8.44) provided that one uses the *time-dependent Riccati matrix of the output feedback control*, i.e., matrix $\mathbf{P}'(t) \in \mathrm{C}^{2n\times 2n}$ in place of the Riccati matrix $\mathbf{P}(t) \in \mathrm{C}^{2n\times 2n}$appearing in Eq.(8.44). Matrix $\mathbf{P}'(t)$ is the solution of the differential matrix Riccati equation and its terminal condition for output feedback control:

$$\dot{\mathbf{P}}' = \mathbf{P}'\grave{\mathbf{B}}\mathbf{R}^{-1}\grave{\mathbf{B}}^T\mathbf{P}' - \mathbf{P}'\mathbf{A} - \mathbf{A}^T\mathbf{P}' - \hat{\mathbf{C}}^T\mathbf{Q}\hat{\mathbf{C}}, \text{ for } 0 < t < T \qquad (8.54)$$

and

$$\mathbf{P}'(T) = \hat{\mathbf{C}}^T\mathbf{F}\hat{\mathbf{C}} \qquad (8.55)$$

These are very similar to the ones given by Eq.(8.41 and 8.42) for state feedback control. Matrix $\mathbf{P}'(t)$ behaves similar to $\mathbf{P}(t)$ [see Fig.(8.4)]. If the use of constant optimal gain matrix is permissible (see Subsection 8.5.3), one may use Eq.(8.48), employing the constant Riccati matrix of the output feedback control, i.e., matrix $\mathbf{P}' \in \mathrm{C}^{2n\times 2n}$ in place of matrix $\mathbf{P}$ appearing in that equation. Matrix $\mathbf{P}'$ is the solution of the algebraic Riccati equation for output feedback control:

$$\mathbf{P}'\grave{\mathbf{B}}\mathbf{R}^{-1}\grave{\mathbf{B}}^T\mathbf{P}'-\mathbf{P}'\mathbf{A}-\mathbf{A}^T\mathbf{P}'-\hat{\mathbf{C}}^T\mathbf{Q}\hat{\mathbf{C}}=\mathbf{0} \tag{8.56}$$

which is similar to the one given in Eq.(8.47) for state feedback control.

In output feedback control, although one is relieved from observing all of the state variables, one is still required to obtain the state at each control cycle. In practice this is done by means of state estimators as shown in Fig. (8.5). State estimators are costly electronics.

For large discrete parameter systems, there are very serious problems not only with system identification but also with multi-input multi-output control electronics when the order of output vector p and the number of actuators q go beyond a couple dozen. Looking at the advances in the electronics industry, it appears that the latter problem may cease to be a major hurdle, but the former is intrinsic as explained next.

In Subsection 6.1.2, during the discussion of inertial forces, the difficulty in finding a method for the lumping of structural mass at the nodes that meets the needs of both high frequency and low frequency response simulations simultaneously is indicated. It is known that the computed free vibration frequencies and modes shapes, i.e., the square root of eigenvalues and eigenvectors respectively, of $\mathbf{K}'\mathbf{x}=\lambda\mathbf{M}\mathbf{x}$ are very sensitive not only to the way mass matrix $\mathbf{M}$ is defined but also to the precision of the computer arithmetic used during the calculations. The implication of the latter is very serious, since, in practical problems, the relative error in the quantified matrix entries due to the nature of associated physical parameters far exceed the increased machine precision one may use. The conclusion one may drive from this observation is that the computed high natural vibration frequencies and mode shapes may not be used for practical control applications where the aim is in controlling low frequency response.

By eliminating high natural frequencies and mode shapes from consideration one may effectively decrease the number of degrees of freedom in the motion of the nodes, thus dealing with a smaller size control problem where the optimal control theory may be applied. Observing and controlling only a few mode shapes without ever considering the others appears very attractive. This prompts the idea of *direct output control*. This is discussed next.

8.7 Optimal Direct Output Feedback Control

Instead of full state $\zeta \in R^{2n\times 1}$, if a subset of nodal motion is feedback in the control, it is called *direct output control*. The subset may be defined in the actual state space by partitioning ζ as

$$\zeta=\left\{\begin{array}{c}\zeta_1\\ \zeta_2\end{array}\right\} \tag{8.57}$$

where the subset is $\zeta_1 \in R^{2n_1 \times 1}$ with $n = n_1 + n_2$, or it may be defined in the modal space as

$$\grave{\eta} = \left\{ \begin{array}{c} \grave{\eta}_1 \\ \grave{\eta}_2 \end{array} \right\} \tag{8.58}$$

where the subset is $\grave{\eta}_1 \in C^{2n_1 \times 1}$, and $\grave{\eta} \in C^{2n \times 1}$ are the participation factors of the $2n$ eigenvectors $\mathbf{Z} \in C^{2n \times 2n}$ of the first order system as studied in Subsection 7.8.1 where the relationship between the eigenpairs (i.e., eigenvalue - eigenvector pairs) of $\mathbf{A}\zeta = \mu\zeta$ and the corresponding vibration frequencies and modes shapes of $\mathbf{K}'\mathbf{x} = \lambda\mathbf{M}\mathbf{x}$ are given with reference to modal damping matrix $\grave{\mathbf{C}}$ which is also given. So long as one uses an even number for the vector order of $\grave{\eta}_1$, dealing with the complex numbers poses no problem, since they appear as complex conjugate pairs (see Subsection 7.8.1). Using Eq.(7.107), one may relate the participation factors $\grave{\eta}_1$ and $\grave{\eta}_2$ to the nodal state as

$$\zeta = \zeta' + \zeta'' \tag{8.59}$$

where

$$\zeta = \sum_{k=1}^{2n} \mathbf{z}_k \grave{\eta}_k \tag{8.60}$$

$$\zeta' = \sum_{k=1}^{2n_1} \mathbf{z}_k \grave{\eta}_k = \mathbf{Z}_1 \grave{\eta}_1 \tag{8.61}$$

and

$$\zeta'' = \sum_{k=2n_1+1}^{2n} \mathbf{z}_k \grave{\eta}_k = \mathbf{Z}_2 \grave{\eta}_2 \tag{8.62}$$

Suppose $\grave{\eta}_1 \in C^{2n_1 \times 1}$, rather than $\zeta^1 \in R^{2n_1 \times 1}$, is the part of the nodal response that is to be controlled.

With the terminology of control discipline, $\grave{\eta}_1 \in C^{2n_1 \times 1}$is the *output* to be controlled. The term output actually refers to what is coming out of the sensors and dependent electronics. Suppose that the sensors and dependent electronics are capable of observing the $2n_1$ participation factors listed in $\grave{\eta}_1 \in C^{2n_1 \times 1}$. If all of the participation factors listed in $\grave{\eta} \in C^{2n \times 1}$were observed, it would lead to a *state feedback control.* Since $n_1 < n$, the control is called *output control.* In addition, if one eliminates the state estimator block in the feedback cycle [see Fig.(8.5)], the feedback control is called *direct output control.* In Fig.(8.6) a sketch of the direct output control is shown. Note that only the output is feedback.

Substituting $\zeta = \mathbf{Z}\grave{\eta}$ and premultiplying both sides with $\mathbf{Z}^{-1}$, the equations of motion in the modal subspace may be obtained from Eqs.(8.18 and 8.19) as

$$\frac{d}{dt}\grave{\eta} = \mathbf{A}^{\eta}\grave{\eta} + \grave{\mathbf{B}}^{\eta}\mathbf{u}, \ \text{for } 0 \le t \le T \tag{8.63}$$

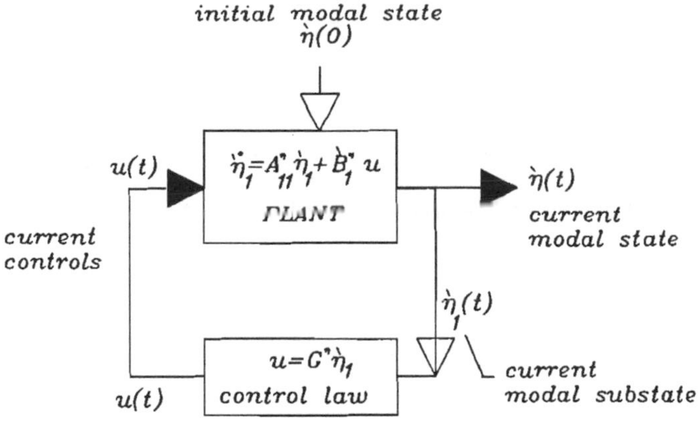

FIGURE 8.6. Direct output feedback control in modal space

where

$$\grave{\eta} = \left\{ \begin{array}{c} \grave{\eta}_1 \\ \grave{\eta}_2 \end{array} \right\} \tag{8.64}$$

$$\mathbf{A}^{\eta} = \mathbf{Z}^{-1}\mathbf{A}\mathbf{Z} = \left[\begin{array}{cc} \mathbf{A}_{11}^{\eta} & \mathbf{A}_{12}^{\eta} \\ \mathbf{A}_{21}^{\eta} & \mathbf{A}_{22}^{\eta} \end{array} \right] \tag{8.65}$$

$$\grave{\mathbf{B}}^{\eta} = \mathbf{Z}^{-1}\grave{\mathbf{B}} = \left[\begin{array}{c} \grave{\mathbf{B}}_1^{\eta} \\ \grave{\mathbf{B}}_2^{\eta} \end{array} \right] \tag{8.66}$$

and

$$\grave{\eta}(0) = \mathbf{Z}^{-1}\zeta_o = \left\{ \begin{array}{c} \grave{\eta}_1(0) \\ \grave{\eta}_2(0) \end{array} \right\} \tag{8.67}$$

$$\grave{\eta}(T) = small = \left\{ \begin{array}{c} \grave{\eta}_1(T) \\ \grave{\eta}_2(T) \end{array} \right\}$$

Eqs.(8.63 and 8.67) may be rewritten in partitioned form as

$$\begin{array}{l} \frac{d}{dt}\grave{\eta}_1 = \mathbf{A}_{11}^{\eta}\grave{\eta}_1 + \grave{\mathbf{B}}_1^{\eta}\mathbf{u} + \mathbf{e}_1 \quad \text{for } 0 \le t \le T \\ \grave{\eta}_1(0), \grave{\eta}_1(T) \quad given \end{array} \tag{8.68}$$

where

$$\mathbf{e}_1(t) = \mathbf{A}_{12}^{\eta}\grave{\eta}_2(t) \tag{8.69}$$

Vector $\mathbf{e}_1(t) \in C^{2n_1 \times 1}$ is called the *spill-over vector* which represents the contribution of unobserved modes on the controlled ones.

In optimal output control, if one can justify that $\|\mathbf{e}_1(t)\|$ is sufficiently small, then one drops $\mathbf{e}_1(t)$ from Eq.(8.68) and rewrites it as

$$\begin{array}{l} \frac{d}{dt}\grave{\eta}_1 \approx \mathbf{A}_{11}^{\eta}\grave{\eta}_1 + \grave{\mathbf{B}}_1^{\eta}\mathbf{u} \quad \text{for } 0 \le t \le T \\ \grave{\eta}_1(0), \grave{\eta}_1(T) \quad \text{given} \end{array} \tag{8.70}$$

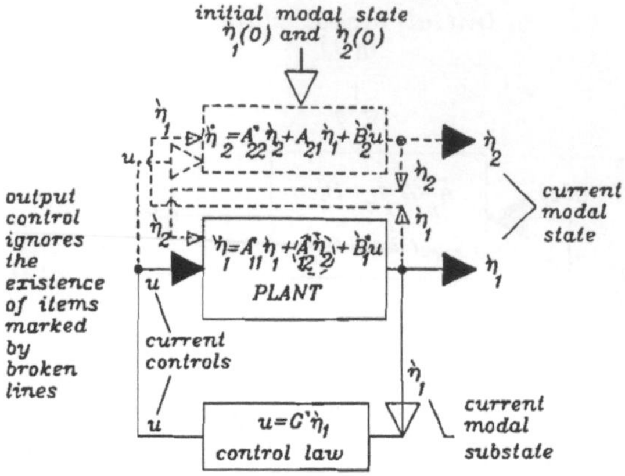

FIGURE 8.7. Effect of uncontrolled modes on direct output feedback control

Then these equations approximately represent the original $(2n)$th order system; yet they are of much smaller order of $(2n_1)$. Applying the optimal control theory to this reduced system, one may find its Riccati matrix $\mathbf{P}^\eta \in R^{2n_1 \times 2n_1}$ and the optimal control law as

$$\mathbf{u}^*(t) = \mathbf{G}^{\eta *}(t)\grave{\eta}_1^*(t) \tag{8.71}$$

where

$$\mathbf{G}^{\eta *}(t) = -\mathbf{R}^{-1}\grave{\mathbf{B}}_1^{\eta T}\mathbf{P}^\eta(t) \tag{8.72}$$

If the control time T is large enough, one may replace $\mathbf{P}^\eta(t)$ with its time-invariant counterpart as explained in the previous section.

If one cannot justify that $\|\mathbf{e}_1(t)\|$ is sufficiently small relative to the norms of the first two terms on the right hand side of Eq.(8.68), then one cannot use optimal output control for controlling the nodal motion, because of the interaction between the controlled and uncontrolled modes. This is sketched in Fig. (8.7).

Considering the fact that all discrete parameter structures are finite degrees of freedom idealizations of their infinite degrees of freedom distributed parameter counterparts, one may conclude that there truly is no state control for flexible bodies. With differing magnitudes in each system, spill-over $\mathbf{e}_1(t)$ always exists. Therefore, one may state that *in the feedback control of flexible bodies, one always has direct output control.*

The current trend in the feedback control of flexible bodies is to include in the feedback cycle the estimate of $\mathbf{e}_1(t)$, i.e., contributions of all uncontrolled modes, by real time on-line measurements. In the planning phase of this, the role of the structural engineer is indispensable.

8.8 Alternatives to Optimal Control

In the feedback control of autonomous motion of nodes in state space, one has to select a control law in order to define controls in real time as a function of observed state. As discussed in Section 8.4, in linear feedback control it is customary to have a linear control law as

$$\mathbf{u} = \mathbf{G}\boldsymbol{\zeta} \tag{8.73}$$

where $\mathbf{G} \in R^{q\times 2n}$ is the gain matrix. The gain matrix can be any convenient matrix provided that it causes the system energy $E(t)$ to monotonically decrease with increasing time, i.e., Eq.(8.21) is satisfied. Such $\mathbf{G}$ matrices are called *admissible gain matrices.* The discussions in Section 8.4 imply that there are infinitely many admissible gain matrix choices. All of these admissible choices are related to the same energy $E(0) - E(T)$ in order to bring the structure from initial state $\boldsymbol{\zeta}(0)$ to terminal state $\boldsymbol{\zeta}(T)$.

Optimal control identifies one of the admissible gain matrices, i.e., the optimal gain matrix $\mathbf{G}^*(t)$ of Eq.(8.44), as a result of minimization of the performance functional defined in Eq.(8.25). In the performance functional positive weight matrices $\mathbf{F}$, $\mathbf{Q}$, and $\mathbf{R}$ are completely arbitrary and yet very useful for meeting the needs of a designer in deciding the control energy consumption rates. Regardless of the values a designer uses for $\mathbf{F}$, $\mathbf{Q}$, and $\mathbf{R}$, the total energy demand of $E(0) - E(T)$ does not change.

There are other possibilities in choosing a gain matrix among infinitely many admissible ones. Some of these are discussed in the following subsections.

8.8.1 Choosing Gain Matrix by Trial and Error

This appears to be an acceptable approach for some. Considering the need for user-friendly and nature-friendly intelligent products on one hand, and the economical availability of sensors, actuators, and microprocessors on the other hand, it is not surprising that locating a working gain matrix among infinitely many admissible ones by trial and error appears to some as an economically viable alternative. The availability of reliable simulation packages is making the approach even more attractive. Until the basic logic of optimal control theory is understood by engineers, it is likely that this trend will continue.

Gain matrix that minimizes the performance functional in Eq.(8.25) is expressed in Eq.(8.44) as $\mathbf{G}^*(t) = -\mathbf{R}^{-1}\grave{\mathbf{B}}^T\mathbf{P}(t)$ where the only quantity not available at the beginning of the analysis is the time-dependent Riccati matrix $\mathbf{P}(t)$. Although it is no problem in modern digital control to handle time varying gain matrices,[5] present day cost considerations and re-

[5] Adaptive control field deals with such problems.

liability may make it advantageous to use constant gain matrices when the possibility exists. As discussed in Subsection 8.5.3, if the control time T is sufficiently large, one may use the asymptotic value of the time-dependent Riccati matrix, i.e., matrix $\mathbf{P}$, in the gain expression. Hence, for situations when this approximation is feasible, reproducing from Eq.(8.48), the optimal constant gain matrix becomes $\mathbf{G} = -\mathbf{R}^{-1}\grave{\mathbf{B}}^T\mathbf{P}$. Using the definition of $\grave{\mathbf{B}} \in \mathrm{R}^{2n\times q}$ from Eq.(6.34), one may rewrite the expression for the optimal constant gain matrix as

$$\mathbf{G} = -\mathbf{R}^{-1}\begin{bmatrix} -\mathbf{I}_\beta^T\vec{\mathbf{K}}\mathbf{B}^T\mathbf{M}^{-1} & \mathbf{0} \end{bmatrix}\begin{bmatrix} \mathbf{P}_1 \\ \mathbf{P}_2 \end{bmatrix} \tag{8.74}$$

or

$$\mathbf{G} = \mathbf{R}^{-1}\mathbf{I}_\beta^T\vec{\mathbf{K}}\mathbf{B}^T\mathbf{M}^{-1}\mathbf{P}_1 \tag{8.75}$$

where $\mathbf{P}_1 \in \mathrm{R}^{n\times 2n}$ and $\mathbf{P}_2 \in \mathrm{R}^{n\times 2n}$ are partitions $\mathbf{P}$, matrix $\mathbf{M} \in \mathrm{R}^{n\times n}$ is the mass matrix, $\mathbf{B}^T \in \mathrm{R}^{m\times n}$ is the matrix transforming nodal deflections into element deformations, $\vec{\mathbf{K}} \in \mathrm{R}^{m\times m}$ is the block diagonal matrix of element stiffnesses, $\mathbf{I}_\beta \in (\mathrm{Int})^{n\times q}$ is the actuator placement matrix, and $\mathbf{R} \in \mathrm{R}^{q\times q}$ is the arbitrary positive definite matrix that provides a handle in deciding the control energy consumption rate. Clearly the actuator placement matrix, like other plant parameters, plays an important role in the definition of the gain matrix. Only matrices $\mathbf{R}$ and $\mathbf{P}_1$ are related to the optimal control theory. If gain matrix is determined by means other than the optimal control theory, one must show that the controlled system energy monotonically decreases from $E(0)$ to $E(T)$. This may be done by trial-and-error and simulation, or better by the following observation.

Let $\mathbf{G} \in \mathrm{R}^{q\times 2n}$ denote the gain matrix obtained without the use of optimal control theory. Using the related control law $\mathbf{u} = \mathbf{G}\boldsymbol{\zeta}$ in the equations of motion in the state space, i.e., in Eq.(8.18), one obtains the formulation in Eq.(8.23), which, by defining

$$\grave{\mathbf{A}} = \mathbf{A} + \grave{\mathbf{B}}\mathbf{G} \tag{8.76}$$

may be rewritten as

$$\begin{gathered} \dot{\boldsymbol{\zeta}} = \grave{\mathbf{A}}\boldsymbol{\zeta}\ , \text{ for } 0 \leq t \leq T \\ \boldsymbol{\zeta}(0) = \boldsymbol{\zeta}_o \\ \boldsymbol{\zeta}(T) = \mathbf{o} \end{gathered} \tag{8.77}$$

The solution of these equations is obtained with the help from Eq.(7.125) as

$$\boldsymbol{\zeta}(t) = e^{\grave{\mathbf{A}}t}\boldsymbol{\zeta}_o \tag{8.78}$$

Matrix $e^{\grave{\mathbf{A}}t} \in \mathrm{R}^{2n\times 2n}$ is the transition matrix which may be expressed, with the help of Eq.(7.122), as

$$e^{\grave{\mathbf{A}}t} = \grave{\mathbf{Z}}diag(e^{\mu_k t})\grave{\mathbf{Z}}^{-1} \tag{8.79}$$

where $\grave{\mathbf{Z}} \in C^{2n\times 2n}$ are the eigenvectors and $\grave{\mu}_k, k = 1,2,\cdots,2n$ are the eigenvalues of $\grave{\mathbf{A}}$. The system energy $E(t)$, as defined in Eq.(7.85), is

$$E(t) = \frac{1}{2}\boldsymbol{\zeta}(t)^T diag(\mathbf{M}, \mathbf{K}')\boldsymbol{\zeta}(t) \tag{8.80}$$

It may be observed from Eq.(8.79) that the energy can decrease monotonically from $E(0)$ to $E(T)$ only if the eigenvalues are all with negative real parts. Therefore, it suffices to check the eigenvalues of $\grave{\mathbf{A}}$ in order to ensure that the gain matrix $\mathbf{G}$ is admissible.

The assessment of gain matrix $\mathbf{G}$ without using the optimal control theory is easier if one starts with the equations of dynamic equilibrium due to the actuator induced deformations studied in Section 7.7. With the help of Eq.(7.78), one may write

$$\mathbf{M}\ddot{\boldsymbol{\xi}} + \grave{\mathbf{C}}\dot{\boldsymbol{\xi}} + \mathbf{K}'\boldsymbol{\xi} = -\mathbf{B}\vec{\mathbf{K}}\mathbf{I}_\beta\mathbf{u} \tag{8.81}$$

Recalling that the state vector consists of nodal velocities and nodal deflections, i.e., $\boldsymbol{\zeta} = [\dot{\boldsymbol{\xi}}^T, \boldsymbol{\xi}^T]^T$, the control law $\mathbf{u} = \mathbf{G}\boldsymbol{\zeta}$ may be rewritten as

$$\mathbf{u} = \mathbf{G}_1\dot{\boldsymbol{\xi}} + \mathbf{G}_2\boldsymbol{\xi} \tag{8.82}$$

where $\mathbf{G}_1 \in R^{q\times n}$ and $\mathbf{G}_2 \in R^{q\times n}$ are partitions of gain matrix $\mathbf{G} \in R^{q\times 2n}$. The control law is called a *velocity control law* if $\mathbf{G}_2 = \mathbf{0}$. It is called a *deflection control law* if $\mathbf{G}_1 = \mathbf{0}$. No qualification is necessary if neither submatrices are zero.

Using $\mathbf{u}$ from the control law of Eq.(8.82) in Eq.(8.81) the latter may be written as

$$\mathbf{M}\ddot{\boldsymbol{\xi}} + (\grave{\mathbf{C}} + \mathbf{B}\vec{\mathbf{K}}\mathbf{I}_\beta\mathbf{G}_1)\dot{\boldsymbol{\xi}} + (\mathbf{K}' + \mathbf{B}\vec{\mathbf{K}}\mathbf{I}_\beta\mathbf{G}_2)\boldsymbol{\xi} = \mathbf{o} \tag{8.83}$$

It may be observed that the control law modifies the dynamic characteristics of the structure by altering the damping and stiffness matrices of the system. The following are very important conclusions:

- A velocity control law alters only damping (since $\mathbf{G}_2 = 0$).
- A deflection control law alters only the stiffness(since $\mathbf{G}_1 = 0$).
- A control law with nonzero $\mathbf{G}_1, \mathbf{G}_2$ alters both stiffness and damping.

If one uses $\mathbf{u}$ from the time-invariant optimal control law of Eqs.(8.48 and 8.49) in Eq.(8.81), one may obtain

$$\mathbf{M}\ddot{\boldsymbol{\xi}} + (\grave{\mathbf{C}} + \mathbf{W}\mathbf{M}^{-1}\mathbf{P}_{11})\dot{\boldsymbol{\xi}} + (\mathbf{K}' + \mathbf{W}\mathbf{M}^{-1}\mathbf{P}_{12})\boldsymbol{\xi} = \mathbf{o} \tag{8.84}$$

where

$$\mathbf{W} = \mathbf{B}\vec{\mathbf{K}}\mathbf{I}_\beta\mathbf{R}^{-1}\mathbf{I}_\beta^T\vec{\mathbf{K}}\mathbf{B}^T \tag{8.85}$$

and the nth order partitions of time-invariant Riccati matrix are displayed as:

$$\mathbf{P} = \begin{bmatrix} \mathbf{P}_{11} & \mathbf{P}_{12} \\ \mathbf{P}_{21} & \mathbf{P}_{22} \end{bmatrix} \tag{8.86}$$

Clearly optimal control law alters both stiffness and damping matrices of the structure during the control time.

A velocity control law is a very safe way of controlling the nodal motion if changes in stiffness characteristics are not desirable. A *velocity control law* may be obtained from Eq.(8.82) with $\mathbf{G}_2 = \mathbf{0}$:

$$\mathbf{u} = \mathbf{G}_1 \dot{\boldsymbol{\xi}} \tag{8.87}$$

and the dynamic equilibrium equation with this velocity control become

$$\mathbf{M}\ddot{\boldsymbol{\xi}} + (\grave{\mathbf{C}} + \mathbf{B}\vec{\mathbf{K}}\mathbf{I}_\beta \mathbf{G}_1)\dot{\boldsymbol{\xi}} + \mathbf{K}'\boldsymbol{\xi} = \mathbf{o} \tag{8.88}$$

If one is interested in a class of non-optimal yet admissible velocity gain matrices $\mathbf{G}_1$ to experiment with, one may choose them to ensure that the $\mathbf{B}\vec{\mathbf{K}}\mathbf{I}_\beta \mathbf{G}_1$ term next to $\grave{\mathbf{C}}$ is at least positive.

For example, let $\grave{\mathbf{R}} \in R^{q \times q}$ be any positive definite matrix. Then one can show that

$$\mathbf{G}_1 = \grave{\mathbf{R}}\mathbf{I}_\beta^T \vec{\mathbf{K}}\mathbf{B}^T \tag{8.89}$$

is a class of admissible non-optimal velocity gain matrices with arbitrary parameter $\grave{\mathbf{R}} \in R^{q \times q}$ that can be a qth order diagonal matrix with arbitrary positive diagonal elements. Note that the substitution of $\mathbf{G}_1$ from Eq.(8.89) into Eq.(8.88) yields a positive coefficient matrix[6] for $\dot{\boldsymbol{\xi}}$ which is required for continuous power leak from the system (see Subsection 6.1.3).

The use of non-optimal admissible gain matrices discussed above does not guarantee that at the end of the control time T, the nodal motion will become acceptably small. However, by trial and error, changing control time T and/or gain matrix $\mathbf{G}$, one may converge to an acceptable set of control time T, gain matrix $\mathbf{G}$, and terminal error $\boldsymbol{\zeta}(T)$.

8.8.2 Choosing Gain Matrix by Eigenvalue Assignment

The discussions in Section 8.6 lead to the conclusion that in the feedback control of flexible bodies one always has *output control* even if one thinks that one has *state control*. This is because of the fact that the flexible bodies, being distributed parameter systems, are always with infinite degrees of freedom, and we define their state with a small subset of them. A discrete parameter structure is an idealization of the actual distributed parameter

---

[6] Since $\mathbf{B}\vec{\mathbf{K}}\mathbf{I}_\beta \grave{\mathbf{R}}\mathbf{I}_\beta^T \vec{\mathbf{K}}\mathbf{B}^T$ is a congruent transform of real, symmetric, and positive definite arbitrary matrix $\grave{\mathbf{R}} \in R^{q \times q}$.

structure. As a consequence, there is always a spill-over, and the feedback control of the structure is always of direct output type.

In Subsection 7.8.3, the dynamics of controlled autonomous systems is briefly discussed without prescribing any control law. In the previous subsection, substituting $\mathbf{u}$ from a general linear control law of

$$\mathbf{u} = \mathbf{G}\zeta \tag{8.90}$$

in the state equations of nodal motion

$$\begin{aligned} \dot{\zeta} &= \mathbf{A}\zeta + \grave{\mathbf{B}}\mathbf{u} \quad \text{for } 0 < t \leq T \\ \zeta(0) &= \zeta_o \end{aligned} \tag{8.91}$$

The latter was rewritten as

$$\begin{aligned} \dot{\zeta} &= \grave{\mathbf{A}}\zeta \quad \text{for } 0 < t \leq T \\ \zeta(0) &= \zeta_o \end{aligned} \tag{8.92}$$

where

$$\grave{\mathbf{A}} = \mathbf{A} + \grave{\mathbf{B}}\mathbf{G} \tag{8.93}$$

With the help of discussions in Subsection 7.8.2, the solution of the problem in Eq.(8.92) may be obtained as

$$\zeta = e^{\grave{\mathbf{A}}t}\zeta_o \tag{8.94}$$

Using Eq.(7.123), the transition matrix $e^{\grave{\mathbf{A}}t} \in \mathrm{C}^{2n\times 2n}$ may be expressed as in Eq.(8.79); hence, the solution becomes

$$\zeta = \grave{\mathbf{Z}}\, diag(e^{\grave{\mu}_k t})\, \grave{\mathbf{Z}}^{-1}\zeta_o \tag{8.95}$$

where $\grave{\mathbf{Z}} \in \mathrm{C}^{2n\times 2n}$ is the matrix of eigenvectors and $\grave{\mu}_k, k = 1, 2, \cdots, 2n$ are the eigenvalues of the controlled system matrix $\grave{\mathbf{A}} \in \mathrm{R}^{2n\times 2n}$. The distribution of the eigenvalues in the complex plane and their relationship with the vibration frequencies of structure are shown in Fig. (7.2).

If the structure were not controlled, i.e., if $\mathbf{u} = \mathbf{o}$, then the equations of motion in Eq.(8.91) would be

$$\begin{aligned} \dot{\zeta}' &= \mathbf{A}\zeta' \quad \text{for } 0 < t \leq T \\ \zeta'(0) &= \zeta_o \end{aligned} \tag{8.96}$$

and the solution

$$\zeta' = \mathbf{Z}\, diag(e^{\mu_k t})\, \mathbf{Z}^{-1}\zeta_o \tag{8.97}$$

where $\mathbf{Z} \in \mathrm{C}^{2n\times 2n}$ is the matrix of eigenvectors and $\mu_k, k = 1, 2, \cdots, 2n$ are the eigenvalues of the uncontrolled system matrix $\mathbf{A} \in \mathrm{R}^{2n\times 2n}$.

The time dependence of nodal motions of controlled and uncontrolled systems differs from each other only in eigenvalues $\grave{\mu}_k$ and μ_k, and eigenvectors $\grave{\mathbf{z}}_k$ and $\mathbf{z}_k$, $k = 1, 2, \cdots, 2p$. In the eigenvalue assignment method,

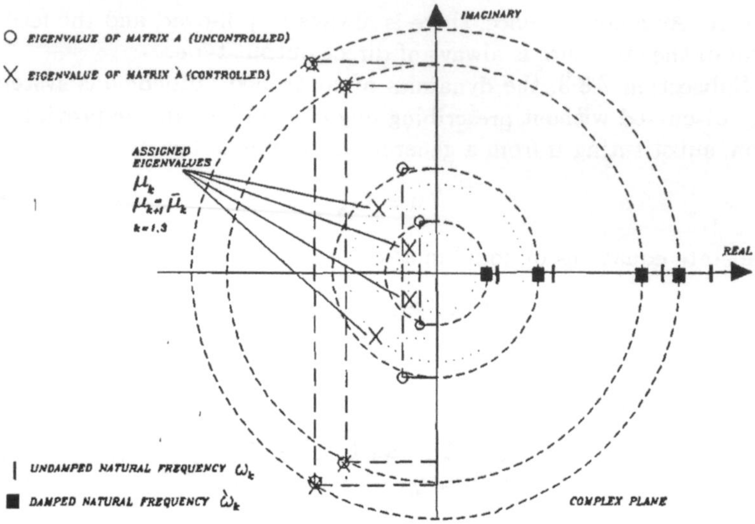

FIGURE 8.8. Eigenvalues of uncontrolled system and those of the controlled system when first four are assigned

one strives to identify a gain matrix **G** such that the selected eigenvalue pairs of the controlled system take the assigned values and the remaining eigenvalues and eigenvectors stay the same as the corresponding ones in the uncontrolled system. A case where only the first four eigenvalues (i.e., $p = 2$) of the uncontrolled system would take the assigned values in the controlled system is sketched in Fig. (8.8). Note that, other than the first four, all the remaining eigenvalues are the same in both systems. Not sketched in the figure is the additive advantage that, excluding the first four, the eigenvectors of both systems would be the same, if one could find a gain matrix to materialize this expectation. A possible method for identifying such a gain matrix is explained next.

Suppose one desires to find a gain matrix **G** such that the first $2p$ eigenvalues of the controlled system take assigned values and the others do not change. For this purpose one may write from Eq.(7.122)

$$\mathbf{Z}^{-1}\mathbf{A}\mathbf{Z} = diag(\mu_k) \tag{8.98}$$

and using matrix **Z** one may obtain a *similarity transform* of $\grave{\mathbf{A}}$, i.e., $\hat{\mathbf{A}} = \mathbf{Z}^{-1}\grave{\mathbf{A}}\mathbf{Z}$. Using Eq.(8.93 and 8.98)

$$\hat{\mathbf{A}} = diag(\mu_k) + \mathbf{P}\mathbf{Q} \tag{8.99}$$

where

$$\begin{aligned} \mathbf{P} &= \mathbf{Z}^{-1}\grave{\mathbf{B}} \\ \mathbf{Q} &= \mathbf{G}\mathbf{Z} \end{aligned} \tag{8.100}$$

or using the normalized eigenvectors $\hat{\mathbf{Z}}$ of $\hat{\mathbf{A}}$ and noting that $\grave{\mathbf{A}}$ and $\hat{\mathbf{A}}$ are *similar*, hence have the same eigenvalues:

$$[diag(\mu_k)+\mathbf{PQ}]\hat{\mathbf{Z}} = \hat{\mathbf{Z}}diag(\grave{\mu}_k) \tag{8.101}$$

If one selects the gain matrix rows in the subspace of the first $2p$ eigenvectors of $\mathbf{A}$, then from the second of Eqs.(8.100) one may observe that the last $2n-2p$ columns of $\mathbf{Q}$ will be zero

$$\mathbf{Q} = \mathbf{GZ} = [\mathbf{q}_1, \mathbf{q}_2, \cdots, \mathbf{q}_{2p}, \mathbf{o}, \cdots, \mathbf{o}] = [\mathbf{Q}_1, \mathbf{0}] \tag{8.102}$$

where $\mathbf{Q}_1 \in C^{q\times 2p}$. Assuming that

$$q = 2p \tag{8.103}$$

and

$$\mathbf{Q}_1\hat{\mathbf{Z}}_{11} = \mathbf{I} \tag{8.104}$$

where $\hat{\mathbf{Z}}_{11} \in C^{2p\times 2p}$ is the leading principal submatrix of $\hat{\mathbf{Z}}$, by inspection one may show that

$$\left.\begin{array}{l} \hat{\mathbf{z}}_k = diag(\frac{1}{\mu_k-\mu_1}, \cdots, \frac{1}{\mu_k-\mu_{2n}})\mathbf{p}_k \\ \grave{\mu}_k = \mu_k \end{array}\right\} \text{for } k = 1, \cdots, 2p$$
$$\left.\begin{array}{l} \hat{\mathbf{z}}_k = \mathbf{i}_k \\ \grave{\mu}_k = \mu_k \end{array}\right\} \text{for } k = 2p+1, \cdots, 2n \tag{8.105}$$

where $\mathbf{i}_k$ and $\mathbf{p}_k$ are the kth columns of identity matrix $\mathbf{I} \in (\mathrm{Int})^{2n\times 2n}$ and matrix $\mathbf{P} \in C^{2n\times q}$, respectively. Assuming that the inverse of $\hat{\mathbf{Z}}_{11}$ exists, from Eq. (8.104) one may obtain

$$\mathbf{Q}_1 = \hat{\mathbf{Z}}_{11}^{-1} \tag{8.106}$$

and using this in Eq.(8.102), the gain matrix $\mathbf{G}$ may be expressed as

$$\mathbf{G} = [\mathbf{Q}_1, \mathbf{0}]\mathbf{Z}^{-1} \tag{8.107}$$

Although the columns of $\hat{\mathbf{Z}}$ are linearly independent, there is no guarantee that $\hat{\mathbf{Z}}_{11}^{-1}$ exists. In case of failure, one may have to repeat the process by slightly altering the quantities used for the assigned eigenvalues.[7]

Even if one is successful in computing the gain matrix $\mathbf{G}$ that will alter only $2p$ eigenvalues and eigenvectors of the uncontrolled system, there is no guarantee that the motion associated with those altered eigenvalues will cease to exist at the end of control time T. One has to properly choose the values for the assigned eigenvalues, and then verify by

---

[7]When $q < 2p$, the assumption of $q = 2p$ may be achieved by splitting certain components of output vector as explained in Section 3.3 of L.-Y. Lu's doctoral dissertation: *On the Placement of Active Members in Adaptive Truss Structures*, 1991, Civil and Environmental Engineering Department, Duke University, Durham, NC.

computation that $e^{\mu_k T}, k = 1, 2, \cdots, 2p$ are sufficiently small in magnitude. Since the eigenvalues are p pairs of complex conjugate numbers [see Eq.(7.106)], using the Euler's formula, one may prefer to show that $e^{-\nu_k \omega_k T} Cos\grave{\omega}_k T$, $k = 1, 2, \cdots, p$ are sufficiently small in magnitude, instead of $e^{\mu_k T}$, $k = 1, 2, \cdots, 2p$. In Fig.(8.8), ω_k and $\grave{\omega}_k$ values are shown on the positive real axis with vertical line segments and solid squares, respectively.

8.9 Actuator Placement

The location of actuators in adaptive structures is as important as any other physical parameter identifying the structure. In this text it is represented by the binary matrix $\mathbf{I}_\beta$ and it is part of the definition of $\grave{\mathbf{B}}$ matrix. From Eq.(6.34) one may obtain the definition of $\grave{\mathbf{B}} \in R^{2n \times q}$ appearing in equations of motion of nodes in state space, i.e., in $\dot{\zeta} = \mathbf{A}\zeta + \grave{\mathbf{B}}\mathbf{u}$. For convenience it is reproduced below:

$$\grave{\mathbf{B}} = \begin{bmatrix} -\mathbf{M}^{-1}\mathbf{B}\vec{\mathbf{K}}\mathbf{I}_\beta \\ 0 \end{bmatrix} \tag{8.108}$$

where $\mathbf{M} \in R^{n \times n}$ is the mass matrix studied in Subsection 6.1.2, $\mathbf{B} \in R^{n \times m}$ is the matrix studied in Subsection 2.2.2 for transforming element forces into the global description of internal nodal forces, $\vec{\mathbf{K}} \in R^{m \times m}$ is the block diagonal matrix of element stiffnesses studied in Section 2.4, $\mathbf{0} \in R^{n \times q}$ is the zero matrix, and $\mathbf{I}_\beta \in (Int)^{n \times q}$ is the actuator placement matrix. Chapters 3, 4, and 5 discuss how to determine the location of the actuators in order to minimize the control energy for cases where the inertial forces are small and there are no preexisting element forces in the structural elements.

In this section the actuator placement problem is discussed for controlling the autonomous motion of nodes. The trajectory of nodes in state space for a controlled autonomous system is sketched in Fig.(8.1). The total energy to be removed from the system is the work done in the actuators. In autonomous systems this is the initial energy $E(0)$

$$E(0) = \frac{1}{2}\zeta_o^T diag(\mathbf{M}, \mathbf{K}')\zeta_o \tag{8.109}$$

i.e., a fixed quantity identified by the initial state ζ_o. The driver of the actuators needs this much energy in order to stop the motion, since when the motion stops at terminal time T, the system energy becomes zero:

$$E(T) = 0 \tag{8.110}$$

Since $E(0)$ is a fixed quantity it cannot be minimized by actuator placement. One has to find other common sense criteria in order to place the actuators rationally. Some of such criteria are discussed in the following subsections.

8.9.1 Placement Criteria to Minimize Control Time

As discussed in Section 4.4, in the static control of nodal deflections, controllability is an important issue. In static control the actuators are placed in order to maximize the controllability. However, as explained in conjunction with Eq.(8.45), as far as nodal vibrations are concerned, an adaptive structure is always controllable, regardless of the physical location of actuators. One may not use controllability as a criterion for actuator placement in vibration control. Among other common sense criteria that one may use in actuator placement is effectiveness of a placement in reducing the system's vibrational energy. Actuator placement with this criterion is discussed in this subsection.

A non-equilibrium initial state ζ_o determines the initial energy $E(0)$ by Eq. (8.109); the rate of energy consumption at the actuators defines the duration of the control, i.e., the control time T. The faster the energy consumed at the actuators, the smaller is the duration of control. Hence, in order to decrease the control duration, one should place the actuators at structural elements where the motion-induced element forces are the largest in magnitude.

The motion-induced element forces are represented by the first term on the right hand side of Eq.(8.9),[8] or by Eq.(8.14). We assume in this section that there are no preexisting stresses in the structure, hence $\grave{\mathbf{K}} = 0$. With this and recalling that $\dot{\boldsymbol{\zeta}} = [\ddot{\boldsymbol{\xi}}^T, \dot{\boldsymbol{\xi}}^T]^T = \grave{\mathbf{A}}\boldsymbol{\zeta}$, $\boldsymbol{\zeta} = e^{\grave{\mathbf{A}}t}\boldsymbol{\zeta}_o$, one may rewrite Eq.(8.14) as

$$\mathbf{s}(t) = -(\mathbf{I} - \mathbf{K}_c\vec{\mathbf{K}}^{-1})\mathbf{C}'[\mathbf{M}, \grave{\mathbf{C}}]\grave{\mathbf{A}}e^{\grave{\mathbf{A}}t}\boldsymbol{\zeta}_o \tag{8.111}$$

where $\mathbf{K}_c \in \mathrm{R}^{m\times m}$ is as defined in Eq.(8.8), $\mathbf{C} \in \mathrm{R}^{n\times n}$ is the damping matrix, $\mathbf{C}' \in \mathrm{R}^{m\times n}$ is a generalized inverse of $\mathbf{B} \in \mathrm{R}^{n\times m}$ discussed in Subsection 3.1.2, and $\grave{\mathbf{A}} = \mathbf{A} + \grave{\mathbf{B}}\mathbf{G} \in \mathrm{R}^{2n\times 2n}$ is the controlled system matrix that involves the gain matrix $\mathbf{G} \in \mathrm{R}^{q\times 2n}$. Unless one knows the initial state $\boldsymbol{\zeta}_o$ and the gain matrix $\mathbf{G}$, it is not possible to assess the element stresses beforehand. Initial states, i.e., the states when the controls are turned on, are random, and the determination of $\mathbf{G}$ requires the location of the actuators. Therefore, from the $\mathbf{s}(t)$ expression in Eq.(8.111), a general rule that identifies mostly stressed elements cannot be obtained.

However, if there is a bias in the expected initial states, one may identify the highly stressed elements. For example for a cantilever truss structure if the expected initial states consist of smooth transverse deflections, the highly stressed elements are the chord elements closest to the fixed support. For that structure, as far as the minimization of the control time is

[8] We assume that the element forces represented by the second term of the right-hand side of Eq.(8.9) are zero either because the structure is statically determinate, or the structure is statically indeterminate, but actuator inserted deformations are compatible or made compatible by means of the synchronous secondary actuators (see Subsection 8.2.2, and Chapter 5).

concerned, the best place for an actuator is the chord element closest to the fixed support.

A similar situation arises if one is interested in controlling certain modes. As in the eigenvalue assignment method discussed in the previous section, one can easily identify the highly stressed elements for those modes, and place actuators on the identified elements to control the modes. For this purpose, one may deform the structure according to the mode shape, and compute the element forces. Then one may order the elements with the magnitude of element forces developed in order to find the location of the actuators to control that mode in the shortest time.

For this purpose, instead of using the expressions for element forces $\mathbf{s}(t)$ given by Eqs.(8.5, or 8.9, or 8.111), one may prefer to use the equation in the second row partition of Eq.(8.1). From this equation, the part of the stress that depends on the nodal motion may be obtained as:

$$\mathbf{s}'(t) = \vec{\mathbf{K}}\mathbf{B}^T \boldsymbol{\xi}(t) \tag{8.112}$$

or describing $\boldsymbol{\xi}(t)$ in terms of modal coordinates $\eta(t)$ [see Eq.(7.60)]

$$\mathbf{s}'(t) = \vec{\mathbf{K}}\mathbf{B}^T \mathbf{X}\eta(t) \tag{8.113}$$

Let $\mathbf{s}_k(t)$ denote the element forces related to the kth mode $\mathbf{x}_k$. Then from the last equation one may write

$$\mathbf{s}_k'(t) = \grave{\mathbf{s}}_k \eta_k(t) \tag{8.114}$$

where

$$\grave{\mathbf{s}}_k = \vec{\mathbf{K}}\mathbf{B}^T \mathbf{x}_k \tag{8.115}$$

The element forces $\grave{\mathbf{s}}_k$ are the ones developed when the nodes are displaced according to the kth mode shape $\mathbf{x}_k$. If one actuator is to be used to control the kth mode in the shortest time, then it should be placed on the element identified by that row index of the largest magnitude entry of $\mathbf{s}_k$.

8.9.2 Placement Criteria to Minimize Spill-Over

As discussed in Section 8.6, in the feedback control of flexible bodies one always has output control, since spill-over, i.e., the effect of the unobserved part of the state on the feedback control of the observed part, is always present. The depiction of this in modal space is in Fig. (8.7). The idea of placing the actuators with the criteria of minimizing spill-over may become important if the spill-over is of large magnitude. In order to study the spill-over, it is advantageous to start from the dynamic equilibrium equations of the nodes. Adopting Eqs.(7.78) for the non-equilibrium initial state of $\boldsymbol{\zeta}_o = [\boldsymbol{\xi}_o^T, \boldsymbol{\xi}_o'^T]^T$ one may write these equations for the controlled autonomous case as

$$\mathbf{M}\ddot{\boldsymbol{\xi}} + \grave{\mathbf{C}}\dot{\boldsymbol{\xi}} + \mathbf{K}'\boldsymbol{\xi} = -\mathbf{B}\vec{\mathbf{K}}\mathbf{I}_\beta \mathbf{u} \quad 0 < t \leq\ll T \tag{8.116}$$

and

$$\boldsymbol{\xi}(0) = \boldsymbol{\xi}_o, \ \dot{\boldsymbol{\xi}}(0) = \boldsymbol{\xi}'_o \tag{8.117}$$

where $\mathbf{M}, \grave{\mathbf{C}}, \mathbf{K}', \mathbf{B}, \vec{\mathbf{K}}, \mathbf{I}_\beta$ matrices are all as defined before. Suppose one uses a control law similar to the one in Eq.(8.82):

$$\mathbf{u} = \mathbf{G}_1\dot{\boldsymbol{\xi}} + \mathbf{G}_2\boldsymbol{\xi} \tag{8.118}$$

where $\mathbf{G}_1, \mathbf{G}_2 \in R^{q\times n}$. Using $\mathbf{u}$ from this equation and the transformation to modal coordinates $\boldsymbol{\xi} = \mathbf{X}\boldsymbol{\eta}$ from Eq.(7.30), Eq.(8.116) may be rewritten as

$$\ddot{\boldsymbol{\eta}} + [\mathbf{D}\mathbf{G}_1\mathbf{X} + diag(2\gamma_k\omega_k)]\dot{\boldsymbol{\eta}} + [\mathbf{D}\mathbf{G}_2\mathbf{X} + diag(\omega_k^2)]\boldsymbol{\eta} = \mathbf{0} \tag{8.119}$$

where

$$\mathbf{D} = \mathbf{X}^T\mathbf{B}\vec{\mathbf{K}}\mathbf{I}_\beta \in R^{n\times q} \tag{8.120}$$

Let $\boldsymbol{\eta}_1 \in R^{p\times 1}$ denote the part of the modal coordinates, i.e., mode participation factors, associated with the controlled modes, and $\boldsymbol{\eta}_2 \in R^{(n-p)\times 1}$ denote the remainder such that

$$\boldsymbol{\eta} = \left\{ \begin{array}{c} \boldsymbol{\eta}_1 \\ \boldsymbol{\eta}_2 \end{array} \right\} \tag{8.121}$$

By partitioning Eqs.(8.119) conformably, one may write

$$\begin{array}{l} \ddot{\boldsymbol{\eta}}_1 + \grave{\mathbf{E}}_{11}\dot{\boldsymbol{\eta}}_1 + \mathbf{E}_{11}\boldsymbol{\eta}_1 = \mathbf{e}_1 \\ \ddot{\boldsymbol{\eta}}_2 + \grave{\mathbf{E}}_{21}\dot{\boldsymbol{\eta}}_2 + \mathbf{E}_{21}\boldsymbol{\eta}_2 = \mathbf{e}_2 \end{array} \tag{8.122}$$

where

$$\left. \begin{array}{l} \mathbf{e}_1 = -\grave{\mathbf{E}}_{12}\dot{\boldsymbol{\eta}}_2 - \mathbf{E}_{12}\boldsymbol{\eta}_2 \\ \mathbf{e}_2 = -\grave{\mathbf{E}}_{22}\dot{\boldsymbol{\eta}}_1 - \mathbf{E}_{22}\boldsymbol{\eta}_1 \end{array} \right\} \tag{8.123}$$

and

$$\left. \begin{array}{l} \grave{\mathbf{E}}_{ij} = [\mathbf{D}\mathbf{G}_1\mathbf{X} + diag(2\gamma_k\omega_k)]_{ij} \\ \mathbf{E}_{ij} = [\mathbf{D}\mathbf{G}_2\mathbf{X} + diag(\omega_k^2)]_{ij} \end{array} \right\} \text{ for } i = 1,2 \text{ and } j = 1,2 \tag{8.124}$$

and $\grave{\mathbf{E}}_{11}, \mathbf{E}_{11} \in R^{p\times q}$, $\grave{\mathbf{E}}_{12}, \mathbf{E}_{12} \in R^{p\times(n-p)}$, $\grave{\mathbf{E}}_{21}, \mathbf{E}_{21} \in R^{(n-p)\times q}$, $\grave{\mathbf{E}}_{22}, \mathbf{E}_{22} \in R^{(n-p)\times(n-p)}$. From Eq.(8.122) one may observe that the participation factors of controlled modes $\boldsymbol{\eta}_1$ are excited by the spill-over $\mathbf{e}_1 \in R^{p\times 1}$ which is caused by the unobserved modes, and the participation factors of unobserved $\boldsymbol{\eta}_2$ are excited by the spill-over $\mathbf{e}_2 \in R^{(n-p)\times 1}$ which is caused by the controlled modes.[9]

[9] For a more detailed treatment, see "On the Placement of Active Members in Adaptive Truss Structures for Vibration Control," L.-Y. Lu, S. Utku, and B. K. Wada, *Journal of Smart Materials and Structures*, No. 1, pp. 8-23, 1992.

By proper placement of the actuators, one may be able to minimize the spill-over. Suppose one would like to control only the kth mode by using a single actuator imbedded in the jth element. Therefore, for this case, one has $q = p = 1$, $\mathbf{G}_1, \mathbf{G}_2 \in \mathrm{R}^{1 \times n}$. The expression for $\mathbf{e}_1 = e_1 \in \mathrm{R}^{1\times 1}$ may be obtained as:

$$e_1 = -\grave{s}_{kj}(\mathbf{G}_1\mathbf{X}_2\dot{\eta}_2 + \mathbf{G}_2\mathbf{X}_2\eta_2) \tag{8.125}$$

where

$$\grave{s}_{kj} = \mathbf{x}_k^T \mathbf{B}\vec{\mathbf{K}}\mathbf{1}_j \tag{8.126}$$

and

$$\mathbf{X}_2 = [\mathbf{x}_1, \cdots, \mathbf{x}_{k-1}, \mathbf{x}_{k+1}, \cdots, \mathbf{x}_n] \tag{8.127}$$

One may interpret $\grave{s}_{kj}$ from its definition in Eq.(8.126) as the stress in the jth entry of element forces $\mathbf{s} \in \mathrm{R}^{m\times 1}$ when the normalized kth mode, i.e., $\mathbf{x}_k$, is the nodal deflections. For small magnitude spill-over e_1, one would like to have both $\grave{s}_{kj}$ and its multiplier in Eq.(8.125) with small magnitude. One may place the actuator with the purpose of making the magnitude of $\grave{s}_{kj}$ small, but this may conflict with the objective of decreasing the control time.[10] The multiplier of $\grave{s}_{kj}$ can be made zero by choosing the rows of $\mathbf{G}_1$ and $\mathbf{G}_2$ in the subspace of controlled modes, i.e., controlled eigenvectors.

8.10 Time Lag in Feedback Control

It may be observed from output feedback control diagram of Fig. (8.6) that lots of operations are being done during a control cycle. If the sampling rate of the control system is 50 Hz, then the control cycle time is $\Delta_c = 1/50 = 20$ milliseconds. During this time interval, according to the figure, one has to

- obtain the sensor outputs $\mathbf{y} \in \mathrm{R}^{p\times 1}$,
- compute the estimate of state $\boldsymbol{\zeta} \in \mathrm{R}^{2n\times 1}$,
- compute the controls $\mathbf{u} \in \mathrm{R}^{q\times 1}$,
- and physically implement them through the actuators.

Since most of the modern control is done digitally, one has to include the time for converting analogue output of sensors into digital data of $\mathbf{y}$, and the time for converting the digital data of $\mathbf{u}$ into analogue signals for actuating the actuators. Let Δt denote the total time required to do all

[10] An interesting treatment of actuator placement by minimizing spill-over may be found in "On the Placement of Active Members in Adaptive Truss Structures for Vibration Control," Lyan-Ywan Lu, Doctoral Dissertation, Department of Civil and Environmental Engineering, Duke University, September 1991.

these operations. The time duration Δt is called *time lag*. It is the time interval between observing the deviation of the system from the desired state and implementing the necessary actuations to correct it. Depending upon the characteristics of structure and its loading, and system parameters $p, q, 2n$, the control scheme may become unstable, even if the $\frac{\Delta t}{\Delta_c}$ ratio is small. There are many studies for the negative effects of time lag in feedback controls.[11] The time lag problem may become more pronounced in controlling transients that are not repeating themselves indefinitely.[12]

In feedback control of flexible mechanical systems, control cycle time Δ_c is usually taken as one order of magnitude smaller than the fundamental period, and lag time Δt is expected to be a fraction of Δ_c. This puts definite limits to system parameters n, p, and q. The limits may be exceeded by faster microprocessors and communication lines of the on-board electronics, and by simple and yet efficient control algorithms.

It is a real challenge to modern engineers to develop very efficient computational control algorithms that take full advantage of *concurrent processing*.[13] Once the reliability and efficiency of such algorithms are established, they may be converted into *firmware* for general use.

8.11 Recapitulation, Autonomous Case

In this chapter the linear feedback control of time-invariant discrete parameter adaptive structures with linear excitation-response relations is discussed for the autonomous case. It is shown that the control of any response quantity can be done by controlling all or some attributes of nodal motion. Since control by deformation inserting actuators may be done efficiently by minimizing the resistance to insertion, it is shown that the resistance is minimum if inserted deformations are compatible. The optimal state feedback control with a quadratic performance functional is discussed for adaptive structures in order to highlight the limitations and the power of

---

[11] For example, "On Vibration Control of Building Structures Subjected to Seismic Excitation," MS Thesis, Ahmet Suner, Department of Civil and Environmental Engineering, Duke University, 1995 (a summary may be found in "Effect of Time Lag and Use of Compensators in the Active Control of Buildings Subjected to Earthquake Excitations" by Ahmet Suner, Abhijit Nagchaudhuri, and Senol Utku, *Journal of Structural Control*, Vol.2, No.2, pp.79-91, December 1995).

[12] See, for example, "Active-Passive Base Isolation System for Seismic Response Controlled Structures", Murat Sener, S. Utku, *Proceedings of 36th Structures, Structural Dynamics and Materials Conference*, paper AIAA-95-1088, New Orleans, LA, April 1995.

[13] For example, "Real-Time Computation of Control Torques for Mechanical Manipulators using Concurrent Processors" by A. V. Ramesh, S. K. Das, S. Utku, L.-Y. Lu, M. Salama, *Proceedings of International Conference of Computational Engineering Science*, Georgia Tech., Atlanta, GA, pp 43.vii.1-43.vii.6, April 1988.

the theory. With the same performance functional, output feedback control and its relation to state feedback control, direct output feedback control, and other non-optimal yet admissible controls are studied. The actuator placement problem to meet various objectives is discussed.

9

Active Control of Response, Non-Autonomous Case

In this chapter active control of response in discrete parameter adaptive structures is studied for cases where there are excitations on the structure in addition to the actuator induced element deformations during the control. With the notation of Section 7.8, $\boldsymbol{\zeta}_o \neq \mathbf{o}$, and $\mathbf{u} \neq \mathbf{o}$, $\mathbf{g} \neq \mathbf{o}$, for $0 \leq t \leq T$. The equations of motion in state space are given in Subsection 7.8.4, and a sketch of the trajectory of nodes is given in Fig. (7.5). This chapter deals with the real time determination of $\mathbf{u}(t)$.

9.1 Total Response Including Control Excitations

This is discussed in detail in Chapter 7. The important results of Chapter 7 for treatments in n-space and also in state space are summarized in the following subsections for quick reference.

9.1.1 Treatment in n-Space

As explained in Section 7.1, using a displacement method approach, one may obtain the nodal motion $\boldsymbol{\xi}(t)$ from the initial state and the dynamic equilibrium equations of nodes, expressed in terms of $\boldsymbol{\xi}$ and its time derivatives:

$$\begin{aligned} &\mathbf{M}\ddot{\boldsymbol{\xi}} + \dot{\mathbf{C}}\dot{\boldsymbol{\xi}} + \mathbf{K}'\boldsymbol{\xi} = \mathbf{p} - \mathbf{B}\vec{\mathbf{K}}\grave{\mathbf{v}}_e - \mathbf{B}\vec{\mathbf{K}}\mathbf{B}_s^T\boldsymbol{\xi}^s - \mathbf{B}\vec{\mathbf{K}}\mathbf{I}_\beta\mathbf{u} \quad \text{for } t > 0\,, \\ &\boldsymbol{\xi}(0) = \boldsymbol{\xi}_o \text{ and } \dot{\boldsymbol{\xi}}(0) = \boldsymbol{\xi}'_o \quad \text{at } t = 0 \end{aligned} \tag{9.1}$$

where

$$\mathbf{K}' = \mathbf{B}\vec{\mathbf{K}}\mathbf{B}^T + \grave{\mathbf{K}} \tag{9.2}$$

Then having $\boldsymbol{\xi}(t)$, one may express the element elongations $\mathbf{v}$ from the geometric relations as

$$\mathbf{v} = \mathbf{B}^T\boldsymbol{\xi} + (\grave{\mathbf{v}}_e + \mathbf{B}_s^T\boldsymbol{\xi}^s + \mathbf{I}_\beta\mathbf{u}) \tag{9.3}$$

and finally, using the stiffness relations, the element forces $\mathbf{s}$ may be expressed as

$$\mathbf{s} = \vec{\mathbf{K}}\mathbf{v} \tag{9.4}$$

In these equations the response quantities are

$\boldsymbol{\xi}$	$\in \mathrm{R}^{n\times 1}$	nodal deflections
$\mathbf{s}$	$\in \mathrm{R}^{m\times 1}$	element forces
$\mathbf{v}$	$\in \mathrm{R}^{m\times 1}$	element deformations

and the system related parameters are

$\mathbf{B}$	$\in \mathrm{R}^{n\times m}$	matrix defining internal nodal forces as $\mathbf{Bs}$
$\mathbf{B}_s^T$	$\in \mathrm{R}^{m\times b}$	matrix relating $\boldsymbol{\xi}^s$ to element deformations
$\grave{\mathbf{C}}$	$\in \mathrm{R}^{n\times n}$	damping matrix
$\grave{\mathbf{K}}$	$\in \mathrm{R}^{n\times n}$	stiffness matrix due to preexisting stresses
$\vec{\mathbf{K}}$	$\in \mathrm{R}^{m\times m}$	block diagonal matrix of element stiffnesses
$\mathbf{I}_\beta$	$\in (\mathrm{Int})^{m\times q}$	actuator placement matrix

The quantities

$\boldsymbol{\xi}_o$	$\in \mathrm{R}^{n\times 1}$	initial deflections of the nodes
$\boldsymbol{\xi}_o'$	$\in \mathrm{R}^{n\times 1}$	initial velocities of the nodes
$\mathbf{p}$	$\in \mathrm{R}^{n\times 1}$	nodal loads
$\grave{\mathbf{v}}_e$	$\in \mathrm{R}^{m\times 1}$	load type prescribed element deformations
$\boldsymbol{\xi}^s$	$\in \mathrm{R}^{b\times 1}$	support movements
$\mathbf{u}$	$\in \mathrm{R}^{q\times 1}$	actuator induced element deformations

are the excitation quantities. Although the initial conditions $\boldsymbol{\xi}_o$ and $\boldsymbol{\xi}_o'$ are constant vectors, the nodal loads $\mathbf{p}$, the load type prescribed element deformations $\grave{\mathbf{v}}_e$, support movements $\boldsymbol{\xi}^s$, and the actuator induced element deformations $\mathbf{u}$ are all functions of time. Sometimes it may be more useful to separately designate the time dependency of vectors $\mathbf{p}$, $\grave{\mathbf{v}}_e$, $\boldsymbol{\xi}^s$, and $\mathbf{u}$. As discussed in Section 7.1, denoting time dependencies by diagonal matrices, and largest magnitude values by primed symbols, one may write for the excitations

$$\left.\begin{aligned} \mathbf{p} &= diag(\hat{p}_k)\ \mathbf{p}' \\ \grave{\mathbf{v}}_e &= diag(\hat{v}_e^k)\grave{\mathbf{v}}_e' \\ \boldsymbol{\xi}^s &= diag(\hat{\xi}_k^s)\boldsymbol{\xi}^{s\prime} \\ \mathbf{u} &= diag\ (\hat{u}_k)\ \mathbf{u}' \end{aligned}\right\} \tag{9.5}$$

where the time-dependent matrices $diag\ (\hat{p}_k) \in R^{n\times n}$, $diag\ (\hat{v}_e^k) \in R^{m\times m}$, $diag\ (\hat{\xi}_k^s) \in R^{b\times b}$, and $diag(\hat{u}_k) \in R^{q\times q}$ are with no physical dimension, and constant vectors $\mathbf{p}' \in R^{n\times 1}$, $\grave{\mathbf{v}}_e' \in R^{m\times 1}$, $\boldsymbol{\xi}^{s\prime} \in R^{b\times 1}$ and $\mathbf{u}' \in R^{q\times 1}$ are with physical dimension.

Using superposition, the solution of the linear second order coupled ordinary differential equations with constant coefficients appearing in Eq.(9.1) may be expressed as the sum of the individual contributions of

$\boldsymbol{\xi}_o$	$\in R^{n\times 1}$	initial deflections of the nodes
$\boldsymbol{\xi}_o'$	$\in R^{n\times 1}$	initial velocities of the nodes
$\mathbf{p} = diag\ (\hat{p}_k)\ \mathbf{p}'$	$\in R^{n\times 1}$	nodal loads
$\grave{\mathbf{v}}_e = diag\ (\hat{v}_e^k)\ \grave{\mathbf{v}}_e'$	$\in R^{m\times 1}$	load type prescribed element deformations
$\boldsymbol{\xi}^s = diag\ (\hat{\xi}_k^s)\boldsymbol{\xi}^{s\prime}$	$\in R^{b\times 1}$	support movements
$\mathbf{u} = diag\ (\hat{u}_k)\ \mathbf{u}'$	$\in R^{q\times 1}$	actuator induced element deformations

in the form of

$$\dot{\boldsymbol{\xi}}(t) = \begin{bmatrix} \mathbf{A}_{\xi_o} & \mathbf{A}_{\xi_o'} & \mathbf{A}_{p'} & \mathbf{A}_{\grave{v}_e'} & \mathbf{A}_{\xi^{s\prime}} & \mathbf{A}_{u'} \end{bmatrix} \begin{Bmatrix} \boldsymbol{\xi}_o \\ \boldsymbol{\xi}_o' \\ \mathbf{p}' \\ \grave{\mathbf{v}}_e' \\ \boldsymbol{\xi}^{s\prime} \\ \mathbf{u}' \end{Bmatrix} \tag{9.6}$$

where the time effects are included in matrices $\mathbf{A}_{\xi_o} \in R^{n\times n}$, $\mathbf{A}_{\xi_o'} \in R^{n\times n}$, $\mathbf{A}_{p'} \in R^{n\times n}$, $\mathbf{A}_{\grave{v}_e'} \in R^{n\times m}$, $\mathbf{A}_{\xi^{s\prime}} \in R^{n\times b}$, and $\mathbf{A}_{u'} \in R^{n\times q}$ which are available from Eqs. (7.55), (7.56), (7.68), (7.72), (7.76), and (7.80), respectively. Note that the multipliers $\boldsymbol{\xi}_o, \boldsymbol{\xi}_o', \mathbf{p}', \grave{\mathbf{v}}_e', \boldsymbol{\xi}^{s\prime}$, and $\mathbf{u}'$ in the superposition equation are all constant vectors.

Using Eqs.(9.3, and 9.4), one may write

$$\begin{Bmatrix} \boldsymbol{\xi} \\ \mathbf{s} \\ \mathbf{v} \end{Bmatrix} = \begin{bmatrix} \cdot \\ \vec{\mathbf{K}} \\ \mathbf{I} \end{bmatrix} (\grave{\mathbf{v}}_e + \mathbf{B}_s^T \boldsymbol{\xi}^s + \mathbf{I}_\beta \mathbf{u}) + \begin{bmatrix} \mathbf{I} \\ \vec{\mathbf{K}}\mathbf{B}^T \\ \mathbf{B}^T \end{bmatrix} \boldsymbol{\xi}(t) \tag{9.7}$$

The substitution of $\boldsymbol{\xi}(t)$ from Eq.(9.6) into this equation leads to the inverse relations as

$$\begin{Bmatrix} \boldsymbol{\xi} \\ \mathbf{s} \\ \mathbf{v} \end{Bmatrix} = \grave{\mathbf{D}}(t) \begin{Bmatrix} \boldsymbol{\xi}_o \\ \boldsymbol{\xi}_o' \\ \mathbf{p}' \\ \grave{\mathbf{v}}_e' \\ \boldsymbol{\xi}^{s\prime} \\ \mathbf{u}' \end{Bmatrix} \tag{9.8}$$

where

$$\dot{\mathbf{D}}(t) = \begin{bmatrix} \mathbf{A}_{\xi_o} & \mathbf{A}_{\xi'_o} & \mathbf{A}_{p'} & \mathbf{A}_{\dot{v}'_e} \\ \vec{\mathbf{K}}\mathbf{B}^T\mathbf{A}_{\xi_o} & \vec{\mathbf{K}}\mathbf{B}^T\mathbf{A}_{\xi'_o} & \vec{\mathbf{K}}\mathbf{B}^T\mathbf{A}_{p'} & \vec{\mathbf{K}}[diag(\hat{v}^k_e)+\mathbf{B}^T\mathbf{A}_{\dot{v}'_e}] \\ \mathbf{B}^T\mathbf{A}_{\xi_o} & \mathbf{B}^T\mathbf{A}_{\xi'_o} & \mathbf{B}^T\mathbf{A}_{p'} & diag(\hat{v}^k_e)+\mathbf{B}^T\mathbf{A}_{\dot{v}'_e} \end{bmatrix}$$

$$\begin{matrix} \mathbf{A}_{\xi^{s\prime}} & \mathbf{A}_{u'} \\ \vec{\mathbf{K}}[\mathbf{B}^T_s diag(\hat{\xi}^s_k)+\mathbf{B}^T\mathbf{A}_{\xi^{s\prime}}] & \vec{\mathbf{K}}[\mathbf{I}_\beta diag(\hat{u}_k)+\mathbf{B}^T\mathbf{A}_{u'}] \\ \mathbf{B}^T_s diag(\hat{\xi}^s_k)+\mathbf{B}^T\mathbf{A}_{\xi^{s\prime}} & \mathbf{I}_\beta diag(\hat{u}_k)+\mathbf{B}^T\mathbf{A}_{u'} \end{matrix} \tag{9.9}$$

Matrix $\dot{\mathbf{D}}(t)$ is the *dynamic load factor matrix.*

9.1.2 Treatment in State Space

It is more informative to study the response in state space. Defining the state vector $\boldsymbol{\zeta}(t)$ as $[\dot{\boldsymbol{\xi}}(t)^T, \boldsymbol{\xi}(t)^T]^T$, and the initial state vector $\boldsymbol{\zeta}(0)$ as $\boldsymbol{\zeta}_o = [\boldsymbol{\xi}_o^T, \boldsymbol{\xi}_o^{'T}]^T$, one may obtain the first order equations of nodal motion studied in Subsection 6.4.2 as:

$$\begin{aligned} \dot{\boldsymbol{\zeta}} &= \mathbf{A}\boldsymbol{\zeta} + \dot{\mathbf{B}}\mathbf{u} + \mathbf{g} \qquad \text{for } t > 0 \\ \boldsymbol{\zeta}(0) &= \boldsymbol{\zeta}_o \qquad \text{at } t = 0 \end{aligned} \tag{9.10}$$

where $\mathbf{A} \in \mathrm{R}^{2n\times 2n}$ and $\dot{\mathbf{B}} \in \mathrm{R}^{2n\times q}$ represent system parameters, $\boldsymbol{\zeta}_o \in \mathrm{R}^{2n\times 1}$ is the initial state, $\mathbf{g} \in \mathrm{R}^{2n\times 1}$is the loading type excitations, and $\mathbf{u} \in \mathrm{R}^{q\times 1}$ is the control excitations. The definition of matrices $\mathbf{A}, \dot{\mathbf{B}}$, and $\mathbf{g}$ are given in Subsection 6.4.2 and repeated below:

$$\mathbf{A} = \begin{bmatrix} -\mathbf{M}^{-1}\dot{\mathbf{C}} & -\mathbf{M}^{-1}\mathbf{K}' \\ \mathbf{I} & \cdot \end{bmatrix} \tag{9.11}$$

$$\dot{\mathbf{B}} = \begin{bmatrix} -\mathbf{M}^{-1}\mathbf{B}\vec{\mathbf{K}}\mathbf{I}_\beta \\ \cdot \end{bmatrix} \tag{9.12}$$

$$\mathbf{g} = \begin{Bmatrix} \mathbf{M}^{-1}\mathbf{p} - \mathbf{M}^{-1}\vec{\mathbf{K}}(\dot{\mathbf{v}}_e + \mathbf{B}^T_s\boldsymbol{\xi}_s) \\ \cdot \end{Bmatrix} \tag{9.13}$$

where all symbols are as defined in the previous subsection. Since $\mathbf{g} \neq \mathbf{o}$, the system is called non-autonomous.

By using the transition matrix $e^{\mathbf{A}t} \in \mathrm{R}^{2n\times 2n}$ of the system, the solution $\boldsymbol{\zeta}(t)$ may be expressed as

$$\boldsymbol{\zeta}(t)=e^{\mathbf{A}t}\boldsymbol{\zeta}_o + \int_0^t e^{\mathbf{A}(\tau-t)}[\dot{\mathbf{B}}\mathbf{u}(\tau-t) + \mathbf{g}(\tau-t)]d\tau, \quad \text{for } 0 \le t \tag{9.14}$$

where the definition of $e^{\mathbf{A}t}$ may be obtained from Eq.(7.123) as

$$e^{\mathbf{A}t} = \mathbf{Z}diag(e^{\mu_k t})\mathbf{Z}^{-1} \tag{9.15}$$

where the kth column of $\mathbf{Z} \in C^{2n \times 2n}$, $\mathbf{z}_k$ is the kth eigenvector and μ_k is the kth eigenvalue of $\mathbf{A}$ as discussed in Subsection 7.8.1 where it is shown that the eigenvalues and the eigenvectors of $\mathbf{A}$ are n complex conjugate pairs.

9.2 Response Control

Although it is possible to control the response of a structural system by altering some of its parameters,[1] in adaptive structures this is done by inserting deformations. The inserted deformations are the control variables. This is similar to what was adopted in the biological world where change in geometric configurations, motility, and motion takes place by deformation insertion in the cells. The deformation insertion in adaptive structures is done in real time by actuators that can be of thermal, electro-mechanical, hydraulic, piezo-electric, shape-memory alloy, and other types.

Deformation insertion of an actuator may be resisted by the pre-existing stresses, and also by the structural system itself. As discussed in Section 8.2, and earlier for static case in Sections 3.8 and 3.9, the resistance from the structural system may not take place if the inserted deformations are *compatible.* If the structural system is statically determinate, i.e., if $m = n$, inserted deformations are always compatible. In this case no special arrangement is necessary for eliminating the structure's resistance to deformation insertion. However, if the structure is statically indeterminate, i.e., if $m > n$, and the inserted deformations are not compatible, then there will be structural resistance to deformation insertion. To be compatible, the inserted deformations, i.e., the control variables $\mathbf{u}$:

$$\mathbf{u}(t) = diag[\hat{u}_k(t)]\mathbf{u}' \tag{9.16}$$

must satisfy the compatibility requirement of

$$\mathbf{C}\mathbf{I}_\beta \mathbf{u}(t) = \mathbf{o} \tag{9.17}$$

where the columns of $\mathbf{C} \in R^{m \times (m-n)}$ establish a basis in the null space of matrix $\mathbf{B} \in R^{n \times m}$ (see Subsection 3.1.2) and $\mathbf{I}_\beta \in (\mathrm{Int})^{m \times q}$ is the actuator placement matrix (see Sections 3.4, 4.3, and 5.6). If the compatibility is not satisfied, by using synchronized secondary actuators, the controls can be made compatible (see Section 5.5). We assume that either the structure is statically determinate, or it is statically indeterminate but the controls are compatible or made compatible by synchronized secondary actuators.

---

[1] For example, by using electro- or magneto-rheological fluids in structural junctions, one can alter the shearing strength of the connection instantaneously by changing the electrical or the magnetic field, thus altering the response.

From Eq.(9.7) one may observe that the response may be obtained with the knowledge of nodal motion $\boldsymbol{\xi}$, loads $\dot{\mathbf{v}}_e$ and $\boldsymbol{\xi}^s$, and controls $\mathbf{u}$. With the knowledge of loads, one may control the response by controlling nodal motion. In this chapter, the control of nodal motion in non-autonomous systems is discussed. The only difference from the discussions of the previous chapter is in the presence of active excitations on the structure during the control time. The control of nodal motion of a building frame during an earthquake, or the control of nodal motion of a paraboloidal antenna structure during a wind storm, are examples of non-autonomous control.

9.3 Energy Considerations

The energy imparted from excitations to a structure during the control plays a very important role in the control of adaptive structures. In the autonomous control discussed in the previous chapter, there is no energy flow into the structure during the control. The only energy flow is from the structure into the energy sinks provided by the actuators. However in non-autonomous control, one has to deal also with the concurrent energy flow into a structure from excitations.

The trajectory of nodes in state space, sketched in Fig. (7.5) for a non-autonomous structural system, is for the case with a known non-equilibrium initial state of $\boldsymbol{\zeta}_o$, where controls are turned on at $t = t'$ and they are turned off at $t = t''$. This figure is expanded to contain both controlled and uncontrolled trajectories in Fig.(9.1). For the present case the controls are turned on at $t = 0$, and they are turned off at $t = T$ to produce the controlled trajectory. When the controls are never turned on, the uncontrolled trajectory is produced.

Depending upon the energy imparted from the excitations, the uncontrolled trajectory can spiral out and be unbounded. However, if the imparted energy is finite, the uncontrolled trajectory may go out to a maximum energy level of $E_u(t')$ and eventually settle on a hypersphere with energy level $E_u(T)$ (not shown in the figure) such that $E_u(t') \geq E_u(T)$. The controlled trajectory starts from the initial state with an energy level of $E(0)$, and, depending upon the on-going excitations and control stratagem, may go out to a maximum energy level of $E_c(t'') \geq E(0)$ and hopefully end at the zero energy level of at-rest state at point O, i.e., at the point where $E_c(T) = 0$. The difference between the two trajectories is due to the energy sink at the controls. The points A and A' in the sketch show the states at time t of controlled and uncontrolled structure, respectively. According to the sketch the controlled state A is at a lower energy state than that of the uncontrolled state A'. This is due to the assumption that the objective of controls is to bring the structure to its nominal equilibrium state which is presumably of lower energy. If we knew the energy level $E_u(t')$ of the

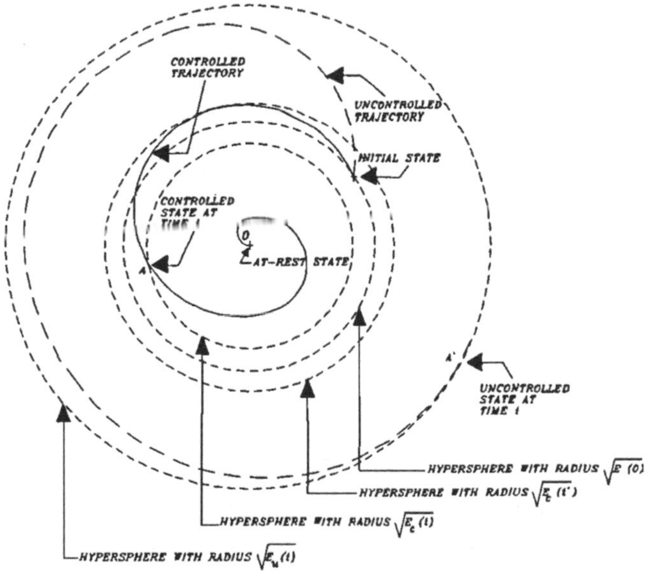

FIGURE 9.1. Controlled and uncontrolled nodal trajectories, non-autonomous system

hypersphere that envelops the uncontrolled trajectory at time t', then the cumulative energy required by the actuators at time t' would be $E_u(t')$, since we assumed that $E_c(T) = 0$. Unfortunately, for naturally occurring excitations, only a crude estimate of the energy level of the hypersphere that envelops the uncontrolled trajectory may be available as a result of past experience.

It is possible that the terminal point of the controlled trajectory may not be point O, but another point O' (not shown in the sketch). For example, the uncontrolled nodal trajectory of a building frame constructed near an active fault line and subjected to an earthquake may start from point O of the state space, but may end at point O' due to a permanent ground movement. In this case the controlled trajectory would also start at point O of the state space, ideally would remain in the close neighborhood of O, and end at point O'. The energy requirement of the actuators in this example would not be less than

$$E_u(t') - \left\|\overrightarrow{OO'}\right\|^2_{energy}$$

where $E_u(t')$ is proportional to the radius-square of the hypersphere that envelops the uncontrolled trajectory (see Section 7.8 for energy norm).

It is also possible that the terminal state of controlled trajectory may be some settled or unsettled point on a hypersphere with energy level $E_c(T)$. For example, when the nominal velocity of an aircraft changed from v_o to

$3v_o$, the uncontrolled trajectory of its nodes may move from the hypersphere with energy level $E(0)$ corresponding to the steady state vibrations at velocity v_o to an enveloping hypersphere of energy $E_u(t')$ and settle down to a hypersphere of energy $E_u(T')$, such that

$$E(0) < E_u(T') << E_u(t')$$

The controlled trajectory for the same scenario ideally may leave the hypersphere of energy $E(0)$ and go to the hypersphere of energy $E_c(T)$ with monotonically increasing energy levels, such that

$$E(0) < E_c(T) = E_u(T')$$

In this case, the energy required by the actuators may be much less than

$$E_u(t') - E_c(T)$$

However, the vibration control of mechanical systems as they move from one known steady state of vibrations to another is not within the scope of this text.

The examples above indicate that the energy requirement of the actuators is a function of the uncontrolled response of the structure which cannot be determined unless the excitations are known *a priori* and exactly. In other words, in Eq.(9.10), one rarely knows the excitation term $\mathbf{g}(t)$ quantitatively for most of the interesting problems. The *a priori* knowledge of $\mathbf{g}(t)$ is assumed not available for problems discussed in Chapters 10 and 11 where one has to be content with the knowledge of a crude upper bound for excitations.

9.4 Optimal State Feedback Control

Using performance functional J defined in Eq.(8.25), optimal state feedback control is discussed in Section 8.5 for autonomous systems where excitations $\mathbf{g}(t)$ are nonexistent. In this section the feedback control in the presence of excitations $\mathbf{g}(t)$ for bringing the trajectory of nodes, optimally with respect to some criterion, from a known initial state of $\boldsymbol{\zeta}_o$ to a known terminal state of $\boldsymbol{\zeta}(T)$ is studied.

In the following subsections it is shown that optimal control by minimizing performance functional J requires *a priori* knowledge of excitation $\mathbf{g}(t)$ or an equivalent attribute of it.

9.4.1 Case When Excitation is Known a Priori

For non-autonomous systems where $\mathbf{g}(t)$ is known *a priori*, the optimal state feedback control with the performance functional J of Eq.(8.25) is possible. Referring to the formulation in Eq.(9.10), let $\boldsymbol{\zeta}(t)$ denote the controlled trajectory with excitations $\mathbf{g}(t)$, controls $\mathbf{u}(t)$, and initial state $\boldsymbol{\zeta}_o$, i.e., the solution of

$$\begin{aligned} \tfrac{d}{dt}\boldsymbol{\zeta} &= \mathbf{A}\boldsymbol{\zeta} + \grave{\mathbf{B}}\mathbf{u} + \mathbf{g} && \text{for } t > 0 \\ \boldsymbol{\zeta}(0) &= \boldsymbol{\zeta}_o && \text{at } t = 0 \end{aligned} \tag{9.18}$$

Let $\grave{\boldsymbol{\zeta}}(t)$ denote the uncontrolled trajectory with excitations $-\mathbf{g}(t)$, controls $\mathbf{u}(t) = \mathbf{o}$, and initial state $\grave{\boldsymbol{\zeta}}(0) = \mathbf{o}$, that is, the solution of

$$\begin{aligned} \tfrac{d}{dt}\grave{\boldsymbol{\zeta}} &= \mathbf{A}\grave{\boldsymbol{\zeta}} - \mathbf{g} && \text{for } t > 0 \\ \grave{\boldsymbol{\zeta}}(0) &= \mathbf{o} && \text{at } t = 0 \end{aligned} \tag{9.19}$$

Defining error vector $\mathbf{e}(t)$ as the sum of these two trajectories, i.e.,

$$\begin{aligned} \mathbf{e}(t) &= \boldsymbol{\zeta}(t) + \grave{\boldsymbol{\zeta}}(t) \\ \mathbf{e}(0) &= \mathbf{e}_o = \boldsymbol{\zeta}(0) + \grave{\boldsymbol{\zeta}}(0) = \boldsymbol{\zeta}_o \end{aligned} \tag{9.20}$$

and using them in the expressions that one may obtain by adding Eqs.(9.19) to Eqs.(9.18), one may arrive at the non-autonomous problem of:

$$\begin{aligned} \tfrac{d}{dt}\mathbf{e} &= \mathbf{A}\mathbf{e} + \grave{\mathbf{B}}\mathbf{u} && \text{for } t > 0 \\ \mathbf{e}(0) &= \boldsymbol{\zeta}_o && \text{at } t = 0 \end{aligned} \tag{9.21}$$

A performance functional J for this problem may be obtained from Eq.(8.25) as

$$J = \frac{1}{2}\mathbf{e}(T)^T\mathbf{F}\ \mathbf{e}(T) + \frac{1}{2}\int_0^T [\mathbf{e}(t)^T\mathbf{Q}\ \mathbf{e}(t) + \mathbf{u}(t)^T\mathbf{R}\ \mathbf{u}(T)]dt \tag{9.22}$$

One may apply the optimal control theory of Section 8.5 to find the optimal controls $\mathbf{u}^*(t)$, which make J a minimum.[2] This corresponds to bringing e-trajectory from initial state $\mathbf{e}(0) = \boldsymbol{\zeta}_o$ to terminal state $\mathbf{e}(T) = \boldsymbol{\zeta}(T) + \grave{\boldsymbol{\zeta}}(T) \approx \mathbf{o}$ optimally. Note that, by changing the name of the state variable from $\mathbf{e}$ to $\boldsymbol{\zeta}$, the present problem reduces to the one studied in Section 8.5. The differential matrix Riccati equation for this case is exactly the same as the one given in Eqs.(8.41 and 8.42). One may use its solution $\mathbf{P}(t)$ for the optimal controls $\mathbf{u}^*(t)$, and with the help of Eqs.(8.43 and 8.44) write:

$$\mathbf{u}^*(t) = \mathbf{G}^*(t)\mathbf{e}(t) \tag{9.23}$$

where

$$\mathbf{G}^*(t) = -\mathbf{R}^{-1}\grave{\mathbf{B}}^T\mathbf{P}(t) \tag{9.24}$$

[2] This idea was mentioned in the paper "Direct Computation of Optimal Control of Forced Linear Systems" by S. Utku, C. P. Kuo, and M. Salama in *Proceedings of AIAA/ASME/ASCE 26th SDM Conference, 4/15/1985.*

Using the definition of $\mathbf{e}(t)$ from Eqs.(9.20), one may rewrite Eq.(9.23) as

$$\mathbf{u}^*(t) = \mathbf{G}^*(t)\boldsymbol{\zeta}(t) + \mathbf{h}^*(t) \tag{9.25}$$

where

$$\mathbf{h}^*(t) = \mathbf{G}^*(t)\grave{\boldsymbol{\zeta}}(t) \tag{9.26}$$

The optimal control law in Eq.(9.25) of the non-autonomous case is quite similar to the one given in Eq.(8.43) except for the presence of time-dependent $\mathbf{h}^*(t)$.

Observe that vector $\mathbf{h}^*(t)$ can be available at time t, if $\grave{\boldsymbol{\zeta}}(t)$ is available, and $\grave{\boldsymbol{\zeta}}(t)$ is available by solving Eqs.(9.19), either by real time or by *a priori* knowledge of $\mathbf{g}$. However, under very unusual circumstances this may be possible for the real time optimal state feedback control of nodal trajectories in non-autonomous systems.

9.4.2 Case When Tracked Trajectory is Known a Priori

Sometimes, instead of excitations $\mathbf{g}$, one may know *a priori* the intent of the excitations, i.e., the trajectory $\grave{\boldsymbol{\zeta}}(t)$ that the nodes should follow from the known initial state $\boldsymbol{\zeta}_o$ to the terminal state $\boldsymbol{\zeta}(T)$. This is called a tracking problem.[3] Since the system is observable, one may assume an output feedback control where sensor output $\mathbf{y}(t)$ is as descriptive as the state $\boldsymbol{\zeta}(t)$. Defining $\mathbf{e}(t)$ as the difference between the output of tracked trajectory $\grave{\boldsymbol{\zeta}}(t)$, i.e., $\hat{\mathbf{C}}\grave{\boldsymbol{\zeta}}(t)$, and the actual output $\mathbf{y}(t)$, i.e.

$$\mathbf{e}(t) = \hat{\mathbf{C}}\grave{\boldsymbol{\zeta}}(t) - \mathbf{y}(t) \tag{9.27}$$

and using the process outlined in Section 8.5 in the minimization of the performance functional J in Eq.(9.22) subject to the constraints

$$\left.\begin{array}{c}\dot{\boldsymbol{\zeta}}(t) = \mathbf{A}\boldsymbol{\zeta}(t) + \grave{\mathbf{B}}\mathbf{u}(t) \text{ for } 0 < t < T \\ \boldsymbol{\zeta}(0) = \boldsymbol{\zeta}_o\end{array}\right\} \tag{9.28}$$

the optimal controls $\mathbf{u}^*(t)$ may be obtained as

$$\mathbf{u}^*(t) = \mathbf{G}^*(t)\boldsymbol{\zeta}(t) + \mathbf{h}(t) \tag{9.29}$$

where $\mathbf{G}^*(t)$ is as defined in Eq.(9.24), and $\mathbf{h}(t)$ is the solution of the following terminal value problem

$$\left.\begin{array}{c}\dot{\mathbf{h}}(t) = -[\mathbf{A} - \grave{\mathbf{B}}\mathbf{R}^{-1}\grave{\mathbf{B}}^T\mathbf{P}]^T\mathbf{h}(t) - \hat{\mathbf{C}}^T\mathbf{Q}\grave{\boldsymbol{\zeta}}(t) \\ \mathbf{h}(T) = \hat{\mathbf{C}}^T\mathbf{F}\grave{\boldsymbol{\zeta}}(T)\end{array}\right\} \tag{9.30}$$

---

[3]See, for example, "Art. 9-9 The Tracking Problem" of *Optimal Control* by Michael Athans and Peter L. Falb, McGraw-Hill, 1966. In the article the authors refer to R. E. Kalman's work "The Theory of Optimal Control and the Calculus of Variations" in *Mathematical Optimization Techniques*, R. Bellman (ed), University of California Press, Berkeley, CA, 1963.

which depends on the knowledge of the tracked trajectory $\dot{\zeta}(t)$. The expression in Eq.(9.29) is the control law. Unfortunately it depends on the prior knowledge of tracked trajectory and the solution of Eq.(9.30).

9.4.3 Case When Excitation is not Known a Priori

In the presence of excitations $\mathbf{g}(t)$ that cannot be known *a priori*, to bring the trajectory of nodes, optimally with respect to some criterion, from a known state ζ_o to another known state $\zeta(T)$ is a very difficult task. This is mainly due to the unknown nature of the excitations $\mathbf{g}(t)$. In the previous two subsections, it is shown that the optimal feedback control by minimizing the performance functional J of Eq.(9.22) is possible for non-autonomous systems where either excitations $\mathbf{g}(t)$ or a tracked trajectory $\dot{\zeta}(t)$ is known *a priori.*

How would one control optimally the motion of the nodes of a building frame during an earthquake of completely unknown nature? From a mathematical point of view, the only way out of this difficult problem may be to find an optimality criterion that would make a mathematical solution possible. From a practical point of view, the solution needs to be obtained by non-optimal strategies. Once such non-optimal but admissible class of solutions are found, one may try to construct an optimality criterion that may help to identify one of the admissible solutions as the optimal one.

In summary, in non-autonomous systems when the excitations are of unknown and transient type, as in seismic excitations, one may be better off if one looks for non-optimal but admissible solutions.

9.5 Non-Optimal Control Possibilities

This subject is discussed for autonomous systems in Section 8.8., where, by modeling a feedback control law after the optimal control law, a large class of admissible gain matrices that monotonically decrease system energy is given. In this section non-optimal feedback control of non-autonomous systems is discussed from an energy point of view.

As shown in Eq.(9.25) and Eq.(9.29), the linear feedback control law in non-autonomous systems, optimal with respect to performance functional J of Eq.(9.22), is in the form of:

$$\mathbf{u}(t) = \mathbf{G}(t)\zeta(t) + \mathbf{h}(t) \tag{9.31}$$

where gain matrix $\mathbf{G} \in C^{q\times 2n}$ depends on the knowledge of system matrices $\mathbf{A}$ and $\dot{\mathbf{B}}$ and vector $\mathbf{h} \in R^{q\times 1}$ depends, in addition, on the *a priori* knowledge of excitations. In selecting non-optimal control laws of this type, the following may be useful:

- The practical considerations, such as time lag, stability, etc., may make it desirable to have a constant **G** matrix in the control law.

- As discussed in the previous section, gain matrix **G** is the same for autonomous and non-autonomous cases of the same structure. Therefore non-optimal yet admissible gain matrices discussed in Section 8.6 for autonomous cases can be used also for non-autonomous cases of the same structure.

- It may be rarely possible to have a constant vector for $\mathbf{h}(t)$ to represent the trajectory caused by excitations and without controls.

- For vector $\mathbf{h}(t)$, modeling after Eq.(9.26), in the absence of any other mathematical help, one may use

$$\mathbf{h}(t) = \mathbf{G}\grave{\zeta}_e(t) \tag{9.32}$$

  where **G** is the selected non-optimal yet admissible gain matrix and $\grave{\zeta}_e(t)$ is the expected trajectory under the negatives of excitations without controls.

- If there is not enough confidence in the validity of $\grave{\zeta}_e(t)$, one may use a zero vector instead and the non-optimal yet admissible control law becomes

$$\mathbf{u} = \mathbf{G}\zeta \tag{9.33}$$

  which implies that to control the nodal trajectory by ignoring the ongoing excitations is also an option.

Other non-optimal control strategies are included in the discussions of next two chapters for the real time feedback control of adaptive structures under low power excitations (e.g., adaptive building frames subjected to wind forces) and under high power excitations (e.g., adaptive building frames subjected to seismic ground motions).

9.6 Actuator Placement

The actuator placement problem in the feedback control of non-autonomous systems depends not only on the control objectives but also on the nature of excitations. The actuator placement problem studied in Chapters 4 and 5 for static control is based on the robustness of control, since control energy requirements in these problems can be minimized by guaranteeing the compatibility of actuator inserted deformations. However, in the feedback control of non-autonomous and autonomous systems, the placement problem is, in general, based on the rate of energy dissipation in the actuators.

From this point of view, in the feedback control of non-autonomous systems, the excitations which are effective during the control play a dominant role in determining the locations of the actuators. The excitation attributes that may play an important role include

1. type (deformations, deflection, force),

2. *a priori* knowledge,

3. transitoriness, and

4. power.

Types of excitations were briefly discussed in Subsection 2.1.2. For example, wind forces are of force type, seismic effects are of deflection type, and thermal loads are of deformation type excitations.

A priori knowledge about the excitations can alter not only the control stratagem (see Section 9.4), but also location of the actuators. The excitations induced by a truck of known axle weight and speed on a bridge structure is quite different than the excitations induced by a tornado of unknown class on the same bridge structure.

By comparing Section 7.9 with the rest of Chapter 7, one may observe that the transient or steady state excitations are handled differently even in analysis. The actuator placement problem is likewise affected by the transitoriness of the excitations.

The term power here refers to the energy imparted to the structure by the excitations in unit time. For example, in building structures, power input from winds is of much smaller magnitude than power input from seismic ground motions. The actuator placement depends highly on the power input from excitations.

From these attributes, only the type of excitations and the level of power input from excitations are further discussed in this book (see next two chapters).

In absence of knowledge about the attributes of excitations, one may try to place the actuators from the standpoint of their effectiveness in bringing initial state of the nodes to their terminal state. If one assumes that the terminal state is a lower energy state than the initial state, then the actuators may be considered as energy sinks. Since the actuators of adaptive structures are of deformation inducing type, one may conclude that they may be placed on elements that are highly stressed by the nodal motion. The actuator placement problem is discussed from this point of view in Section 8.9.

9.7 Recapitulation, Non-Autonomous Case

In this chapter the linear feedback control of time-invariant discrete parameter adaptive structures with linear excitation-response relations is discussed for the non-autonomous case. It is shown that in this case the presence of excitations during the control time makes the feedback control very difficult, especially if the loads are not known *a priori*. The linear optimal feedback control with quadratic performance functional is discussed for cases where either excitations are known *a priori* or a trajectory to be tracked is prescribed. Selection from non-optimal yet admissible linear control laws and actuator placement problems are also discussed.

10
Active Control Against Wind

Due to the increase in the ultimate strength of engineering materials, many modern earthbound structures in conventional geometric configurations that are designed with the ultimate strength paradigm are becoming too flexible. According to the numbers given in a recent book on structural control,[1] on a windy day one may get seasick at the top of New York's World Trade Center Tower (built in 1973), whereas nothing of the sort happens at the top of the Empire State Building (built there in 1932). Modern structures designed with modern design paradigms that minimize material usage may need active vibration control during their use. In fact by employing tuned mass damper systems, this is being done almost routinely in many tall buildings under horizontal excitations all over the world, following the initial successful experience in the John Hancock Tower of Boston.[2] How could this be done if one were to use the concepts of adaptive structures in the tower? This and other related questions are dealt with in this chapter as applications of the material discussed earlier.

---

[1] *Introduction to Motion Based Design* by Jerome J. Connor & Boutros S. A. Klink, Computational Mechanics Publications, Ashurst, Southampton SO40 7AA, UK, or Billerica, MA 01821, 1966.

[2] John Hancock Tower, *Engineering News Record*, Oct. 1075. However, Prof. Robert J. Hansen of M.I.T. (member of the committee related to the activities that led ultimately to the installation of the tuned mass damper system in the tower) reports that the first case was probably the Citicorp Building in New York City.

10.1 State Equations for Wind Type Excitations

Compared to seismic excitations, wind excitations are generally of much less power. Moreover, the wind excitations are time dependent forces acting on the nodes, whereas seismic excitations are ground motions acting on the supports. In both cases, the structure is at-rest at $t = 0$, and, except for some crude upper bounds of some of the attributes, nothing is known *a priori* about the excitations. In this setting, one would like to keep the nodal trajectory within an acceptable neighborhood of the at-rest state throughout many randomly encountered loading episodes, by means of active control.

The state equations of nodal motion due to wind type excitations and control deformations may be stated as in Eq.(7.134):

$$\left.\begin{array}{c}\dot{\boldsymbol{\zeta}} = \mathbf{A}\boldsymbol{\zeta} + \grave{\mathbf{B}}\mathbf{u} + \mathbf{g} \quad \text{for } t > 0 \\ \boldsymbol{\zeta}(0) = \boldsymbol{\zeta}_o = \mathbf{o}\end{array}\right\} \tag{10.1}$$

where

$$\boldsymbol{\zeta}(t) = \left\{\begin{array}{c}\dot{\boldsymbol{\xi}}(t) \\ \boldsymbol{\xi}(t)\end{array}\right\}$$

and, with the help of Eqs.(7.88, 7.89, and 7.90),

$$\mathbf{A} = \left[\begin{array}{cc}-\mathbf{M}^{-1}\grave{\mathbf{C}} & -\mathbf{M}^{-1}\mathbf{K}' \\ \mathbf{I} & \cdot\end{array}\right] \tag{10.2}$$

$$\grave{\mathbf{B}} = \left[\begin{array}{c}-\mathbf{M}^{-1}\mathbf{B}\vec{\mathbf{K}}\mathbf{I}_\beta \\ \cdot\end{array}\right] \tag{10.3}$$

and

$$\mathbf{g}(t) = \left\{\begin{array}{c}\mathbf{M}^{-1}\mathbf{p}(t) \\ \cdot\end{array}\right\} \tag{10.4}$$

In these equations, $\boldsymbol{\xi}(t) \in R^{n\times1}$ is independent nodal deflections (see Subsection 2.1.3), $\dot{\boldsymbol{\xi}}(t) \in R^{n\times1}$ is independent nodal velocities, $\mathbf{M} \in R^{n\times n}$ is positive definite mass matrix (see Subsection 6.1.2), $\grave{\mathbf{C}} \in R^{n\times n}$ is positive definite modal damping matrix (see Section 7.3), $\mathbf{K}' \in R^{n\times n}$ is positive definite global stiffness matrix of the structure (see Section 3.1.1), $\mathbf{I}_\beta \in (\text{Int})^{m\times q}$ is the actuator placement matrix (see Sections 3.4, 3.9, 4.3, 5.3, and 5.4), $\mathbf{u}(t) \in R^{q\times1}$ is the actuator induced element deformation representing controls (see Sections 3.4, 4.3, and 5.6), $\vec{\mathbf{K}} \in R^{m\times m}$ is positive definite block diagonal matrix of element stiffnesses (see Section 2.4), $\mathbf{B} \in R^{n\times m}$ is the matrix transforming description of element forces into the global description of internal nodal forces (see Subsection 2.2.2), and $\mathbf{p}(t) \in R^{n\times1}$ is the nodal wind loads (see Subsection 2.1.2). As discussed in Section 7.1, nodal wind loads may be expressed as

$$\mathbf{p}(t) = diag[\hat{p}_k(t)]\mathbf{p}' \tag{10.5}$$

where $\mathbf{p}' \in R^{n\times 1}$ is the maximum magnitudes (or estimates of maximum magnitudes) of nodal components of $\mathbf{p}$ and $diag[\hat{p}_k(t)] \in R^{n\times n}$ physically non-dimensional time functions showing time variations of nodal loads. The solution of Eqs.(10.1), i.e., the nodal trajectory, could be expressed as in Eq.(7.135) if one knew nodal wind loads $\mathbf{p}(t)$ and controls $\mathbf{u}(t)$. Unfortunately in the real time control of nodal motions, one does not know $\mathbf{p}(t)$ *a priori*.

10.2 Control Possibilities of Nodal Motion

One may observe from the right-hand side of Eqs.(10.1) that the motion is caused by the existence of the $\grave{\mathbf{B}}\mathbf{u} + \mathbf{g}$ term, where $\mathbf{g}$ represents nodal wind forces, and $\mathbf{u}$ represents the controls to eliminate the motion caused by $\mathbf{g}$. Since the objective is to ensure that $\boldsymbol{\zeta} \approx \mathbf{o}$ at all times, the following possibilities exist.

10.2.1 Insulate Structure Against Wind Forces

This method of control may be used when appropriate. For example, ray domes, among other things, protect the communication antennas from wind induced deformations, and thus ensure uninterrupted communications. The presence of an independent spherical dome around an antenna structure makes $\mathbf{g} = \mathbf{o}$ for the antenna structure, and hence $\mathbf{u} = \mathbf{o}$, in Eqs.(10.1), thus yielding $\boldsymbol{\zeta} = \mathbf{o}$ at all times. However, such solutions may not be appropriate for large structures.

10.2.2 Determine Controls by Measuring Nodal Wind Forces

The non-autonomous problem in Eq.(10.1) would become autonomous if one could find $\mathbf{u}(t)$ such that

$$\grave{\mathbf{B}}\mathbf{u}(t)+\mathbf{g}(t)= \mathbf{o} \tag{10.6}$$

for all times. This would mean $\boldsymbol{\zeta} = \mathbf{o}$ at all times since $\boldsymbol{\zeta}_o = \mathbf{o}$ [see Eq.(10.1)]. Using the definitions of $\grave{\mathbf{B}}$ and $\mathbf{g}$ from Eqs.(10.3 and 10.4), this may be reduced to

$$\mathbf{B}\vec{\mathbf{K}}\mathbf{I}_\beta\mathbf{u}(t) = \mathbf{p}(t) \tag{10.7}$$

from where one may attempt to solve for $\mathbf{u}(t)$. Unfortunately, since in general

$$m \geq n > q \tag{10.8}$$

it may not be possible to solve $\mathbf{u}(t)$ from Eq.(10.7).

In the case of statically determinate structures, if one is allowed to have as many actuators as the number of element forces, one has

$$m = n = q \tag{10.9}$$

In this case $\mathbf{I}_\beta = \mathbf{I}$, and from Eq.(10.7) one may obtain

$$\mathbf{u}(t) = \vec{\mathbf{K}}^{-1}\mathbf{B}^{-1}\mathbf{p}(t) \tag{10.10}$$

since $\vec{\mathbf{K}}^{-1} \in \mathrm{R}^{n\times n}$ and $\mathbf{B}^{-1} \in \mathrm{R}^{n\times n}$ both exist (see Section 2.4 and Subsection 2.2.2). The conclusion that may be deduced from this equation is that by appropriate deformation insertion in structural elements, one may keep the nodes stationary, by measuring the nodal wind forces in real time, and then computing controls from Eq.(10.10) and implementing them without delay. Of course, with present day technology, the realization of this idea is remote even for statically determinate earthbound structures.

10.2.3 Determine Controls by Measuring the State

Thanks to modern technology, instead of measuring nodal loads $\mathbf{p}(t)$, one may measure state $\boldsymbol{\zeta}(t)$ in real time[3] [i.e., the deviation of nodes from at-rest state (say, about 50 times a second)], and using an appropriate feedback control law, one may compute controls $\mathbf{u}(t)$ and implement them by means of the actuators (again, say, about 50 times a second) in order to keep the nodes at immediate proximity of at-rest state $\boldsymbol{\zeta}_o = \mathbf{o}$. Since we want to control the nodal motion at the same time when the wind is blowing, this is a non-autonomous feedback control problem which is studied in the previous chapter.

According to the discussions of Section 9.4, optimal state or output feedback control, using performance functional J from Eq.(8.25) or Eq.(8.53), appears not useful, since *a priori* knowledge of $\mathbf{p}(t)$ for $0 \le t \le T$ is not available. Non-optimal feedback control possibilities are briefly discussed in Section 9.5 where, in Eq.(9.31), a linear control law of $\mathbf{u} = \mathbf{G}\boldsymbol{\zeta} + \mathbf{h}$ with an admissible gain matrix $\mathbf{G}$ and some guessed vector $\mathbf{h}$ to compensate the ongoing excitations is suggested.

If one uses a control time T which is sufficiently small relative to the time it takes for a structure to reach a predetermined energy level that triggers the controls, then the presence of nodal wind forces may be ignored during the controls. Such an assumption enables one to use $\mathbf{h} = \mathbf{o}$ in the control law, and treat the control problem as an autonomous one. This is explained in the next section.

---

[3]Since adaptive structures are *observable* (see Section 8.6), one measures only a few components of the state continuously. Using that data, one may estimate the state.

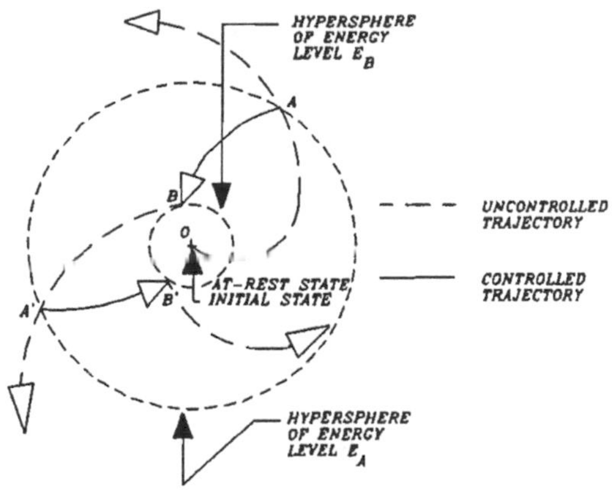

FIGURE 10.1. Sketch of controlled and uncontrolled nodal trajectories in state space for wind induced vibration control using intermittent control episodes

10.3 Excitation Power

In Fig.(10.1) hypothetical uncontrolled and controlled nodal trajectories of an adaptive structure under wind loading are sketched in state space with broken and solid lines, respectively. In the uncontrolled case the nodal trajectory starts from the zero energy point O, and gradually increases its energy. The controlled trajectory starts at point A (on hypersphere with energy level E_A) where the controls are turned on. From this point the controlled trajectory sharply loses energy up to point B (on hypersphere with energy level E_B). At point B the controls are turned off and the trajectory resumes to increase its energy gradually due to wind loads until it reaches point A' on the hypersphere with energy E_A that triggers the controls. The energy level E_A is a design parameter indicating the upper bound of tolerable vibrations. Likewise E_B is a design parameter representing the vibrational energy of *almost at-rest* states. The difference between these two bounds, ΔE, may be expressed as

$$\Delta E = |E_A - E_B| \tag{10.11}$$

Let t_A, t_B, and $t_{A'}$ denote the times when the controlled trajectory is at points A, B, and A', respectively. Control duration T of one control episode may be expressed as

$$T = t_B - t_A \tag{10.12}$$

and the succeeding uncontrolled wind loading duration $\hat{T}$ is

$$\hat{T} = t_{A'} - t_B \tag{10.13}$$

If the magnitude of the average energy dissipation rate during a control episode is at least one order of magnitude larger than the magnitude of the average energy build-up rate during uncontrolled wind loading, i.e., if

$$\frac{\Delta E}{T} >> \frac{\Delta E}{\hat{T}} \tag{10.14}$$

or if

$$\hat{T} >> T \tag{10.15}$$

then one may ignore the energy build-up by the wind loading during a control episode. One may interpret Eq.(10.14) that the power of actuators is much larger than the power of excitations. This means that the non-autonomous control problem of *control in the presence of wind excitations* may be approximated by the autonomous control problem of *control by ignoring the wind excitations.* In this chapter it is assumed that power of actuators is much larger than power of excitations.

10.4 What to Control

Problems discussed in this chapter and the next deal with excitations that cannot be known *a priori.* By past observations, very rough upper bound estimates of some of their attributes may be available. As opposed to this, the dynamic nodal response of a structure with linear excitation-response relations materializes at known discrete frequencies, i.e., natural vibration frequencies, with known patterns, i.e., free vibration mode shapes (see Eq. 7.30 or 7.107 and the related discussions in Chapter 7). Depending on the purpose of structure and the *frequency content* of the expected excitations,[4] many of the mode shapes need not be controlled; only the ones that interfere with the integrity and the functionality of the structure should be controlled.

Wind records indicate that the frequency content of wind loads lies in the frequency range of $\omega_L \cong 0.0$ Hz and $\omega_U \cong 5$ Hz. Since the excitation-response relations of the structure are linear, the integrity of the structure may require the control of the vibration modes with frequencies falling in this range. In tall tower-like structures under wind excitations this means the control of only a few mode shapes. For example, for the tall planar

[4]This may be obtained from the "power spectral density versus frequency" graphs of wind loads. They are generated from the time records of wind loads by Fourier transforms. For natural excitations, graphs in frequency versus power reference frames start from the origin, climb to some maximum, and curve down to zero. According to Parceval's theorem, the area under the graph is the root-mean-square of the excitations in time domain; therefore it is always finite. The frequency interval between a lower bound frequency ω_L and an upper bound frequency ω_U contains most of the graph, and the interval $< \omega_L, \omega_U >$ is called the "frequency content of excitations." For further information, the reader may refer to *Dynamics of Structures* by Walter C. Hurty and Moshe F. Rubinstein, Prentice-Hall, Inc., Englewood Cliffs, NJ, 1964.

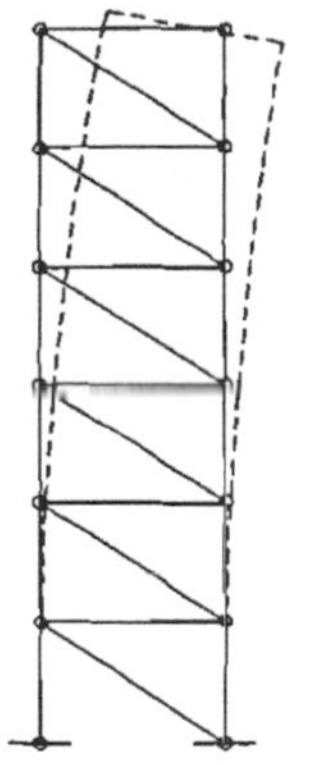

FIGURE 10.2. A statically determinate planar truss tower structure and sketch of its fundamental mode shape

truss tower structure shown in Fig. (10.2), the fundamental vibration mode shape, which is associated with the lowest natural vibration frequency, may be the only one that falls in the frequency range. The functionality of structure may also require the control of the same mode, since that is the one that may cause sea sickness at the top floor.

Let p denote the number of mode shapes to be controlled. One may obtain p as the count of natural vibration frequencies falling in the frequency range of $< \omega_L, \omega_U >$.[5] In order to apply a feedback control technique one has to observe p number of participation factors of the vibration mode shapes, $\mathbf{x}_i$, $i = 1, \cdots, p$. With the notation of Section 7.2, these are η_i, $i = 1, \cdots, p$. Alternately, with the state space notation of Subsection 7.8.1, one may observe $2p$ number of participation factors

$$\hat{\eta}_i \quad i = 1, \cdots, 2p \tag{10.16}$$

of eigenvectors

$$\mathbf{z}_i \quad i = 1, \cdots, 2p \tag{10.17}$$

of matrix $\mathbf{A} \in C^{2n \times 2n}$ of Eq.(10.1). Note that the $2p$ eigenvectors of $\mathbf{A}$ are p pairs of complex conjugate vectors, and these are related to the vibration mode shapes as discussed in Section 7.8.1.

Since instead of complete state $\boldsymbol{\zeta} \in R^{2n \times 1}$ some of its attributes $\mathbf{y} \in R^{2p \times 1}$ are observed, one needs matrix $\hat{\mathbf{C}}$ in order to relate $\boldsymbol{\zeta}$ to $\mathbf{y}$ as in Eq.(8.50),

[5] One may obtain p from [(number of negative diagonal elements of matrix $\mathbf{D}_U$) minus (number of negative diagonal elements of matrix $\mathbf{D}_L$)] where $\mathbf{D}_U$ and $\mathbf{D}_L$ are the nth order diagonal matrices appearing in the $U^T DU$ factorizations of $[\mathbf{K}' - \omega_U^2 \mathbf{M}]$ and $[\mathbf{K}' - \omega_L^2 \mathbf{M}]$, respectively. For details, the reader may refer, for example, to the technical note "Sturm Sequences or Law of Inertia of Quadratic Forms?" by Senol Utku, vol.3, pp.419-420, *Computers and Structures*, 1973.

if one is to use the optimal output feedback control explained in Section 8.6. In the present case, the observed attributes are displayed in Eq.(10.16). Assuming that the sensor output consists of the participation factors listed in Eq.(10.16) (a good challenge for sensor industry) and with the help of Eq.(7.107), one may write

$$\mathbf{y} = \mathbf{I}_\alpha^T \mathbf{Z}^{-1} \boldsymbol{\zeta} \tag{10.18}$$

The binary matrix $\mathbf{I}_\alpha \in (\mathrm{Int})^{2n\times 2p}$ is defined as

$$\mathbf{I}_\alpha = [\mathbf{i}_1, \mathbf{i}_2, \cdots, \mathbf{i}_{2p-1}, \mathbf{i}_{2p}] \tag{10.19}$$

where $\mathbf{i}_k$ is the kth column of $2n$th order identity matrix. Matrix $\mathbf{I}_\alpha$ is the *observed component identification matrix* (see Section 3.5). Comparing Eq.(10.18) with Eq.(8.50) one may obtain

$$\hat{\mathbf{C}} = \mathbf{I}_\alpha^T \mathbf{Z}^{-1} \tag{10.20}$$

With this preparation, one may use any of the autonomous feedback control techniques discussed in Chapter 8, provided that the number and the locations of the actuators, i.e., the quantity q and the actuator placement matrix $\mathbf{I}_\beta \in (\mathrm{Int})^{2n\times 2q}$, are available. This is discussed in the next section.

10.5 Actuator Placement

In principle, actuators are placed on places where they are most effective in fulfilling their purpose. If the control goal is to dissipate the p number of lowest vibration mode shapes, then for each mode shape, they should be placed in those structural elements where they are challenged most, as discussed in Section 8.9.

For the statically determinate planar truss tower shown in Fig. (10.2), suppose $p = 1$ and the goal of control is to dissipate vibrations associated with the fundamental mode $\mathbf{x}_1$ which is sketched in the figure by broken lines. Using Eq.(8.115), one may compute bar forces $\grave{\mathbf{s}}_1$ caused by $\mathbf{x}_1$. Ordering the entries of $\grave{\mathbf{s}}_1$ in decreasing magnitudes, one may observe that the chord elements closer to the supports are stressed more than the others. Suppose $q = 2$, then the locations of the two actuators are the two chord elements next to the supports, as shown in Fig. (10.3a). Unfortunately, these are the members that carry a large preexisting stress due to the dead load of the structure.

The adverse effect of preexisting element forces on the energy needs of the actuators is discussed in Section 3.8 and 8.2. From this point of view, the actuator placement in Fig. (10.3a) is not acceptable. Noting that the diagonal member in the ground floor has no preexisting element forces due to dead load, it is an ideal place for one actuator. Considering that the actuator induced deformations can be positive or negative, one may use two

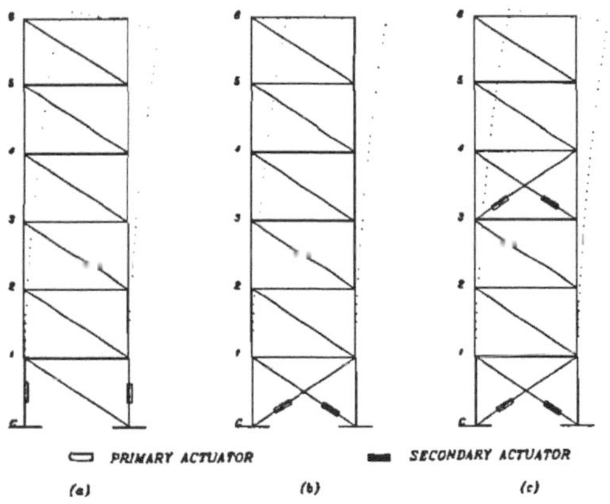

FIGURE 10.3. Actuator placement for controlling the first vibration mode of a planar truss tower structure

diagonals in the ground floor in order not to limit the capacity of actuators by buckling considerations. Then one may place the two actuators on the ground floor diagonals as shown in Fig. (10.3*b*). Unfortunately the additional diagonal makes the planar truss tower statically indeterminate to the first degree. As discussed in Sections 5.4 and 5.5, in order to satisfy compatibility [see Eq.(8.13)], one may use one of the two actuators as a secondary actuator whose synchronized deformation insertion becomes subservient to that of the primary one in order to satisfy the single compatibility relation in this case. Any of the two may be considered as primary.

According to the discussions of Section 10.3, the non-autonomous control problem can be treated as an autonomous one, provided that the power consumed at the actuators is much larger than the power input of the excitations. If one wishes to treat the present control problem as an autonomous control problem, one may have to increase the power consumed by the actuators. One way of doing this is to increase the number of actuators. In Fig. (10.3*c*) a placement of actuators when $q = 4$ is sketched and primary and secondary actuators are identified.

If the statically determinate planar truss tower structure shown in Fig. (10.2) were a frame structure as shown in Fig. (10.4*a*), one might be tempted to place two actuators in place of the two dampers shown in (*a*). This solution would have two major drawbacks:

1. During insertion of positive element deformations, actuators will have to work against the large preexisting column forces due to the deadweight of structure.

2. Since the connection between the floors is statically indeterminate, the actuator's deformation insertions will be resisted by the structure.

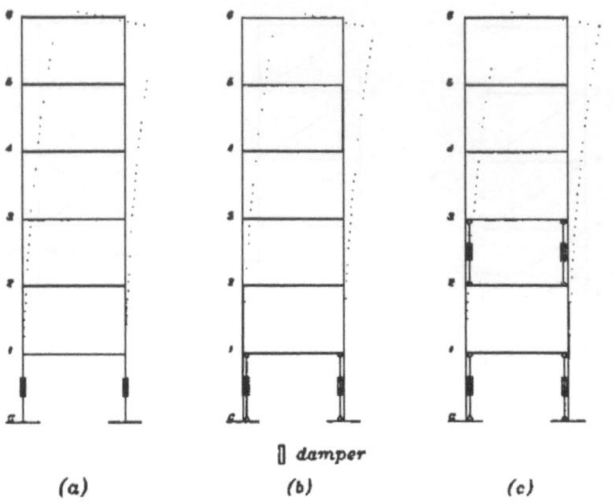

FIGURE 10.4. Passive damper placement for damping the first vibration mode of a planar frame tower structure

The solution shown in Fig. (10.4*b*) eliminates the first drawback if the elements carrying the actuators are retrofitted. The actuator carrying members will not have any of the large column forces. However, because of the statically indeterminate connection of the floors, the second drawback will persist, i.e., the deformation insertion by the actuator will be resisted by the structure. The elimination of the second drawback requires many synchronized secondary actuators (not shown in the figure).

Instead of active control of the fundamental mode of vibration, if one considers passive control, then as shown in Fig. (10.4) one may use dampers in lieu of actuators. Fig.(10.4*c*) shows the placement of additional dampers to increase the rate of dissipation of the fundamental mode of vibration of the frame tower.

10.6 Control Law

After determining the attributes of state that are to be observed [i.e., after determining matrix $\hat{\mathbf{C}}$ in Eq.(8.50)], and identifying the number and the locations of actuators that satisfy compatibility [i.e., defining matrix $\grave{\mathbf{B}}$ of Eq.(10.1) that ensures Eq.(8.13) by appropriately selected actuator placement matrix $\mathbf{I}_\beta$ (see Sections 5.5 and 5.6)], one has to find a suitable control law that makes it possible to drive the structure from energy level E_A to energy level E_B in control time T as discussed in Section 10.3. The discussions of Chapter 8 may be used as a guide in selecting a control law suitable for the problem. Since the problem is being treated as an autonomous problem, in an output feedback control scheme one may use

$$\mathbf{u} = \mathbf{G}\boldsymbol{\zeta} \tag{10.21}$$

where gain matrix $\mathbf{G} \in R^{q \times 2n}$ is preferably optimal but necessarily admissible. In this case, a state estimator (see Section 8.6) may be needed. In a direct output feedback scheme, one may use

$$\mathbf{u} = \mathbf{G}\mathbf{y} \tag{10.22}$$

where gain matrix $\mathbf{G} \in R^{q \times 2p}$ is preferably optimal but always admissible. The feedback control diagram is similar to the one shown in Fig. (8.5) for the output feedback control and it is similar to the one shown in Fig. (8.6) for the direct output feedback control.

10.7 Recapitulation of Active Control Against Wind

In this chapter active control of wind induced vibrations of adaptive tower structures is discussed. The non-autonomous control problem is reduced to intermittent autonomous controls by assuming that the power dissipated through the actuators is much larger than the power input from wind forces. From the facts that wind excitations have low frequency content and functionality of tower structures requires small amplitude vibrations, it is concluded that the fundamental vibration mode shape and its few neighbors may be controlled. On this basis, the attributes of state that need to be observed are identified (i.e., matrix $\hat{\mathbf{C}}$ is identified). The number and the place of actuators are determined (i.e., matrix $\grave{\mathbf{B}}$ is determined) so as to fulfill the control objectives effectively and by requiring least energy. Chapter 8 discussions are referred to for determination of the control law for the output feedback control that can be optimal and/or direct.

11

Active Control Against Seismic Loads

In the previous chapter, as an example of non-autonomous control, active control of adaptive structures against wind loads was studied. In this chapter another important non-autonomous control problem is studied: active control of adaptive structures against seismic loads. Compared to wind excitations, seismic excitations are generally of much more power. Moreover, the wind excitations are time dependent forces acting on the nodes, whereas seismic excitations are the prescribed and yet totally unknown motions of the supports. In both cases, the structure is in an at-rest state at $t = 0$, and, except for some crude upper bounds of some of the attributes, nothing is known *a priori* about the excitations. In this setting, one would like to keep the nodal trajectory within an acceptable neighborhood of the at-rest state throughout the loading episode, by means of active control.

11.1 State Equations for Seismic Excitations

The state equations of nodal motion due to seismic excitations and control deformations may be stated as in Eq.(7.134):

$$\left.\begin{array}{c} \dot{\boldsymbol{\zeta}} = \mathbf{A}\boldsymbol{\zeta} + \dot{\mathbf{B}}\mathbf{u} + \mathbf{g} \quad \text{for } t > 0 \\ \boldsymbol{\zeta}(0) = \boldsymbol{\zeta}_o = \mathbf{o} \end{array}\right\} \tag{11.1}$$

where

$$\boldsymbol{\zeta}(t) = \left\{ \begin{array}{c} \dot{\boldsymbol{\xi}}(t) \\ \boldsymbol{\xi}(t) \end{array} \right\} \tag{11.2}$$

and, with the help of Eqs.(7.88, 7.89, and 7.90),

$$\mathbf{A} = \begin{bmatrix} -\mathbf{M}^{-1}\mathbf{\grave{C}} & -\mathbf{M}^{-1}\mathbf{K}' \\ \mathbf{I} & \cdot \end{bmatrix} \tag{11.3}$$

$$\mathbf{\grave{B}} = \begin{bmatrix} -\mathbf{M}^{-1}\mathbf{B}\vec{\mathbf{K}}\mathbf{I}_\beta \\ \cdot \end{bmatrix} \tag{11.4}$$

and

$$\mathbf{g}(t) = \begin{Bmatrix} -\mathbf{M}^{-1}\mathbf{B}\vec{\mathbf{K}}\mathbf{B}_s^T\boldsymbol{\xi}^s(t) \\ \cdot \end{Bmatrix} \tag{11.5}$$

In these equations, $\boldsymbol{\xi}(t) \in \mathrm{R}^{n\times 1}$ is independent nodal deflections (see Subsection 2.1.3), $\dot{\boldsymbol{\xi}}(t) \in \mathrm{R}^{n\times 1}$ is independent nodal velocities, $\mathbf{M} \in \mathrm{R}^{n\times n}$ is positive definite mass matrix (see Subsection 6.1.2), $\grave{\mathbf{C}} \in \mathrm{R}^{n\times n}$ is positive definite modal damping matrix (see Section 7.3), $\mathbf{K}' \in \mathrm{R}^{n\times n}$ is positive definite global stiffness matrix of the structure (see Section 3.1.1), $\mathbf{I}_\beta \in (\mathrm{Int})^{m\times q}$ is the actuator placement matrix (see Sections 3.4, 3.9, 4.3, 5.3, and 5.4), $\mathbf{u}(t) \in \mathrm{R}^{q\times 1}$ is the actuator induced element deformation representing controls (see Sections 3.4, 4.3, and 5.6), $\vec{\mathbf{K}} \in \mathrm{R}^{m\times m}$ is positive definite block diagonal matrix of element stiffnesses (see Section 2.4), $\mathbf{B} \in \mathrm{R}^{n\times m}$ is the matrix transforming description of element forces into global descriptions of internal nodal forces (see Subsection 2.2.2), and $\mathbf{B}_s^T \in \mathrm{R}^{m\times b}$ is matrix relating support motions $\boldsymbol{\xi}^s(t) \in \mathrm{R}^{b\times 1}$ to element deformations (see Subsections 2.1.2 and 6.2.1). As discussed in Section 7.1, support motions may be expressed as

$$\boldsymbol{\xi}^s(t) = diag[\hat{\xi}_k^s(t)]\boldsymbol{\xi}^{s\prime} \tag{11.6}$$

where $\boldsymbol{\xi}^{s\prime} \in R^{b\times 1}$ is maximum magnitudes (or estimates of maximum magnitudes) of components of $\boldsymbol{\xi}^s$ and $diag[\hat{\xi}_k(t)] \in \mathrm{R}^{b\times b}$ physically non-dimensional time functions showing time variations of support motion. The solution of Eqs.(11.1), i.e., the nodal trajectory, could be expressed as in Eq.(7.135) if one knew support motions $\boldsymbol{\xi}^s(t)$ and controls $\mathbf{u}(t)$. Unfortunately in the real-time control of nodal motions, one does not know $\boldsymbol{\xi}^s(t)$ *a priori*.

11.2 Control Possibilities of Nodal Motion

One may observe from the right-hand side of Eqs.(11.1) that the motion is caused by the existence of the $\grave{\mathbf{B}}\mathbf{u} + \mathbf{g}$ term where $\mathbf{g}$ represents support motion, and $\mathbf{u}$ represents the controls to eliminate the motion caused by $\mathbf{g}$. Since the objective is to ensure that $\boldsymbol{\zeta} \approx \mathbf{o}$ at all times, the following possibilities exist.

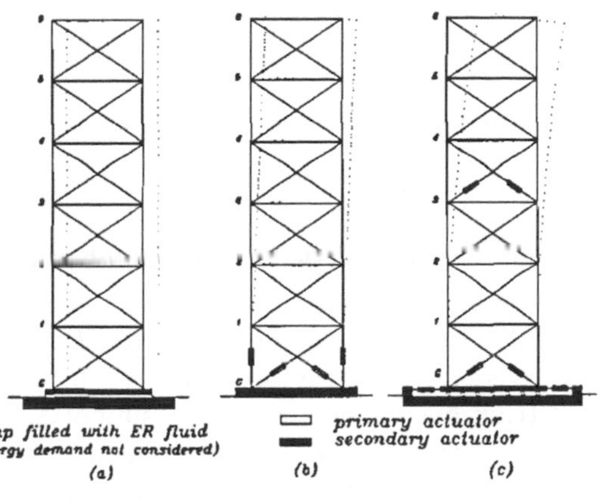

FIGURE 11.1. Possibilities of controlling nodal motion due to seismic excitations

11.2.1 Insulate Structure Against Support Motions

Although insulating a structure against wind requires another structure of a ray dome type, it is a much more realizable proposition to insulate a structure against the support motions, at least in the horizontal direction, during the seismic events. Electro-rheological (ER) materials[1] appear to have great potential in many applications in addition to their potential in transforming instantaneously fixed supports into roller supports practically with no friction [see Fig. (11.1*a*)]. When **g** in Eq.(11.1) represents horizontal ground motion, such a facility may be used to cut the transmission of ground motions to the structure. The present day realizable version of this idea is the passive base isolation pads. Due to their low stiffness and high damping characteristics, the pads decrease effectively power input to the structure from horizontal seismic excitations. They are being used widely in areas of high seismic activity when it is technically and economically feasible.

11.2.2 Determine Controls by Sensing Support Motion

The non-autonomous problem in Eq.(11.1) would become autonomous if one could find $\mathbf{u}(t)$ such that

$$\grave{\mathbf{B}}\mathbf{u}(t)+\mathbf{g}(t)=\mathbf{o} \tag{11.7}$$

for all times. This would mean $\boldsymbol{\zeta} = \mathbf{o}$ at all times since $\boldsymbol{\zeta}_o = \mathbf{o}$ [see Eq.(11.1)]. Using the definitions of $\grave{\mathbf{B}}$ and $\mathbf{g}$ from Eqs.(11.4 and 11.5), Eq.(11.1) may

[1] See, for example, "ER Material Models and Vibration Control," Henri P. Gavin, *Proceedings of 11th Symposium on Structural Dynamics and Control*, May 12-14, 1997, VPI&SU, Blacksburg, VA.

be reduced to

$$\mathbf{I}_\beta \mathbf{u}(t) = \mathbf{B}_s^T \boldsymbol{\xi}^s(t) \tag{11.8}$$

From this equation one may solve $\mathbf{u}(t)$ as

$$\mathbf{u}(t) = \mathbf{I}_\beta^T \mathbf{B}_s^T \boldsymbol{\xi}^s(t) \tag{11.9}$$

since $\mathbf{I}_\beta^T \mathbf{I}_\beta = \mathbf{I} \in (\text{Int})^{q \times q}$. Let q' denote the number of nonzero rows of $\mathbf{B}_s^T \in \text{R}^{m \times b}$ with labels $\beta_1, \beta_2, \cdots \beta_{q'}$. As explained in Subsection 6.2.1, this information is available from the way one generates $\mathbf{B}_s^T \in \text{R}^{m \times b}$. For example, if the structure were a truss structure as shown in Fig. (11.1b), q' would be the number of truss bar connected to the supports, and $\beta_1, \beta_2, \cdots \beta_{q'}$ would be the labels of these bars. Hence if one chooses the number of actuators q to be equal to q', and the actuator placement matrix $\mathbf{I}_\beta$ as

$$\mathbf{I}_\beta = [\mathbf{i}_{\beta_1}, \mathbf{i}_{\beta_2}, \cdots, \mathbf{i}_{\beta_{q'}}] \tag{11.10}$$

where $\mathbf{i}_k$, $k = \beta_1, \beta_2, \cdots \beta_{q'}$, is the kth column of the mth order identity matrix, then one may see from Eq.(11.9) that one may determine the controls $\mathbf{u}(t)$ by the real-time measurements of support motions $\boldsymbol{\xi}^s(t)$. This is similar to our efforts to remain upright during ground motion by adjusting our leg and foot muscles. The conclusion that may be deduced from Eq.(11.9) and Eq.(11.1) is that by appropriate deformation insertion in the structural elements that connect the structure to the supports, one may keep the nodes stationary, i.e., $\boldsymbol{\zeta} = \mathbf{o}$. This requires the sensing of support motion $\boldsymbol{\xi}^s(t)$, and then determining the controls $\mathbf{u}(t)$ by Eq.(11.9) and implementing them in real time without any delay.

11.2.3 Determine Controls by Sensing the State

Thanks to modern technology, one may measure in real time the state[2] $\boldsymbol{\zeta}(t)$ [i.e., the deviation of nodes from at-rest state (say, about 50 times a second)], and using an appropriate feedback control law, one may compute controls $\mathbf{u}(t)$ and implement them by means of actuators (again, say, about 50 times a second) in order to keep the nodes in immediate proximity of at-rest state, without ever sensing $\boldsymbol{\xi}^s(t)$. Since we want to control the nodal motion at the same time when the ground is moving, this is a non-autonomous feedback control problem which is studied in Chapter 9.

According to the discussions of Section 9.4, optimal state or output feedback control, using performance functional J from Eq.(8.25) or Eq.(8.53), appears not useful since *a priori* knowledge of $\boldsymbol{\xi}^s(t)$ for $0 \le t \le T$ is not available. Non-optimal feedback control possibilities are briefly discussed

[2] As mentioned before, since adaptive structures are *observable* (see Section 8.6), one senses only a few components of the state continuously. Using that data, one may estimate the state.

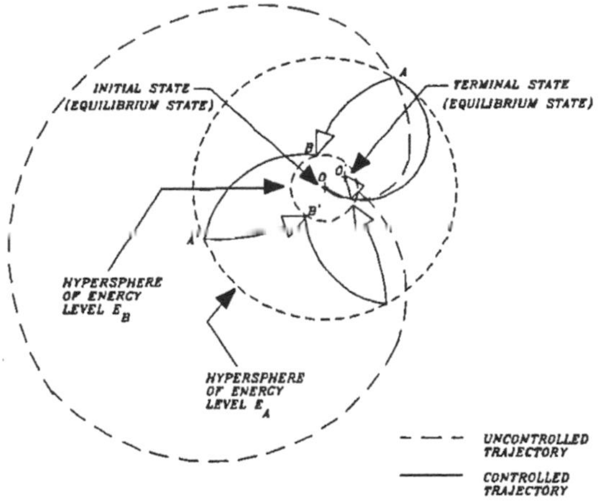

FIGURE 11.2. Controlled and uncontrolled nodal trajectories in state space for active base isolated adaptive structure under seismic excitations

in Section 9.5 where, in Eq.(9.31), a linear control law of $\mathbf{u} = \mathbf{G}\boldsymbol{\zeta} + \mathbf{h}$ with an admissible gain matrix $\mathbf{G}$ and some guessed vector $\mathbf{h}$ to compensate the on-going excitations is suggested. In the seismic excitations case, it is very difficult to come up with an acceptable vector $\mathbf{h}$.

If one uses an active base isolation system, such as the one outlined in the previous subsection and sketched in Fig. (11.1c), the power input to structure from ground motion may be reduced to low levels comparable to that of wind excitations. If this can be achieved, then the low power inputs can be handled with the autonomous feedback control methods explained in the previous chapter.

11.3 Managing Excitation Power

In Fig.(11.2) hypothetical uncontrolled and controlled nodal trajectories of an adaptive structure under seismic loading are sketched in state space with broken and solid lines, respectively.

In the uncontrolled case the nodal trajectory starts from the at-rest equilibrium state of zero energy point O, and after passing from a maximum energy level, it exhausts its energy, possibly by destroying the integrity of the structure on the way, and arrives at the terminal state of O' soon after the seismic episode ends. This is shown by the broken line in the figure. As mentioned in Section 9.3, if there is a permanent ground dislocation due to slips along the fault lines, then points O and O' are two different points. However, if the seismic activity does not involve any ground dislocation, then points O and O' overlap. For the integrity of structure, the seismic energy transmitted to a structure through foundations should not exceed an allowable energy limit E_A. Depending on the type of structure, by means of passive, active, or hybrid base isolation systems, the seismic energy transmitted to the structure through the foundations can be controlled in many structures. The allowable energy limit E_A is usually much higher than the energy limit E_B representing an upper bound for the serviceability of structure. Hence one may need two independent active control systems: one, an active base isolation system, to keep the energy input to the structure from exceeding E_A, and one, an autonomous output feedback control system, to keep the energy level most of the time much less than E_A, as sketched in Fig. (11.1*c*).

Assuming two independent control systems as shown in Fig. (11.1*c*), the controlled trajectory in Fig. (11.2) starts at point O and climbs to point A on hypersphere with energy level E_A, and, thanks to the active base isolation system, with a much longer path than the uncontrolled case. At point A the second control system, i.e., an autonomous output feedback control system, is turned on. This second system is similar to the one discussed in the previous chapter for wind excitations, and assumes that the base isolation system can retain the power input to the structure at very low levels. Because of this non-autonomous second control system, the controlled trajectory sharply loses energy up to point B on the hypersphere with energy level E_B. At point B the controls are turned off and the trajectory resumes to increase its energy gradually, due to energy seeping through the base isolation system, until it reaches point A' on the hypersphere with energy E_A which triggers the non-autonomous second controls again. These intermittent control episodes continue until the vibrational energy drops to levels E_B or less permanently.

The allowable energy level E_A is a design parameter indicating the upper bound of vibrational energy for the structure to remain undamaged. Likewise E_B is a design parameter representing the upper bound vibrational energy for a serviceable environment within the structure. For the characteristics of non-autonomous control system operating intermittently, one may refer to the discussions of Chapter 10.

After a brief overview of passive base isolation systems in the next section, the remainder of this chapter discusses only the active base isolation systems.

11.4 Passive Base Isolation Systems

In isolating structures against horizontal seismic excitations, seismic base isolation pads are widely used. In areas of high seismic activity, when it is feasible, many older building structures are retrofitted with these pads.

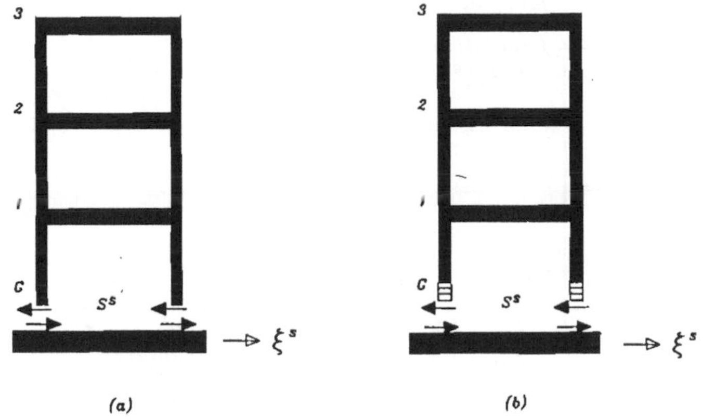

FIGURE 11.3. Foundation level shear forces of a planar frame structure (with and without passive base isolation pads) under horizontal seismic ground movement

11.4.1 Justification

Compared to concrete or steel, the pads are very flexible and with high damping characteristics. They transmit horizontal seismic ground motions with reduced magnitude and delay. They have high damping characteristics. They can be retrofitted to many existing structures, and they are renewable.

In Fig.(11.3) free body diagrams of a planar frame structure under seismic excitations are shown: in (*a*) without pads, and in (*b*) with pads. The energy E transmitted to the structure from the seismic ground motion $\boldsymbol{\xi}^s(t) \in R^{b\times 1}$ is the work done by the element force resultant $\mathbf{s}^s(t) \in R^{b\times 1}$ under $\boldsymbol{\xi}^s$ during the seismic episode, namely,

$$E = \int_0^T \dot{\boldsymbol{\xi}}^s(t)^T \mathbf{s}^s(t)dt \tag{11.11}$$

where $\boldsymbol{\xi}^s$ and $\mathbf{s}^s$ are as shown in the figure. One may express the element force resultants $\mathbf{s}^s$ as

$$\mathbf{s}^s(t) = \mathbf{K}^s \Delta\boldsymbol{\xi}^s(t) \tag{11.12}$$

where $\mathbf{K}^s \in R^{b\times b}$ is the stiffness of structural elements that are connected to the foundation, in the directions of the components of $\boldsymbol{\xi}^s$, and

$$\Delta\boldsymbol{\xi}^s(t) = \boldsymbol{\xi}^s(t) - \boldsymbol{\xi}^{sf}(t) \tag{11.13}$$

where $\boldsymbol{\xi}^{sf}(t)$ represents the movements of the far ends of the elements due to the motion of structure. Using $\mathbf{s}^s(t)$ from Eq.(11.12) in Eq.(11.11), one may write

$$E = \int_0^T \dot{\boldsymbol{\xi}}^s(t)^T \mathbf{K}^s \Delta\boldsymbol{\xi}^s(t)dt \tag{11.14}$$

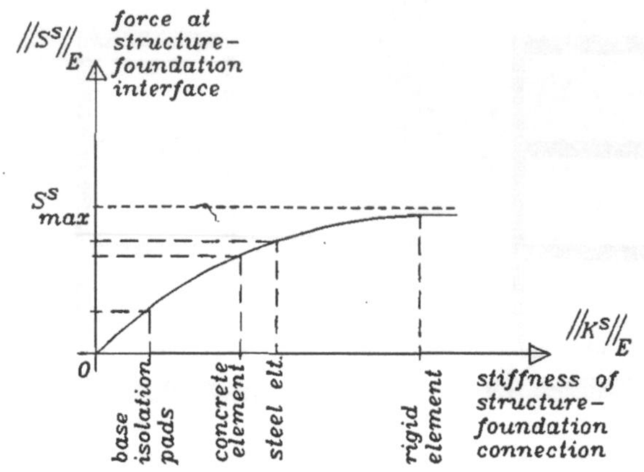

FIGURE 11.4. Shear force at structure-foundation interface as a function of shear stiffness of structure-foundation connection

From these equations one may observe the following:

1. If $\mathbf{K}^s = \mathbf{0}$, then no energy is transmitted to the structure through the supports, as if the structure is on roller supports with zero friction and subjected to horizontal seismic ground motion. In this case, there are no horizontal interface forces at the cross-section over the foundation. When the foundation moves, the structure will remain stationary.

2. If $\mathbf{K}^s = \mathbf{K}^s_p$ ($\mathbf{K}^s_p$ representing the stiffness of the base isolation pads which are of very low shear rigidity), then there is some force $\mathbf{s}^s(t)$ at the interface with the foundation; hence, some energy will be transmitted to the structure. When the foundation moves, the structure will follow it reluctantly.

3. If $\mathbf{K}^s = \mathbf{K}^s_c$ ($\mathbf{K}^s_c$ representing the stiffness of steel or concrete structural elements, connecting the structure to the supports on the foundation, in direction of the components of $\boldsymbol{\xi}^s$), then there will be substantial force $\mathbf{s}^s(t)$ at the interface with the foundation; hence, considerable energy will be transmitted to the structure. When the foundation moves, the structure will follow it rigorously.

4. If the structure were made of extremely stiff material (say, of material mined from a neutrino star), then $\mathbf{K}^s$ would be infinitely large, and $\Delta\boldsymbol{\xi}^s$ would be infinitely small, producing a finite product that would be much larger than that of the previous case. In this case when the foundation moves, the structure would duplicate it.

In Fig. (11.4), $\|\mathbf{s}^s\|_E$, i.e., Euclidean norm (over components and time) of interface forces $\mathbf{s}^s(t)$, is sketched against $\|\mathbf{K}^s\|_E$, i.e., Euclidean norm of

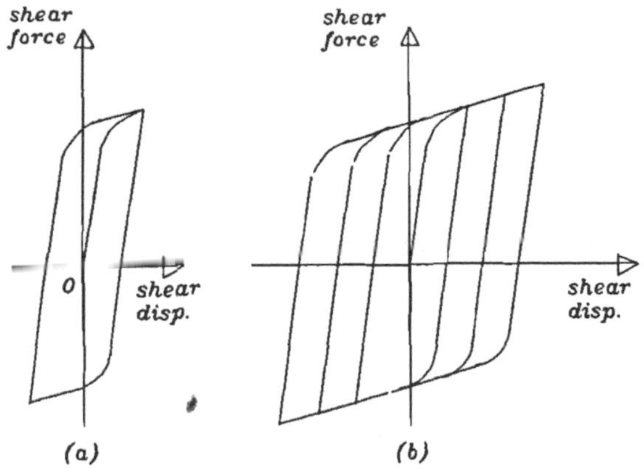

FIGURE 11.5. Typical shear force versus shear displacement curves of a base isolation pad

stiffness $\mathbf{K}^s$. The curve for $\|\mathbf{s}^s\|_E$ starts from the origin, increases monotonically with increasing stiffness of *structure-to-foundation connection*, and approaches to its largest value asymptotically. It follows then, that the motion of structure above foundation decreases its intensity with decreasing stiffness of structure-foundation connection. As stated earlier, for horizontal seismic ground motions, when it is economically and technically feasible, passive seismic base isolation pads with low shear rigidity are preferred to rigid connections.

11.4.2 Analysis

The passive base isolation pads, in addition to their low shear and high axial stiffness, have excellent damping characteristics against shear inducing motions. They are rubber blocks with steel reinforcement. During loading-unloading in shear, they imitate the elastic-rigid plastic behavior, with a large hysteretic loop.[3] A typical shear force versus shear deformation curve of an isolation pad is sketched in Fig. (11.5): for a slightly damaged pad in (*a*), and for an extensively damaged pad in (*b*). Slopes of the curve are proportional to the pad's shearing stiffness, and the area of the hysteretic loops are proportional to the pad's damping constants. Since both are varying with the deformation level, element force versus element deformation relations for the pads are nonlinear.

[0]See, for example, *Seismic Isolation and Response Control for Nuclear and Non-Nuclear Structures*, Takafumi Fujita (ed), Institute of Industrial Science, University of Tokyo, for the 11th International Conference on Structural Mechanics in Reactor Technology (SMIRT 11), Aug. 1991.

For analysis of the passively base isolated structure in the absence of controls, one may rewrite the state equations of nodal motion and initial conditions given in Eqs.(11.1) as

$$\left.\begin{array}{l} \dot{\boldsymbol{\zeta}} = [\mathbf{A}(\boldsymbol{\zeta})]\boldsymbol{\zeta} + \mathbf{g} \\ \boldsymbol{\zeta}(0) = \mathbf{o} \end{array}\right\} \tag{11.15}$$

noting that $\mathbf{u} = \mathbf{o}$ and, from Eq.(11.3),

$$\mathbf{A} = \mathbf{A}(\boldsymbol{\zeta}) = \begin{bmatrix} -\mathbf{M}^{-1}\dot{\mathbf{C}}(\boldsymbol{\zeta}) & -\mathbf{M}^{-1}\mathbf{K}'(\boldsymbol{\zeta}) \\ \mathbf{I} & \cdot \end{bmatrix}$$

where $\boldsymbol{\zeta} \in \mathrm{R}^{2n\times 1}$is the state vector as defined in Eq.(11.2). The nonlinear initial value problem defined by these equations can be integrated numerically only after one knows nonlinear stiffness and damping characteristics of the pads and the histogram of seismic excitations $\mathbf{g}(\boldsymbol{\zeta},t)$. With the help of Eq.(11.5) one may write:

$$\mathbf{g}(\boldsymbol{\zeta},t) = \left\{ \begin{array}{c} -\mathbf{M}^{-1}\mathbf{B}\vec{\mathbf{K}}(\boldsymbol{\zeta})\boldsymbol{\xi}^{s}(t) \\ \cdot \end{array} \right\} \tag{11.16}$$

where seismic ground motion $\boldsymbol{\xi}^s(t) \in \mathrm{R}^{b\times 1}$ must be guessed.

Since n is usually very large [see Eq.(1.5)], the numerical integration of Eq.(11.15) is a very costly process even for one specific case of guessed excitations $\boldsymbol{\xi}^s(t)$. However, this difficulty can be overcome by decreasing the accuracy of analysis.

Since the stiffness of base isolation pads is much less than the stiffness of other members of structure, linear eigenvalue analyses (see Sections 7.2 and 7.3) using various constant pad stiffnesses within the range may show that, due to isolation pads, the structure acquires new low free vibration frequencies associated with mode shapes that resemble rigid body movements representing translation in horizontal direction, bending rotation about the horizontal axis, and/or torsion rotation about the vertical axis. In other words, for a less accurate analysis which is compatible with the guessing involved for $\boldsymbol{\xi}^s(t)$, one may consider the part of structure above the base isolation pads as a rigid body. Such an approximation reduces the order from $2n$ to a few, and, thus, the response of the base isolated structure to various hypothetical seismic ground motions can be obtained very economically by computer simulation.

The real difficulty in this approach is that we can never know seismic ground motion $\boldsymbol{\xi}^s(t)$ until it happens. The current trend is to use a worst case synthetic seismic ground motion $\boldsymbol{\xi}^s(t)$ that may happen in that locality during the design life of the structure, and check if the structure can survive that synthetic ground motion. Since nature does not like to act according to our prescriptions, more often than not our designs fail. Current design paradigms can benefit from new breakthroughs in sensor, actuator, and microprocessor technologies, i.e., from adaptive structures.

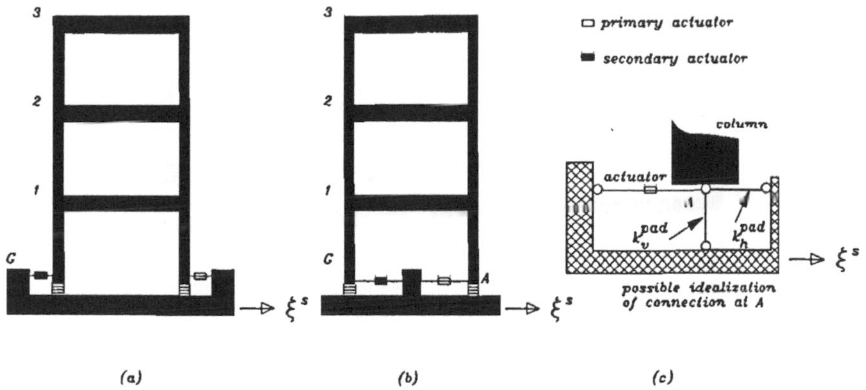

FIGURE 11.6. Actuator placement for active base isolation of a planar frame structure under horizontal seismic ground motion

11.5 Actuator Placement for Active Base Isolation

Active base isolation systems are based on the idea of controlling nodal motion of a structure by sensing in real time seismic ground movements (see Subsection 11.2.2). They can be used in support or in lieu of passive base isolation systems. They can be versatile enough to protect the structure actively against any component of the seismic ground motion.

11.5.1 Planar Frames under Horizontal Ground Motion

In Fig. (11.6*a* and *b*) possible actuator placements for a passively base isolated planar frame structure under horizontal ground motion are shown. They are based on the basic idea depicted in Fig. (11.1*b*). In order to increase actuators' efficiency in preventing the transmission of horizontal ground motion to the structure, and also to protect the actuators from the preexisting element forces due to dead loads, here actuators are placed horizontally and close to the foundation. The compatibility of inserted strains [see Eqs.(8.12 and 8.13) and Sections 5.4 and 5.5] requires that one of the two horizontally placed actuators needs to act as a synchronized secondary actuator such that if the inserted deformation in the primary actuator is u, then the one in the secondary is $-u$ (assuming that the adaptive structure is symmetrical about the vertical axis).

Due to the presence of passive base isolation pads, the actuators' deformation insertion will be resisted by the pads' stiffness in the horizontal direction. In Fig. (11.6*c*) a possible idealization of connection at point A of the frame in Fig. (11.6*b*) is shown. In the figure, axial stiffness of the pad is marked by k_h^{pad} which represents the pad's shearing stiffness. The actuator has to work against this stiffness. In the static case, an actuator

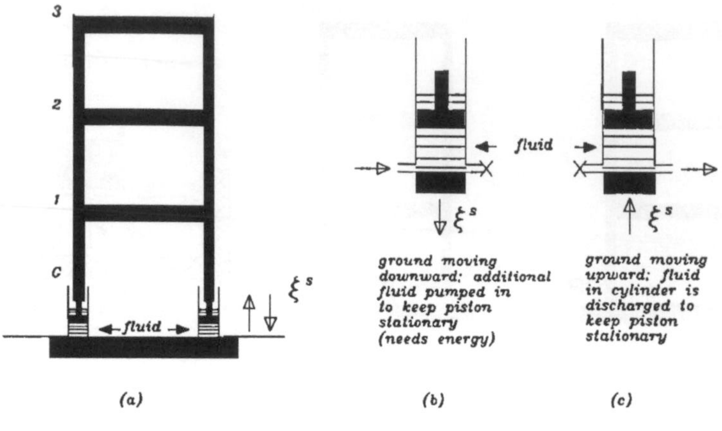

FIGURE 11.7. Actuator placement for active base isolation of a planar frame structure against vertical seismic ground motion

needs energy for overcoming this stiffness. Therefore, passive base isolation pads with smaller shear stiffness are preferable. The actuator stiffness in the vertical direction is marked by k_v^{pad} which is normally at least one order of magnitude larger than k_h^{pad}. As seen from the figure, the vertical stiffness plays no role in the horizontal motion, so long as the deflections are small.

Sensing the horizontal ground motion of the foundation, i.e., $\xi^s(t)$ at time t, one may determine the necessary controls $u(t)$ in order to keep the structure stationary when the ground is moving. This is further discussed for vertical ground motion in the next section.

11.5.2 Planar Frames under Vertical Ground Motion

In Fig.(11.7*a*), placement of piston-cylinder type hydraulic actuators for active base isolation of a planar frame structure against vertical ground motion is sketched. The two actuators need to be synchronized in order to insert the same deformation. This is required not for satisfying compatibility, but for preventing rigid body rotations of the structure [assuming that vertical ground motion $\xi^s(t)$ is uniform].

Since actuators are oriented in the vertical direction, they carry very large preexisting axial forces due to the structure's weight. In order to keep the structure stationary during vertical seismic ground motions, the actuators need to work against these large preexisting forces.

During downward motion of the ground, one needs to pump extra fluid into the cylinder of an hydraulic actuator in order to keep the structure stationary. This is sketched in Fig. (11.7*b*). For this, one may need very large power and energy. If such large power and energy demands cannot be met, then the structure may be allowed to follow the downward motion of the ground, partially.

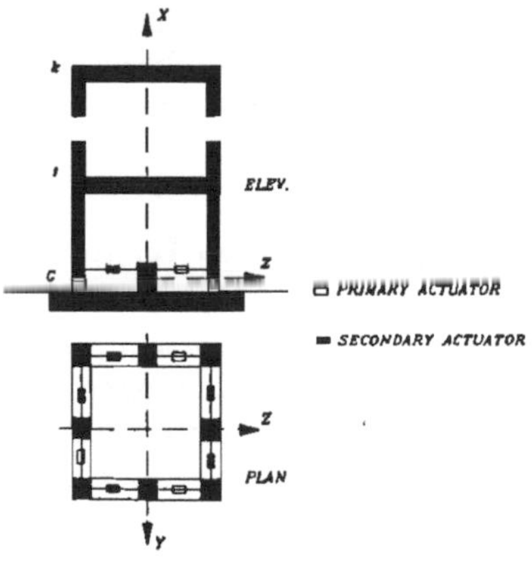

FIGURE 11.8. Actuator placement for active base isolation of a three-dimensional frame structure against seismic ground motions in horizontal plane

During the upward motion of the ground, the fluid in the cylinder is allowed to flow out freely. This is sketched in Fig. (11.7c). In the upward ground motion case, actuators do not need any energy.

By sensing the vertical seismic ground motion, one may compute in real time the necessary control deformations in order to keep the structure stationary.

11.5.3 Space Frames under Horizontal Ground Motion

A possible actuator placement is sketched in Fig. (11.8) for active base isolation of a three-dimensional frame structure against horizontal seismic ground motions. If the ground motion is in the Y direction, then the actuators parallel to the Y direction are activated. On the other hand, if the ground motion is in the Z direction, then the actuators parallel to the Z direction are activated. If the shear center related to the shear forces associated with the passive base isolations pads is not coincident with the centroid of the cross-section at the structure-foundation interface, then the actuators can act to compensate the off-set.

For horizontal excitations, one needs only three linkages (not all parallel) to connect the structure to its foundation in a statically determinate manner. From the actuator placement shown in Fig. (11.8), it may be observed that there are 8 actuators (i.e., $q = 8$); hence there are 8 linkages. This is in addition to the shear connections provided by the passive base isolation pads. Even if one ignores these shear connections, the structure foundation connection in the horizontal plane is statically indeterminate to the degree $q-3 = 8-3 = 5$. In order to ensure compatibility and thus prevent conflicts in the operation of actuators, the control deformations in the 8 actuators need to satisfy geometric compatibility requirements and be synchronized [see Chapter 5]. This can be achieved by using three primary actuators and five synchronized secondary actuators as shown in the figure. The control elongations in the 8 actuators may be obtained by Eq.(11.9).

11.6 Control of Active Base Isolation Systems

In the previous section, the actuator placement problem is discussed for the active base isolation systems for seismic ground motions. Assuming that the control objective is to keep the structure stationary when the ground is moving, actuators are placed on elements that connect the structure to its foundations. These are *connecting active elements.* In the following subsections, control energy, control system, and control law are discussed for active base isolation systems.

11.6.1 Control Energy

In order to minimize the energy needed to run the actuators, the following must be observed:

1. Connecting active elements should not carry any preexisting element forces (see discussions in Section 3.8). In the case of seismic ground motions in the horizontal plane, this could be achieved by creating horizontal connecting elements, as discussed in Subsections 11.5.1 and 11.5.3. In the case of seismic ground motion in the vertical direction, unfortunately, the active connecting elements had to be vertical as discussed in Subsection 11.5.2.

2. Deformation insertion should not be resisted by the structure. This can be achieved by the compatibility of inserted deformations through the use of synchronized secondary actuators (see discussions in Section 3.9 and Chapter 5).

3. If some of the connecting elements are not active elements, then the compatibility equations may never be satisfied. This is the case shown in Figs.(11.6 and 11.8) where passive base isolation pads are non-active connecting elements. When non-active connecting elements are present, primary and secondary actuators may be determined as explained in Chapter 5 by ignoring the non-active connecting elements. This will prevent conflicts in the actions of actuators.

4. Energy required for running the actuators in the presence of non-active connecting elements is proportional to the stiffness of non-active connecting elements; therefore, their stiffness should be as small as possible.

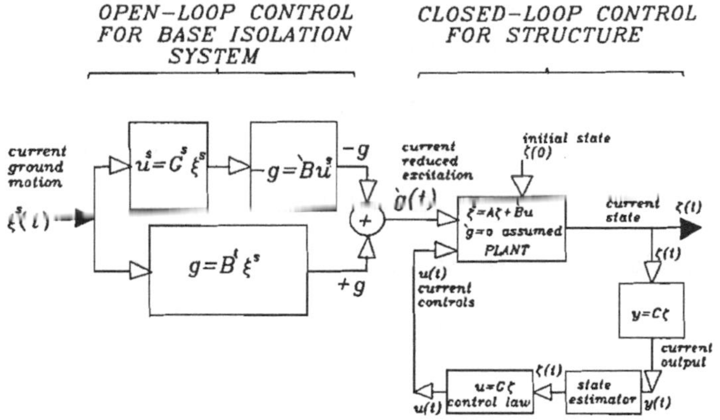

FIGURE 11.9. Open- and closed-loop controls of base isolated adaptive structure against seismic ground motions

With the above observations, one may verify the validity of primary and secondary actuators shown in Figs. (11.6 and 11.8).

11.6.2 Control System

A control system compatible with the nodal trajectories depicted in Fig. (11.2) is shown in Fig. (11.9) for an actively base isolated adaptive structure against seismic ground motions. The control system consists of two parts, an open-loop, one representing a *feedforward control* for the base isolation system, and a closed-loop, one representing a *feedback control* for the structure.[4]

If one ignores the existence of reduced magnitude excitations input, i.e., $\grave{\mathbf{g}}(t) \in \mathrm{R}^{2b\times 1}$, the closed-loop control is an autonomous output feedback control [see Fig. (8.5) and discussions in Section 8.6]. Even if it exists, the reduced magnitude excitation input $\grave{\mathbf{g}}(t)$ is assumed to be a low power one relative to the actuation power of the closed-loop system actuators (see discussions of Sections 10.3 and 11.3).

11.6.3 Control Law

As discussed in Subsection 11.2.2, one may determine the necessary controls with the objective of keeping the structure stationary as the foundations move. Since there are two independent control systems, there are two control vectors. Let $\overset{s}{u} \in R^{q\times 1}$ denote the control vector in the open-loop

[4]See "Active Vibration Control in Buildings with Passive Base Isolation," Purnima Jalihal, Doctoral Dissertation, Department of Civil and Environmental Engineering, Duke University, November 1994.

system of the active base isolation (vector **u** in the figure belongs to the closed-loop control system). Then using Eq.(11.9), one may write

$$\mathbf{u}^s(t) = \mathbf{G}^s \boldsymbol{\xi}^s(t) \tag{11.17}$$

as the control law of open-loop system where the gain matrix $\mathbf{G}^s \in R^{q \times b}$ is:

$$\mathbf{G}^s = \mathbf{I}_\beta \mathbf{B}_s^T \tag{11.18}$$

This control law ensures the vanishing of the $\grave{\mathbf{B}}\mathbf{u}^s + \mathbf{g}$ term on in the state equations

$$\left.\begin{array}{c} \dot{\boldsymbol{\zeta}} = \mathbf{A}\boldsymbol{\zeta} + \grave{\mathbf{B}}\mathbf{u}^s + \mathbf{g} \quad \text{for } 0 < t \leq T \\ \boldsymbol{\zeta}(0) = \mathbf{o} \end{array}\right\} \tag{11.19}$$

According to Eq.(11.5),

$$\mathbf{g}(t) = \mathbf{B}^\xi \boldsymbol{\xi}^s(t) \tag{11.20}$$

and

$$\mathbf{B}^\xi = \begin{bmatrix} -\mathbf{M}^{-1}\mathbf{B}\vec{\mathbf{K}}\mathbf{B}_s^T \\ \cdot \end{bmatrix} \tag{11.21}$$

By substitution, one may verify that indeed

$$\grave{\mathbf{B}}\mathbf{u}^s + \mathbf{g} = \mathbf{o} \quad \text{for } 0 < t \leq T \tag{11.22}$$

which reduces Eqs.(11.19) into

$$\left.\begin{array}{c} \dot{\boldsymbol{\zeta}} = \mathbf{A}\boldsymbol{\zeta} \quad \text{for } 0 < t \leq T \\ \boldsymbol{\zeta}(0) = \mathbf{o} \end{array}\right\} \tag{11.23}$$

From the solution one may observe that $\boldsymbol{\zeta}(t) = \mathbf{o}$ during the control time T.

However, the existence of passive base isolation pads (which are non-active connecting elements in the horizontal plane) and the time delay problem will make Eq.(11.22) only approximately true.[5] There will be some small power seepage to the structure from the active base isolation system. For this reason in addition to the open-loop control system, one may need the closed-loop control system as sketched in Fig. (11.9).

The time delay problem in the open-loop control of the active base isolation system plays a very important role in controlling the amount of energy that may seep through the isolation system. By observing not only the seismic ground motion but also the ground velocities and accelerations, one may be able to decrease the undesirable effects of the time delay.[6]

---

[5]More information may be found in "Adaptive Base Isolation: A System to Control Seismic Energy Flow into Buildings," Murat Sener, Doctoral Dissertation, Department of Civil and Environmental Engineering, Duke University, December 1996.

[6]See "Active-Passive Base Isolation System for Seismic Response Controlled Structures," Murat Sener, S. Utku, *Proceedings of 36th Structures, Structural Dynamics and Materials Conference*, paper AIAA-95-1088, New Orleans, LA, April 1995.

It should be noted that the control criterion used in this chapter is to keep the structure at its rest-state, i.e., at the state before the initiation of the seismic event. If the ground acquires a permanent dislocation, this means that the structure will dislocate from the foundations by the same amount (see Section 11.3). Since this may cause serious problems in structures near active fault lines, one may need to alter the control criterion for such structures.

11.7 Recapitulation of Active Control Against Seismic Excitations

In this chapter active control of adaptive structures against seismic ground motions is discussed. The non-autonomous control problem is reduced to intermittent autonomous controls in the structure by means of active base isolation. Control of an active base isolation system is basically an open-loop control problem. Discussions of the chapter include this problem. Basics of actuator placement and identification of control laws for open-loop control of active base isolation are discussed considering mostly horizontal seismic ground motions. Control against vertical seismic ground motions is also discussed.

12
Distributed Parameter Adaptive Structures

In the previous chapters basic issues related to discrete parameter adaptive structures were discussed. In this last chapter of the book, an introduction to the basic issues related to the distributed parameter adaptive structures is given.

12.1 Incorporating Intelligence

When it was decided in the late Seventies to open a window for distributive computing, everything changed, from the way we think, the way we act, the way we communicate, the way we educate, the way we do sciences, the way we improve and apply our technologies, the way we entertain ourselves, the way we observe our environment, and the way we do our politics. What lay beneath this remarkable decision are the technological breakthroughs,[1] fair business practices, and common sense. Are the changes for the better? From the perspective of involving everyone in our collective life on this planet, this author thinks the answer should be a resounding yes. The changes are continuing to take place in every field, and the field of *mechanics of deformable bodies* is not exempt from them.

[1]See, for example "On Impact of Distributive Computing in Education," S. Utku, J. Lestingi, and M. Salama, Preprint 80-061, *Proceedings of ASCE 1980 National Convention*, Portland, OR, April 14-18, 1980.

How can one incorporate intelligence into engineered products? This question can be better answered by the experts of the product, i.e., engineers and engineering scientists. Sciences, in general, tell us why the things are happening in the way they are. They give us clues as to what may happen if a parameter is changed or an additional player gets involved. However, they are not in the business of taking advantage of what they have found for human advantage. That is the engineers' job. Engineers try to control things under the scrutiny of society. It is a moral obligation for an engineer to sustainably improve human life with user friendly products.

A spider is a wonderfully deformable body with intelligence. Since it knows how to react to its environment in real time, its race is around eons. It has its sensors, its actuators, and its intelligence to coordinate the sensor inputs with its actuators in real time, perhaps with a fixed code, but its success is all too obvious.

Spiders and all might be trickling the minds of engineers all the time. Since the Eighties, interest in actively controlling the behavior of mechanical systems in real time is increasing with increasing rates. This is partly due to the impact of advanced materials technology on sensors and actuator technologies, and partly due to the availability of very powerful and reliable microprocessors at costs that were unimaginably low in the preceding decade. Engineering systems where the mechanics of deformable bodies is either the dominant discipline or the supporting one are orienting themselves to benefit from these developments. Pressures stemming from such reorientations have forced the engineers and engineering scientists to revive and broaden the activities that did not find applications in the domain of large simulations to predict the behavior of various systems under well studied, nevertheless imagined, circumstances.

Thanks to physical sciences, an engineer knows the parameters, the variables, and the laws that determine the behavior of physical systems. By changing the values of the parameters or by assigning values to the variables in real time, he can change the behavior of his product. If he cannot alter the conditions affecting the behavior of his product, he can alter his product to minimize the effects of those conditions. For example, earthquakes affect the behavior of buildings. An earthquake engineer cannot control earthquakes, but he can control his building not to get affected by them through a multitude of ways including designing it as a pyramid, changing support conditions or material in real time, or suspending it in the air. Which one is selected is what the society decides collectively.

The use of recent breakthroughs in sensor, actuator, and microprocessor technologies will not only increase the social choices in engineering products but also will make them intelligent for more beneficial human usage.

12.2 Composite Materials

These are modern materials that are composed of two or more different materials with different physical properties. It is the continuation of remarkable and very successful efforts of civil engineers in the early decades of this century in creating steel-reinforced concrete materials by bringing together steel and concrete. The success of reinforced concrete technology is being a good model for the composite materials technology.

Presently a very effective composites technology exists in bringing together plastics and kevlar fibers, rubber and kevlar fibers, rubber and steel fibers, aluminum and carbon fibers, carbon and boron fibers, etc. to produce composite materials to meet the structural requirements of many diversified engineering products. They are usually produced from laminae with unidirectional fibers. A unidirectional laminate is manufactured as a thin planar sheet or strip by imbedding parallel reinforcing fibers in a matrix of more compliant material. Due to different mechanical properties of the matrix and the fiber, a lamina is an orthotropic material being much stiffer in the fiber direction. As in plywood, by staggering fiber directions of many layers of orthotropic laminae under pressure and high temperature, one may obtain a composite which may be of any desired level of orthotropy, in the plane of the laminate, and isotropic but much softer in the transverse direction. Although it is possible to produce three-dimensional fiber reinforced material blocks with any desired anisotropy, composite materials are usually available commercially in the form of laminate sheets.

12.2.1 Piezo-Electric Composite Materials as Actuators

If one uses piezo-electric material sheets in composites, a very useful material may result. Since tangential strains in a piezo-electric sheet change as a function of voltage gradient in the transverse direction, composite materials using such piezo-electric sheets can be used as distributed actuators. Such actuators use electricity to insert strains into the structure. By controlling voltage gradients in the transverse direction of a composite plate containing piezo-electric sheets, one may control deformations of the plate. This is quite similar to controlling the plate deformations by controlling thermal gradients in the thickness direction. However, high thermal inertia and thermal conduction make control by thermal gradients very inferior to control by voltage gradients.

Behavior control of deformable plates and shells by controlling voltage gradients in the thickness direction is presently confined to distributed parameter structures involving low strain energies due to limitations in actuator power input, control energy efficiency, and implementation difficulties originating from current limitations in related technologies.

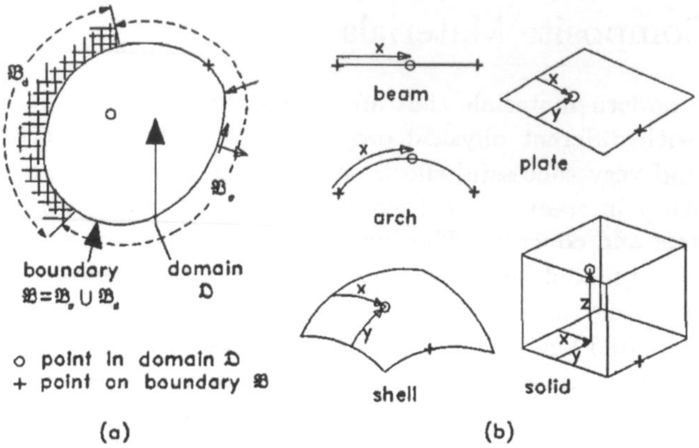

FIGURE 12.1. Definition sketch for domain, boundary, and material points

12.2.2 Piezo-Electric Composite Materials as Sensors

Since strains in the tangential direction of a piezo-electric sheet create a voltage gradient in the transverse direction, they can be used as distributed sensors. Therefore, like strain gages, they are very useful in many structures in observing in real time strains, and stresses. Structural health monitoring systems are using these types of sensors extensively.

There is not a thermal counterpart of piezo-electric sensors, since the creation of strains in most engineering materials does not induce measurable temperature changes.

12.2.3 Distributed Parameter Adaptive Structures

Distributed parameter adaptive structures are distributed parameter structures that are equipped with distributed or discrete parameter sensors, strain inducing actuators, and on-board microprocessors.

Beams, arches, cables, plates, shells, and solids as shown in Fig. (1.1) are examples of distributed parameter structures. In this type of structure one has to distinguish the points inside the boundary with those on the boundary, since the excitation-response relations at a material point of these structures depend on the location of that point relative to the boundary. The points inside the boundary constitute the *domain* which is labelled with $\mathfrak{D}$, and the points on the boundary constitute the *boundary* which is labelled with $\mathfrak{B}$, in Fig. (12.1*a*).

In order to have some ability for extensions to nonlinear problems, the excitation-response relations are described in incremental form. The incremental excitation-response relations at points in domain $\mathfrak{D}$ are of the type of

$$\mathcal{L}\ \Delta\psi = \Delta\mathbf{f} \quad \text{in } \mathfrak{D} \tag{12.1}$$

where $\mathcal{L}$ is a partial differential *real* square matrix operator, operating on the vector of incremental response quantities $\Delta\boldsymbol{\psi}$, and $\Delta\mathbf{f}$ is the vector of incremental excitations that includes incremental control excitations. The operator $\mathcal{L}$ and vectors $\Delta\boldsymbol{\psi}$ and $\Delta\mathbf{f}$ are functions of independent spatial variables $\mathbf{x}$ and temporal variable t. In beam, arch, cable type structures, $\mathbf{x} \in \mathrm{R}^1$ represents a scalar identifying the location of a *cross-section* on the axis of the structure (e.g., in straight axis beams $\mathbf{x} = x$) before the application of incremental loading. In plates and shells, $\mathbf{x} \in \mathrm{R}^2$ represents 2 scalars identifying the location of a *thickness* on the middle surface (e.g., in rectangular plates $\mathbf{x}^T = [x, y]$) before the application of incremental loading. In solid structures, $\mathbf{x} \in \mathrm{R}^3$ represents three scalars identifying the location of a *material point* (e.g., in rectangular prisms, $\mathbf{x}^T = [x, y, z]$) before the incremental loading. These are sketched in Fig. (12.1*b*).

The incremental excitation-response relations at a point on boundary $\mathfrak{B}$ are of the type of

$$\mathcal{B}\ \Delta\boldsymbol{\psi} = \Delta\mathring{\mathbf{f}} \quad \text{on } \mathfrak{B} \tag{12.2}$$

where $\mathcal{B}$ is a partial differential *real* matrix operator, operating on the vector of incremental response quantities $\Delta\boldsymbol{\psi}$, and $\Delta\mathring{\mathbf{f}}$ is the vector of incremental excitations that includes control excitations. Components of $\Delta\mathring{\mathbf{f}}$ may be of stress type or of displacement type. Denoting part of the boundary where stresses are prescribed by $\mathfrak{B}_\sigma$ and the part where displacements are prescribed by $\mathfrak{B}_d$, one may observe that

$$\mathfrak{B} = \mathfrak{B}_\sigma \cup \mathfrak{B}_d \tag{12.3}$$

as sketched in Fig. (12.1*a*).

There are many mathematical distinctions between $\mathcal{L}$ and $\mathcal{B}$ operators. These include the facts that their row and column orders are usually different, and the highest order of differentiation operator in $\mathcal{B}$ is at least one smaller than the highest order differential operator in $\mathcal{L}$. During an incremental loading the system is assumed time invariant, i.e.,

$$\frac{d}{dt}[\mathcal{L}, \mathcal{B}] = 0 \quad \text{for } 0 \le t \le \Delta t \tag{12.4}$$

and $\mathcal{L}$ and $\mathcal{B}$ operators *quasi-linear*, i.e., the highest order differential operators appear linearly. A quasi-linear differential operator $\mathcal{L}$ may be further classified at a point in domain $\mathfrak{D}$ as *elliptic*, *parabolic*, *hyperbolic*, or *purely hyperbolic*, depending upon the number of real directions (i.e., *characteristic directions*) in $(\mathbf{x}, t)$ space, along which discontinuities may exist in the highest order derivative of components $\Delta\boldsymbol{\psi}$ at that point. If no such directions exist, the equation is called elliptic; if only one exists, it is called parabolic; and other cases are called hyperbolic. If the classification at a point remains the same for all points of domain $\mathfrak{D}$, then the excitation-response relations are called accordingly *elliptic*, *parabolic*, *hyperbolic*, or *purely hyperbolic* for that load increment.[2,3] In general, the linear static problems of beams, plates, shells, and solids are elliptic problems; the stress wave propagation problems in linear elasticity are purely hyperbolic problems; and the linear vibration problems of beams, plates, and shells are parabolic problems.

For static control of distributed parameter adaptive structures, the excitation response relations are of *elliptic* type. For dynamic control, the discussions in this chapter are confined to cases where excitation-response relations are of a *parabolic* type.

In elliptic problems, by raising the order of differentiation, it may be possible to reduce Eq.(12.1) into a single scalar differential equation of order $2m$ of a single scalar response quantity $\psi(\mathbf{x})$ as

$$\mathcal{L}_{2\acute{m}}\ \Delta\psi(\mathbf{x}) = \Delta f \text{ in } \mathfrak{D} \tag{12.5}$$

where $\acute{m}$ is a positive integer, and $\mathcal{L}_{2\acute{m}}$ is a quasi-linear elliptic differential operator of order $2\acute{m}$. If the $\mathcal{L}_{2\acute{m}}$ operator is *linear,*[4] *self-adjoint,*[5] *positive definite,*[6] and with *appropriate boundary conditions* on $\mathfrak{B}$ which is not too irregular, then a unique solution exists for the partial differential equation problem defined by Eq.(12.5). Such problems are called *boundary value problems.*

The reduction operations mentioned in the previous paragraph may reduce the boundary conditions in Eq.(12.2) into $\acute{m}$ scalar boundary condi-

---

[2]For a simple and clear treatment of classification of quasi-linear partial differential equations see Art. 6-3, pp.351-371, of *Engineering Analysis, A Survey of Numerical Procedures*, S. H. Crandall, McGraw-Hill, 1956.

[3]For classification of quasi-linear partial differential equation for the cases of "a single nth order equation in R^2," "a single second order equation in R^n," and "n first order equations in R^m," see Arts. 1.6 and 1.7 of *Numerical Solutions of Partial Differential Equations*, Senol Utku, Civil Engineering Department, Duke University, Nov. 1981.

[4]A differential operator $\mathcal{L}$ is *linear* if, for sufficiently smooth arbitrary functions u, v and constants c, d,

$$\mathcal{L}[cu + dv] = c\mathcal{L}[u] + d\mathcal{L}[v] \tag{12.6}$$

holds.

[5]A differential operator $\mathcal{L}$ is *self-adjoint* in domain $\mathfrak{D}$ if, for sufficiently smooth arbitrary functions u, v which satisfy the homogeneous boundary conditions $\mathcal{B}_k[u] = \mathcal{B}_k[v] = 0$, $k = 1, \cdots, \acute{m}$, on $\mathfrak{B}$, the equality

$$\int_{\mathfrak{D}} u\mathcal{L}[v]d\mathfrak{D} = \int_{\mathfrak{D}} v\mathcal{L}[u]d\mathfrak{D} \tag{12.7}$$

holds.

[6]A self-adjoint differential operator $\mathcal{L}$ is *positive-definite* in domain $\mathfrak{D}$ if, for sufficiently smooth arbitrary function u which satisfies the homogeneous boundary conditions $\mathcal{B}_k[u] = 0$, $k = 1, \cdots, \acute{m}$, on $\mathfrak{B}$, the equality

$$\int_{\mathfrak{D}} u\mathcal{L}[u]d\mathfrak{D} > 0 \text{ (zero only if } u = 0) \tag{12.8}$$

holds.

tions that are of order $2\grave{m} - 1$ or less:

$$\mathcal{B}_k \, \Delta\psi(\mathbf{x}) = \Delta\grave{f}_k \quad k = 1, \cdots, \grave{m} \quad \text{on } \mathfrak{B} \tag{12.9}$$

Such boundary conditions are *appropriate boundary conditions* for a well posed boundary value problem.

For example, in linear problems of Kirchhoffian isotropic rectangular plates in bending, the transverse displacement $w(x, y)$ at a middle plane point with Cartesian coordinates $\mathbf{x} = [x, y]^T$ is $\psi(\mathbf{x})$ and the biharmonic operator $\nabla^4$is $\mathcal{L}_{2\grave{m}}$; hence $\grave{m} = 2$. If the plate is clamped all around on unyielding boundaries, then $\Delta\grave{f}_1 = \Delta\grave{f}_2 = 0$, and $\mathcal{B}_1 = 1$, $\mathcal{B}_2 = \frac{\partial}{\partial \grave{n}}$ with $\grave{n}$ being the arch length measured along the normal (in the middle plane) to the boundary. With these, it can be shown that the plate problem is a well-posed boundary value problem, ensuring a unique response to a given excitation.

The linear excitation-response relations of beams, arches, plates, shells, and solids under static loads are well-posed boundary value problems when the boundary conditions are appropriate.

12.3 Static Case

When the excitations are applied on an adaptive distributed parameter adaptive structure slowly such that the material particles gain only negligible inertial forces relative to the other forces, then the problem is called static. Static problems are important not only for their own sake but also for dynamic problems where the static solutions provide the state when all motions cease.

12.3.1 Incremental Linear Excitation-Response Relations

At a material point $\mathbf{x}$ of three-dimensional solid, the incremental response quantities $\Delta\psi$ are

$$\Delta\psi = \begin{Bmatrix} \Delta\mathbf{d} \\ \Delta\boldsymbol{\sigma} \\ \Delta\boldsymbol{\varepsilon} \end{Bmatrix} \tag{12.10}$$

where $\Delta\mathbf{d} \in R^{3\times1}$ represents the incremental displacement vector (i.e., the vector from point $\mathbf{x}$, to corresponding position in the deformed configuration), $\Delta\boldsymbol{\sigma} \in R^{6\times1}$ lists the independent components of the incremental stress tensor, and $\Delta\boldsymbol{\varepsilon} \in R^{6\times1}$ lists the independent components of the incremental strain tensor.

At point $\mathbf{x}$, the incremental excitation quantities for a three-dimensional solid may consist of

$$\Delta\mathbf{f} = \begin{Bmatrix} -\Delta\varkappa \\ -\Delta\boldsymbol{\varepsilon}_o \\ \cdot \end{Bmatrix} \tag{12.11}$$

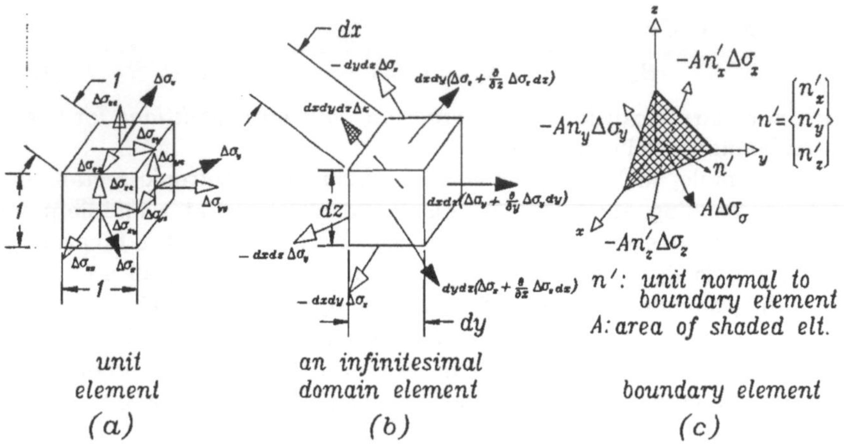

FIGURE 12.2. Definition sketch for incremental stress components and vectors, and free body diagrams of infinitesimal domain and boundary elements

where $\Delta \varkappa \in R^{3\times 1}$ is the incremental body force vector (e.g., change in gravity force acting on material of a unit volume at point $\mathbf{x}$), and $\Delta\boldsymbol{\varepsilon}_o \in R^{6\times 1}$ lists the independent components of the inserted incremental strain tensor (e.g., incremental control strains, thermal strains, shrinkage strains, etc.).

The relations between the incremental response quantities and the incremental excitation quantities are the excitation-response relations for the solid which is a distributed parameter structure. Without the direct or implied knowledge of these relations, it is not possible to obtain the incremental response to a given incremental excitation by any kind of computational or mathematical simulation, unless one uses a physical model and actually measures them.

As in discrete parameter structures, these relations are based on Newton's laws, the rules of Euclidean geometry, and the constitutive laws of structural material. If the excitation-response relations are linear, the boundary conditions are appropriate, boundary $\mathfrak{B}$ is not too irregular, and the constitutive laws of material are positive definite, then they define a well-posed boundary value problem which ensures a unique response to a given excitation.

It is assumed that the incremental excitations-response relations of distributed parameter adaptive structures and appropriate boundary conditions define a well-posed linear boundary value problem.

Equilibrium Equations

In Fig. (12.2a) the free body diagram of a unit cubical material element is shown. The cube is defined by x-constant, y-constant, and z-constant coordinate surfaces of a Cartesian coordinate system (x, y, z). On the positive faces (i.e., the faces where the outer normals are in the same direction as the coordinate axes) the forces acting on the cube are shown by vectors $\Delta\boldsymbol{\sigma}_x, \Delta\boldsymbol{\sigma}_y, \Delta\boldsymbol{\sigma}_z \in R^{3\times1}$ where the subscript indicates direction of the normal of surface on which the force is acting. Since the faces are all of unit area, vectors $\Delta\boldsymbol{\sigma}_x, \Delta\boldsymbol{\sigma}_y, \Delta\boldsymbol{\sigma}_z$ are actually stress vectors. The Cartesian description of these vectors is

$$\left.\begin{aligned} \Delta\boldsymbol{\sigma}_x^T &= [\Delta\sigma_{xx}, \Delta\sigma_{xy}, \Delta\sigma_{xz}] \\ \Delta\boldsymbol{\sigma}_y^T &= [\Delta\sigma_{yx}, \Delta\sigma_{yy}, \Delta\sigma_{yz}] \\ \Delta\boldsymbol{\sigma}_z^T &= [\Delta\sigma_{zx}, \Delta\sigma_{zy}, \Delta\sigma_{zz}] \end{aligned}\right\} \tag{12.12}$$

The stress vectors acting on the negative faces (i.e., the faces where the outer normals are in the opposite direction as the coordinate axes) are $-\Delta\boldsymbol{\sigma}_x$, $-\Delta\boldsymbol{\sigma}_y$, $-\Delta\boldsymbol{\sigma}_z$ since the cube is infinitely small. Vectors $\Delta\boldsymbol{\sigma}_x$, $\Delta\boldsymbol{\sigma}_y$, $\Delta\boldsymbol{\sigma}_z$ together define the Cartesian description of incremental *stress tensor* $\Delta\mathbf{S} \in R^{3\times3}$ at point $\mathbf{x}$:

$$\Delta\mathbf{S} = [\Delta\boldsymbol{\sigma}_x, \Delta\boldsymbol{\sigma}_y, \Delta\boldsymbol{\sigma}_z] = \begin{bmatrix} \Delta\sigma_{xx} & \Delta\sigma_{yx} & \Delta\sigma_{zx} \\ \Delta\sigma_{xy} & \Delta\sigma_{yy} & \Delta\sigma_{zy} \\ \Delta\sigma_{xz} & \Delta\sigma_{yz} & \Delta\sigma_{zz} \end{bmatrix} \tag{12.13}$$

Considering the moment equilibrium equations of the unit cube, one may observe that $\mathbf{S}$ is symmetrical, i.e.,

$$\Delta\mathbf{S} = \Delta\mathbf{S}^T \tag{12.14}$$

Noting that the diagonals and the lower off-diagonals of $\Delta\mathbf{S}$ are sufficient for its definition, a lineal listing of these six components is defined as stress vector $\Delta\boldsymbol{\sigma} \in R^{6\times1}$ at point $\mathbf{x}$:

$$\Delta\boldsymbol{\sigma}^T = [\Delta\sigma_{xx}, \Delta\sigma_{yy}, \Delta\sigma_{zz}, \Delta\sigma_{xy}, \Delta\sigma_{xz}, \Delta\sigma_{yz}] \tag{12.15}$$

where the first three components are normal stresses, and the last three are shearing stresses.

The free body diagram of an infinitesimal rectangular prism material element of a structure, with sides dx, dy, dz, is shown in Fig. (12.2*b*). Since the prism is finite, the forces on negative and positive faces are different in magnitude. Assuming that the deformed geometry of the element can be well-represented by its undeformed geometry, and that there are no preexisting stresses $\mathring{\boldsymbol{\sigma}}_o \in R^{6\times1}$, with the help of the figure, the force equilibrium requirement in x, y, z directions may be obtained as

$$\frac{\partial}{\partial x}\Delta\boldsymbol{\sigma}_x + \frac{\partial}{\partial y}\Delta\boldsymbol{\sigma}_y + \frac{\partial}{\partial z}\Delta\boldsymbol{\sigma}_z = -\Delta\boldsymbol{\varkappa}$$

where $\Delta\boldsymbol{\varkappa}$ is the incremental body force. These equations may be rewritten as

$$\mathcal{L}'\,\Delta\boldsymbol{\sigma} = -\Delta\boldsymbol{\varkappa} \tag{12.16}$$

where the matrix differential operator $\mathcal{L}' \in R^{3\times 6}$ is:

$$\mathcal{L}' = \begin{bmatrix} \frac{\partial}{\partial x} & \cdot & \cdot & \frac{\partial}{\partial y} & \frac{\partial}{\partial z} & \cdot \\ \cdot & \frac{\partial}{\partial y} & \cdot & \frac{\partial}{\partial x} & \cdot & \frac{\partial}{\partial z} \\ \cdot & \cdot & \frac{\partial}{\partial z} & \cdot & \frac{\partial}{\partial x} & \frac{\partial}{\partial y} \end{bmatrix} \tag{12.17}$$

and $\Delta\boldsymbol{\sigma} \in R^{6\times 1}$ is as defined in Eq.(12.15).

If there were large preexisting stresses $\grave{\boldsymbol{\sigma}}_o \in R^{6\times 1}$, due to geometrical change of the element, they would contribute to the equilibrium equations with amounts proportional to the displacement vector $\Delta\mathbf{d} \in R^{3\times 1}$. In this case, the equilibrium equations would become

$$\mathcal{L}'\Delta\boldsymbol{\sigma} + \grave{\mathbf{D}}\Delta\mathbf{d} = -\Delta\varkappa \quad \text{for all } \mathbf{x} \text{ in } \mathfrak{D} \tag{12.18}$$

where $\grave{\mathbf{D}} \in R^{3\times 3}$ is *stiffness of infinitesimal element due to its preexisting stress.* Matrix $\grave{\mathbf{D}}$ is a function of preexisting stresses $\grave{\boldsymbol{\sigma}}_o \in R^{6\times 1}$ and symmetric. Depending upon the values of entries of $\grave{\boldsymbol{\sigma}}_o$, matrix $\grave{\mathbf{D}}$ may be definite, semi-definite, or indefinite. For many problems term $\grave{\mathbf{D}}\Delta\mathbf{d}$ is negligible[7] compared to $\mathcal{L}'\Delta\boldsymbol{\sigma}$.

If material point $\mathbf{x}$ is on boundary $\mathfrak{B}_\sigma$, instead of an infinitesimal cube, equilibrium of an infinitesimal boundary element bordering to $\mathfrak{B}_\sigma$ should be considered. With the help of Fig. (12.2c), the equilibrium equations may be written as

$$-\Delta\boldsymbol{\sigma}_x A n'_x - \Delta\boldsymbol{\sigma}_y A n'_y - \Delta\boldsymbol{\sigma}_z A n'_z + \Delta\boldsymbol{\sigma}_\sigma A = \mathbf{o}$$

where A is the infinitesimal boundary area,

$$\mathbf{n}'^T = [n'_x, n'_y, n'_z] \tag{12.19}$$

is the *unit* vector normal to the boundary area, and $\Delta\boldsymbol{\sigma}_\sigma \in R^{3\times 1}$ is the incremental prescribed *boundary tractions.* After cancellations and using Eq.(12.19), one may rewrite it as

$$\Delta\mathbf{S}\ \mathbf{n}' = \Delta\boldsymbol{\sigma}_\sigma \tag{12.20}$$

where $\Delta\mathbf{S} \in R^{3\times 3}$ is the incremental stress tensor defined in Eq.(12.13), and $\mathbf{n}' \in R^{3\times 1}$ is as defined in Eq.(12.19). Note that $\Delta\mathbf{S}$, $\mathbf{n}'$, and $\Delta\boldsymbol{\sigma}_\sigma \in R^{3\times 1}$ are all described in the same Cartesian coordinate system. By reordering factors, one may rewrite Eq.(12.20) as

$$\mathbf{L}'\ \Delta\boldsymbol{\sigma} = \Delta\boldsymbol{\sigma}_\sigma \quad \text{for all } \mathbf{x} \text{ in } \mathfrak{B}_\sigma \tag{12.21}$$

where

$$\mathbf{L}' = \begin{bmatrix} n'_x & \cdot & \cdot & n'_y & n'_z & \cdot \\ \cdot & n'_y & \cdot & n'_x & \cdot & n'_z \\ \cdot & \cdot & n'_z & \cdot & n'_x & n'_y \end{bmatrix} \tag{12.22}$$

[7] Matrix $\grave{\mathbf{D}}$ is similar to matrix $\grave{\mathbf{K}}$ discussed in Subsection 2.2.3.

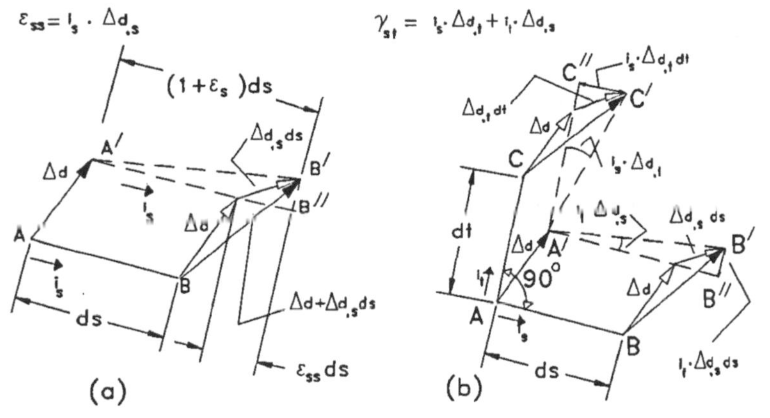

FIGURE 12.3. Definition sketch for lineal and angular incremental strains

and the incremental stress vector $\Delta\boldsymbol{\sigma} \in R^{6\times 1}$ is as defined in Eq.(12.15).

Depending upon the location of the material particle relative to the boundary, one may use either the differential relation in Eq.(12.18) or the algebraic relation in Eq.(12.21), for the equilibrium equations.

Geometric Relations

Under the effect of excitations the material points change position. The vector joining the position of a material point before the incremental excitations applied to its position after they are applied is the *incremental displacement vector* $\Delta\mathbf{d} \in R^{3\times 1}$. The Cartesian description of this vector is:

$$\Delta\mathbf{d}^T = [\Delta d_x, \Delta d_y, \Delta d_z] = [\Delta u, \Delta v, \Delta w] \tag{12.23}$$

Due to the incremental displacements, not only the distances between neighboring material particles change but also particles on orthogonal directions at a point cease to remain in orthogonal directions. The change of a unit distance in a direction is called *lineal strain* in that direction, and the change of the right angle between two orthogonal directions is called *engineering angular strain.* The lineal strains are positive when lengths increase, and angular strains are positive when right angles decrease.

In Fig. (12.3a) the relationship between incremental displacement and lineal strain ε_{ss} in the direction of unit vector $\mathbf{i}_s$ at point A is explained. Let points A and B denote two material points ds distance apart and in the direction $\mathbf{i}_s$, before incremental excitations are applied. Let points A' and B' denote the final positions after the excitations are applied. The lineal strain ε_{ss}, according to its definition

$$\varepsilon_{ss} = \frac{\overline{A'B'} - \overline{AB}}{\overline{AB}} \cong \frac{\overline{A'B''} - \overline{AB}}{\overline{AB}} = \frac{ds + \mathbf{i}_s \cdot \Delta\mathbf{d}_{,s} ds - ds}{ds}$$

may be reduced to

$$\varepsilon_{ss} = \mathbf{i}_s \cdot \Delta \mathbf{d}_{,s} \tag{12.24}$$

where $\Delta \mathbf{d}_{,s}$ is the first partial derivative of $\Delta \mathbf{d}$ with respect to arc length measured in $\mathbf{i}_s$ direction.

In Fig. (12.3*b*) the relationship between incremental displacements and engineering angular strain γ_{st} for orthogonal directions identified by orthogonal unit vectors $\mathbf{i}_s$ and $\mathbf{i}_t$ is explained. Orthogonal line segments $\overrightarrow{AB} = \mathbf{i}_s ds$ and $\overrightarrow{AC} = \mathbf{i}_t dt$ are shown deformed into $\overrightarrow{A'B'}$ and $\overrightarrow{A'C'}$. Decrease of right angle between directions of $\mathbf{i}_s$ and $\mathbf{i}_t$ is the sum of angles $\widehat{C''A'C'} \cong \mathbf{i}_s \cdot \Delta \mathbf{d}_{,t}$ and $\widehat{B''A'B'} \cong \mathbf{i}_t \cdot \Delta \mathbf{d}_{,s}$; hence

$$\gamma_{st} = \gamma_{ts} = \mathbf{i}_s \cdot \Delta \mathbf{d}_{,t} + \mathbf{i}_t \cdot \Delta \mathbf{d}_{,s} \tag{12.25}$$

where $\Delta \mathbf{d}_{,s}$ is as defined before, and $\Delta \mathbf{d}_{,t}$ is the first partial derivative of $\Delta \mathbf{d}$ with respect to arc length measured in the $\mathbf{i}_t$ direction.

With the help of Eqs.(12.23 and 12.24), one may obtain lineal strains $\varepsilon_{xx}, \varepsilon_{yy}$, and ε_{zz} in directions parallel to the Cartesian axes x, y, and z, respectively, as a function of incremental displacement components Δu, Δv, and Δw along the Cartesian axes, as

$$\left.\begin{aligned} \varepsilon_{xx} &= \frac{\partial}{\partial x}\Delta u \\ \varepsilon_{yy} &= \frac{\partial}{\partial y}\Delta v \\ \varepsilon_{zz} &= \frac{\partial}{\partial z}\Delta w \end{aligned}\right\} \tag{12.26}$$

Similarly, referring to the right-handed Cartesian coordinate system (x, y, z), the engineering angular strains $\gamma_{xy} = \gamma_{yx}$, $\gamma_{xz} = \gamma_{zx}$, and $\gamma_{yz} = \gamma_{zy}$, associated with orthogonal directions (x, y), (x, z), and (y, z), respectively, may be obtained from Eqs.(12.23 and 12.25) as

$$\left.\begin{aligned} \gamma_{xy} = \gamma_{yx} &= \frac{\partial}{\partial y}\Delta u + \frac{\partial}{\partial x}\Delta v \\ \gamma_{xz} = \gamma_{zx} &= \frac{\partial}{\partial z}\Delta u + \frac{\partial}{\partial x}\Delta w \\ \gamma_{yz} = \gamma_{zy} &= \frac{\partial}{\partial z}\Delta v + \frac{\partial}{\partial y}\Delta w \end{aligned}\right\} \tag{12.27}$$

One half of an engineering angular strain is called simply *angular strain* and denoted with symbol ε instead of symbol γ:

$$\Delta \varepsilon_{ij} = \frac{1}{2}\Delta \gamma_{ij} \quad \text{for } i = x, y, z \text{ and } j = x, y, z \text{ but } i \neq j \tag{12.28}$$

The nine scalars $\Delta \varepsilon_{ij}$ constitute the components of *incremental strain tensor* $\Delta \mathbf{E} \in \mathrm{R}^{3\times 3}$ at a material point $\mathbf{x}$

$$\Delta \mathbf{E} = \begin{bmatrix} \Delta \varepsilon_{xx} & \Delta \varepsilon_{xy} & \Delta \varepsilon_{xz} \\ \Delta \varepsilon_{yx} & \Delta \varepsilon_{yy} & \Delta \varepsilon_{yz} \\ \Delta \varepsilon_{zx} & \Delta \varepsilon_{zy} & \Delta \varepsilon_{zz} \end{bmatrix} \tag{12.29}$$

which may be also expressed as

$$\Delta\mathbf{E} = \begin{bmatrix} \Delta\varepsilon_{xx} & \frac{1}{2}\Delta\gamma_{xy} & \frac{1}{2}\Delta\gamma_{xz} \\ \frac{1}{2}\Delta\gamma_{yx} & \Delta\varepsilon_{yy} & \frac{1}{2}\Delta\gamma_{yz} \\ \frac{1}{2}\Delta\gamma_{zx} & \frac{1}{2}\Delta\gamma_{zy} & \Delta\varepsilon_{zz} \end{bmatrix} \tag{12.30}$$

Note that $\Delta\mathbf{E}$ is a real symmetric matrix, i.e.,

$$\Delta\mathbf{E} = \Delta\mathbf{E}^T \tag{12.31}$$

One may observe that the diagonal elements of $\Delta\mathbf{E}$ are incremental lineal strains and off-diagonals are incremental angular strains.

The three lineal strains and the three engineering angular strains are listed in a lineal fashion to define *strain vector*. The incremental strain vector at material point $\mathbf{x}$, $\Delta\boldsymbol{\varepsilon}(\mathbf{x}) \in R^{6\times 1}$ may be displayed as

$$\Delta\boldsymbol{\varepsilon}^T = [\Delta\varepsilon_{xx}, \Delta\varepsilon_{yy}, \Delta\varepsilon_{zz}, \Delta\gamma_{xy}, \Delta\gamma_{xz}, \Delta\gamma_{yz}] \tag{12.32}$$

Note that indices of corresponding entries in $\Delta\boldsymbol{\varepsilon}$ and $\Delta\boldsymbol{\sigma}$ are the same [see Eq.(12.15)].

One may rewrite Eqs.(12.26 and 12.27) compactly as

$$\Delta\boldsymbol{\varepsilon} = \mathcal{L}'^T \Delta\mathbf{d} \tag{12.33}$$

where the incremental strain vector $\Delta\boldsymbol{\varepsilon} \in R^{6\times 1}$ is as defined in Eq.(12.32), the incremental displacement vector $\Delta\mathbf{d} \in R^{3\times 1}$ is as defined in Eq.(12.23), and the differential matrix operator $\mathcal{L}'^T \in R^{6\times 3}$ is the transpose of $\mathcal{L}' \in R^{3\times 6}$ defined in Eq.(12.17). The reader may verify this.

If the material element at point $\mathbf{x}$ is subjected to incremental strains $\Delta\boldsymbol{\varepsilon}_o \in R^{6\times 1}$:

$$\Delta\boldsymbol{\varepsilon}_o^T = [\Delta\varepsilon_{o_{xx}}, \Delta\varepsilon_{o_{yy}}, \Delta\varepsilon_{o_{zz}}, \Delta\gamma_{o_{xy}}, \Delta\gamma_{o_{xz}}, \Delta\gamma_{o_{yz}}] \tag{12.34}$$

representing incremental control strains or strains induced by temperature change, then the incremental strain displacement relations of Eq.(12.33) may be restated to show this effect:

$$\Delta\boldsymbol{\varepsilon} = \mathcal{L}'^T \Delta\mathbf{d} + \Delta\boldsymbol{\varepsilon}_o \tag{12.35}$$

The reader may compare this matrix differential relationship, given for a material element at point $\mathbf{x}$ of a distributed parameter adaptive structure, with the algebraic one given in Section 2.3.3 for discrete parameter adaptive structures.

Material's Constitutive Laws

At a domain point $\mathbf{x}$, three equilibrium equations in Eq.(12.18) and six geometric relations in Eq.(12.35) involve fifteen incremental response quantities listed in $\Delta\mathbf{d} \in R^{3\times 1}$, $\Delta\boldsymbol{\sigma} \in R^{6\times 1}$, and $\Delta\boldsymbol{\varepsilon} \in R^{6\times 1}$. One needs six more equations for the determination of a unique incremental response corresponding to a given incremental excitation. The missing equations are provided by the constitutive laws of the material. Confining to linearly elastic materials, according to Hooke's law, one may write

$$\Delta\boldsymbol{\sigma} = \mathbf{D}\Delta\boldsymbol{\varepsilon} \tag{12.36}$$

where matrix $\mathbf{D} \in \mathrm{R}^{6\times 6}$ is a real, symmetric, and positive definite *material matrix.*

For a general anisotropic material, one needs twenty-one laboratory measurements in order to obtain the twenty-one constants defining the material matrix. However, if the material is isotropic, by determining only two constants in the laboratory, i.e., the Young's modulus E, and the Poisson's ratio ν, the material matrix may be expressed as

$$\mathbf{D} = \frac{E}{(1+\nu)(1-2\nu)} \begin{bmatrix} 1-\nu & \nu & \nu & \cdot & \cdot & \cdot \\ \nu & 1-\nu & \nu & \cdot & \cdot & \cdot \\ \nu & \nu & 1-\nu & \cdot & \cdot & \cdot \\ \cdot & \cdot & \cdot & \frac{1-2\nu}{2} & \cdot & \cdot \\ \cdot & \cdot & \cdot & \cdot & \frac{1-2\nu}{2} & \cdot \\ \cdot & \cdot & \cdot & \cdot & \cdot & \frac{1-2\nu}{2} \end{bmatrix} \tag{12.37}$$

The inverse relationship of Eq.(12.36) may be written as

$$\Delta\boldsymbol{\varepsilon} = \mathbf{C}''\Delta\boldsymbol{\sigma} \tag{12.38}$$

where matrix $\mathbf{C}'' \in \mathrm{R}^{6\times 6}$ is a real, symmetric, and positive definite *compliance matrix* which is the inverse of material matrix

$$\mathbf{C}'' = \mathbf{D}^{-1} \tag{12.39}$$

For isotropic materials, by inverting $\mathbf{D}$ in Eq.(12.37), the compliance matrix may be obtained as

$$\mathbf{C}'' = \frac{1}{E} \begin{bmatrix} 1 & -\nu & -\nu & \cdot & \cdot & \cdot \\ -\nu & 1 & -\nu & \cdot & \cdot & \cdot \\ -\nu & -\nu & 1 & \cdot & \cdot & \cdot \\ \cdot & \cdot & \cdot & 2(1+\nu) & \cdot & \cdot \\ \cdot & \cdot & \cdot & \cdot & 2(1+\nu) & \cdot \\ \cdot & \cdot & \cdot & \cdot & \cdot & 2(1+\nu) \end{bmatrix} \tag{12.40}$$

Between isotropic and completely anisotropic linearly elastic materials, there are many other special elastic materials, such as orthotropic, aeolotropic ones, etc. By stacking together highly orthotropic composite laminae with unidirectional fibers in different orientations, modern technology can produce various kinds of linearly elastic materials which may contain piezo-electric sheets or shape memory alloy fibers that may insert strains into the material in a controlled manner. Knowing the material matrix of a unidirectional fiber composite lamina and the relative orientations of such laminae in the stack, one may calculate with a microprocessor the material matrix of the whole stack.[8]

Incremental Excitation Response Relations

By combining equilibrium equations from Eq.(12.18), geometric relations from Eq.(12.35), and constitutive laws of material from Eq.(12.36), one may write for a material point at position **x** in domain $\mathfrak{D}$ the incremental excitation response relations compactly as

$$\begin{bmatrix} \grave{\mathbf{D}} & \mathcal{L}' & \cdot \\ \mathcal{L}'^T & \cdot & -\mathbf{I} \\ \cdot & -\mathbf{I} & \mathbf{D} \end{bmatrix} \begin{Bmatrix} \Delta\mathbf{d} \\ \Delta\boldsymbol{\sigma} \\ \Delta\boldsymbol{\varepsilon} \end{Bmatrix} = \begin{Bmatrix} -\Delta\boldsymbol{\varkappa} \\ -\Delta\boldsymbol{\varepsilon}_o \\ \cdot \end{Bmatrix} \tag{12.41}$$

This may be rewritten as in Eq.(12.1), i.e., as

$$\mathcal{L}\ \Delta\boldsymbol{\psi}(\mathbf{x}) = \Delta\mathbf{f} \quad \text{for } \mathbf{x} \text{ in } \mathfrak{D} \tag{12.42}$$

where $\Delta\boldsymbol{\psi}$ is as in Eq.(12.10), and

$$\mathcal{L} = \begin{bmatrix} \grave{\mathbf{D}} & \mathcal{L}' & \cdot \\ \mathcal{L}'^T & \cdot & -\mathbf{I} \\ \cdot & -\mathbf{I} & \mathbf{D} \end{bmatrix} \tag{12.43}$$

and

$$\Delta\mathbf{f} = \begin{Bmatrix} -\Delta\boldsymbol{\varkappa} \\ -\Delta\boldsymbol{\varepsilon}_o \\ \cdot \end{Bmatrix} \tag{12.44}$$

If the material point is positioned at **x** on boundary $\mathfrak{B}$, by using Eq.(12.21), the incremental excitation relations become

$$\begin{bmatrix} \cdot & \mathbf{L}' & \cdot \\ \mathbf{I} & \cdot & \cdot \end{bmatrix} \begin{Bmatrix} \Delta\mathbf{d} \\ \Delta\boldsymbol{\sigma} \\ \Delta\boldsymbol{\varepsilon} \end{Bmatrix} = \begin{Bmatrix} \Delta\boldsymbol{\sigma}_\sigma \\ \Delta\mathbf{d}_d \end{Bmatrix} \quad \begin{matrix} \mathbf{x} \text{ in } \mathfrak{B}_\sigma \\ \mathbf{x} \text{ in } \mathfrak{B}_d \end{matrix} \tag{12.45}$$

where $\mathbf{I} \in (\text{Int})^{3\times3}$ is the identity matrix, $\mathbf{L}' \in \mathrm{R}^{3\times6}$ is as defined in Eq.(12.22), $\Delta\boldsymbol{\sigma}_\sigma \in \mathrm{R}^{6\times1}$ is the prescribed boundary tractions, if any, and $\Delta\mathbf{d}_d \in \mathrm{R}^{3\times1}$ is the prescribed boundary displacements, if any. This may be rewritten as in Eq.(12.2), i.e., as

$$\mathcal{B}\ \Delta\boldsymbol{\psi}(\mathbf{x}) = \Delta\hat{\mathbf{f}} \quad \mathbf{x} \text{ in } \mathfrak{B} \tag{12.46}$$

---

[8]See, for example, "Design and Material Representation of Lightweight Composite Panels for Large Deployable Reflectors," S. Utku, R. E. Freeland, H. Helms, C. P. Kuo, P. M. McElroy, and C. Porter, JPL-D8912, September 1991, Jet Propulsion Laboratory, Pasadena, CA.

where $\Delta\psi$ is as in Eq.(12.10) and

$$\mathcal{B} = \begin{bmatrix} \cdot & \mathbf{L}' & \cdot \\ \mathbf{I} & \cdot & \cdot \end{bmatrix} \tag{12.47}$$

and

$$\Delta\dot{\mathbf{f}} = \begin{Bmatrix} \Delta\boldsymbol{\sigma}_\sigma \\ \Delta\mathbf{d}_d \end{Bmatrix} \quad \begin{matrix} \mathbf{x} \text{ in } \mathfrak{B}_\sigma \\ \mathbf{x} \text{ in } \mathfrak{B}_d \end{matrix} \tag{12.48}$$

It may be shown that the real self-adjoint operator $\mathcal{L}$ in Eq.(12.43) is positive definite if

$$\left\|\dot{\mathbf{D}}\right\| << \|\mathbf{D}\| \tag{12.49}$$

which is the case if the structure is not in or at the proximity of an imminent buckling state at the beginning of current incremental loading. In fact, for most problems, one may assume

$$\dot{\mathbf{D}} = 0 \tag{12.50}$$

as discussed in Subsection 2.2.3. Assuming that the real linear operator $\mathcal{L}$ is positive definite, the excitation response relations given by Eqs.(12.42 and 12.46) constitute a well-posed boundary value problem provided that boundary $\mathfrak{B}$ is not too irregular (see Subsection 12.3.3). This ensures a unique incremental response $\Delta\psi$ corresponding to given excitations $\Delta\mathbf{f}$ and $\Delta\dot{\mathbf{f}}$. How this unique response may be obtained is discussed in the next subsection when incremental excitations consist of only actuator inserted strains $\Delta\boldsymbol{\varepsilon}_o$.

12.3.2 Inverse Relations

In this chapter, instead of investigating the response for any excitation, only the response due to actuator inserted incremental strains $\Delta\boldsymbol{\varepsilon}_o$ is studied. Although only specific components of $\Delta\boldsymbol{\varepsilon}_o \in R^{6\times 1}$ at point $\mathbf{x}$ of domain $\mathfrak{D}$ may be nonzero, the general treatment here makes no such assumptions.

Response to Controls

Control excitations in the present treatment are the inserted incremental strains $\Delta\boldsymbol{\varepsilon}_o$. For the purpose of the discussions here, how these strains are created is not important. They can be created by changing the temperature of the material, or they may be created by changing the electric field of piezo-electric sheets imbedded in the material, or some other means.

Although thermally induced strains will not dissipate as long as the required temperature is maintained and the structural material is with a large retardation time (i.e., lazy in viscous flow), the same cannot be said for the present day piezo-electric materials. They are rather viscous; therefore, they require much shorter control time.

Pertinent to the discussions of this chapter is the expectation that once the control strains are created they will remain practically constant, at least during the control time.

Present discussions assume the following:

- There are no excitations on the boundary $\mathfrak{B}$, i.e., $\Delta\boldsymbol{\sigma}_\sigma$ and $\Delta\mathbf{d}_d$ are zero, hence, from Eq.(12.48)

$$\Delta\grave{\mathbf{f}} = \mathbf{o} \tag{12.51}$$

- There are no incremental body forces, i.e., $\Delta\varkappa = \mathbf{o}$, hence, from Eq.(12.44)

$$\Delta\mathbf{f} = \begin{Bmatrix} \cdot \\ -\Delta\boldsymbol{\varepsilon}_o \\ \cdot \end{Bmatrix} \tag{12.52}$$

- The effect of preexisting stresses in the equilibrium equations is negligible, i.e.,

$$\dot{\mathbf{D}}\Delta\mathbf{d} \cong \mathbf{o} \tag{12.53}$$

With the above assumptions, the boundary value problem defined by Eqs.(12.42 and 12.46) becomes

$$\left.\begin{array}{ll} \mathcal{L}\,\Delta\boldsymbol{\psi}(\mathbf{x}) = [\,\cdot \quad -\Delta\boldsymbol{\varepsilon}_o^T \quad \cdot\,]^T & \text{for } \mathbf{x} \text{ in } \mathfrak{D} \\ \mathcal{B}\,\Delta\boldsymbol{\psi}(\mathbf{x}) = \mathbf{o} & \text{for } \mathbf{x} \text{ in } \mathfrak{B} \end{array}\right\} \tag{12.54}$$

or using $\mathcal{L}$ and $\mathcal{B}$ operators from Eqs.(12.43 and 12.47), these equations may be restated as:

$$\begin{bmatrix} \cdot & \mathcal{L}' & \cdot \\ \mathcal{L}'^T & \cdot & -\mathbf{I} \\ \cdot & -\mathbf{I} & \mathbf{D} \end{bmatrix} \begin{Bmatrix} \Delta\mathbf{d} \\ \Delta\boldsymbol{\sigma} \\ \Delta\boldsymbol{\varepsilon} \end{Bmatrix} = \begin{Bmatrix} \cdot \\ -\Delta\boldsymbol{\varepsilon}_o \\ \cdot \end{Bmatrix} \quad \text{for } \mathbf{x} \text{ in } \mathfrak{D} \tag{12.55}$$

and

$$\begin{bmatrix} \cdot & \mathbf{L}' & \cdot \\ \mathbf{I} & \cdot & \cdot \end{bmatrix} \begin{Bmatrix} \Delta\mathbf{d} \\ \Delta\boldsymbol{\sigma} \\ \Delta\boldsymbol{\varepsilon} \end{Bmatrix} = \begin{Bmatrix} \mathbf{o} \\ \mathbf{o} \end{Bmatrix} \quad \begin{array}{l} \text{for } \mathbf{x} \text{ in } \mathfrak{B}_\sigma \\ \text{for } \mathbf{x} \text{ in } \mathfrak{B}_d \end{array} \tag{12.56}$$

Note that Eq.(12.53) is used in writing Eq.(12.55).

In order to decrease the number of response quantities during the manipulations, the boundary value problem defined by Eqs.(12.55 and 12.56) may be solved either by displacement method or by force method (see Section 3.1 for the counterparts in discrete parameter structures). If the goal is simulation, because of the economies in computational resources, the displacement method is used after discretizing the boundary value problem by a suitable discretization method such as the finite element method. However, since the goal here is analysis, both methods are briefly studied.

Response by Displacement Method

In this method incremental stresses $\Delta\boldsymbol{\sigma}$ and incremental strains $\Delta\boldsymbol{\varepsilon}$ are eliminated from the equilibrium equations by substitution in order to obtain the governing equations in terms of incremental displacements $\Delta\mathbf{d} \in R^{3\times 1}$ only.

By adding $\mathcal{L}'$ operated equations in the third row partition of Eq.(12.55) to the equations in its first row partition, and then adding $\mathbf{D}\mathcal{L}'$ operated equations in the second row partition to those of the first row partition, one may obtain

$$\mathcal{L}'\mathbf{D}\mathcal{L}'^T\Delta\mathbf{d} = -\mathcal{L}'\mathbf{D}\Delta\boldsymbol{\varepsilon}_o \quad \text{for } \mathbf{x} \text{ in } \mathfrak{D} \qquad (12.57)$$

For isotropic materials these equations become[9]

$$\frac{E}{2(1+\nu)}(\mathbf{I}\ \nabla^2 + \frac{1}{1-2\nu}\nabla\ \nabla^T)\Delta\mathbf{d} = -\mathcal{L}'\mathbf{D}\Delta\boldsymbol{\varepsilon}_o \quad \text{for } \mathbf{x} \text{ in } \mathfrak{D} \qquad (12.58)$$

where $\mathbf{D}$ is as in Eq.(12.37) and

$$\nabla = [\frac{\partial}{\partial x}, \frac{\partial}{\partial y}, \frac{\partial}{\partial z}]^T \qquad (12.59)$$

$$\nabla^2 = \nabla^T\nabla \qquad (12.60)$$

Equations in Eq.(12.57), or Eq.(12.58) for isotropic materials, represent the equilibrium equations which are three second order partial differential equations for the three components of incremental displacement response $\Delta\mathbf{d} \in R^{3\times 1}$. It may be shown that the linear operator $\mathcal{L}'^T\mathbf{D}\mathcal{L}' \in R^{3\times 3}$ is self-adjoint and positive definite.

Since the highest order differentiation is $2\dot{m} = 2$ in all equations, one needs $\dot{m} = 1$ boundary condition at each point on $\mathfrak{B}$. In a well-posed boundary value problem, the highest order differentiation in the boundary conditions is $2\dot{m} - 1 = 1$ or less. These boundary conditions are given in Eq.(12.56) in terms of all the response quantities. Using $\Delta\boldsymbol{\sigma}$ from Eq.(12.36) and $\Delta\boldsymbol{\varepsilon}$ from Eq.(12.35), one may rewrite Eq.(12.56) as

$$\left.\begin{array}{ll} \mathbf{L}'\mathbf{D}\mathcal{L}'^T\Delta\mathbf{d} = -\mathbf{D}\Delta\boldsymbol{\varepsilon}_o & \text{for } \mathbf{x} \text{ in } \mathfrak{B}_\sigma \\ \Delta\mathbf{d} = \mathbf{o} & \text{for } \mathbf{x} \text{ in } \mathfrak{B}_d \end{array}\right\} \qquad (12.61)$$

Although an important control option, in the present treatment we will assume $\Delta\boldsymbol{\varepsilon}_o = \mathbf{o}$ for $\mathbf{x}$ in $\mathfrak{B}_\sigma$ and write

$$\left.\begin{array}{ll} \mathbf{L}'\mathbf{D}\mathcal{L}'^T\Delta\mathbf{d} = \mathbf{o} & \text{for } \mathbf{x} \text{ in } \mathfrak{B}_\sigma \\ \Delta\mathbf{d} = \mathbf{o} & \text{for } \mathbf{x} \text{ in } \mathfrak{B}_d \end{array}\right\} \qquad (12.62)$$

[9] See S. Timoshenko and J. N. Goodier, *Theory of Elasticity*, 2nd ed., 1951, McGraw-Hill Book Co., New York (Art. 79, pp.233-235).

The problem defined by Eqs.(12.57 and 12.62) constitutes a well-posed boundary value problem. The solution of this problem provides the unique values of incremental displacement response $\Delta\mathbf{d}$ corresponding to inserted control strains $\Delta\boldsymbol{\varepsilon}_o$ which are known at all $\mathbf{x}$ in $\mathfrak{D}$. The expressions for $\Delta\mathbf{d}$ in terms of $\Delta\boldsymbol{\varepsilon}_o$ are the inverse relations. For the purpose of actuator placement and actuator selection problems the inverse relations are best expressed in terms of Green's function $\mathbf{G}(\mathbf{x},\mathbf{x}_o)$ which is the replacement of $\Delta\mathbf{d}$ corresponding to the unit value of $\Delta\boldsymbol{\varepsilon}_o$, i.e., to unit pulse strain $\boldsymbol{\delta}(\mathbf{x},\mathbf{x}_o)$ inserted at location $\mathbf{x}_o$ in $\mathfrak{D}$, namely, the solution of the following well-defined boundary value problem

$$\left[\begin{array}{l} \mathcal{L}'\mathbf{D}\mathcal{L}'^T\mathbf{G}(\mathbf{x},\mathbf{x}_o) = -\mathcal{L}'\mathbf{D}\ \boldsymbol{\delta}(\mathbf{x},\mathbf{x}_o) \quad \text{for } \mathbf{x} \text{ in } \mathfrak{D} \\ \begin{bmatrix} \mathbf{L}'\mathbf{D}\mathcal{L}'^T \\ \mathbf{I} \end{bmatrix} \mathbf{G}(\mathbf{x},\mathbf{x}_o) = \mathbf{o} \quad \text{for } \mathbf{x} \text{ on } \mathfrak{B} \end{array}\right] \tag{12.63}$$

which is a restatement of Eqs.(12.57 and 12.62) for

$$\Delta\boldsymbol{\varepsilon}_o = \boldsymbol{\delta}(\mathbf{x},\mathbf{x}_o) \tag{12.64}$$

In these expressions $\boldsymbol{\delta}(\mathbf{x},\mathbf{x}_o)$ is the Dirac delta function with singularity at $\mathbf{x}_o$. Green's function $\mathbf{G}(\mathbf{x},\mathbf{x}_o)$ may be obtained analytically or numerically by solving Eq.(12.63), or experimentally by measuring in laboratory models. Once it is available, $\Delta\mathbf{d}$ for any $\Delta\boldsymbol{\varepsilon}_o$ may be obtained by superposition

$$\Delta\mathbf{d}(\mathbf{x}) = \int_{\mathfrak{D}} \mathbf{G}(\mathbf{x},\mathbf{x}_o)\Delta\boldsymbol{\varepsilon}_o(\mathbf{x}_o)\ d\mathfrak{D}_o \tag{12.65}$$

since the incremental boundary value problem is linear.

Having solved $\Delta\mathbf{d}$ from the boundary value problem of Eqs.(12.57 and 12.62) or from Eq.(12.65), one may obtain the incremental strain response $\Delta\boldsymbol{\varepsilon}$ from equations in the second row partition, and the incremental stress response $\Delta\boldsymbol{\sigma}$ from equations in the third row partition of Eq.(12.55), and write

$$\begin{Bmatrix} \Delta\mathbf{d} \\ \Delta\boldsymbol{\sigma} \\ \Delta\boldsymbol{\varepsilon} \end{Bmatrix} = \begin{bmatrix} \mathbf{I} \\ \mathbf{D}\mathcal{L}'^T \\ \mathcal{L}'^T \end{bmatrix} \Delta\mathbf{d} + \begin{Bmatrix} \cdot \\ \mathbf{D}\Delta\boldsymbol{\varepsilon}_o \\ \Delta\boldsymbol{\varepsilon}_o \end{Bmatrix} \tag{12.66}$$

Although computationally very efficient, the displacement method tells very little if inserted control strains $\Delta\boldsymbol{\varepsilon}_o$ would create stresses $\Delta\boldsymbol{\sigma}$ in the structure. From the control energy standpoint this is very important, since the strain energy ΔU associated with $\Delta\boldsymbol{\sigma}$, i.e.,

$$\Delta U = \int_{\mathfrak{D}} \frac{1}{2}\Delta\boldsymbol{\sigma}^T\mathbf{C}''\Delta\boldsymbol{\sigma}\ d\mathfrak{D} \tag{12.67}$$

where $\mathfrak{D}$ is the material volume of the structure, is met directly from the energy budget of the control system. On this matter a much better insight is obtained by analysis using the force method.

Response by Force Method

In this method, in order to obtain in domain $\mathfrak{D}$ six partial differential equations involving only the six components of incremental stress response $\Delta\boldsymbol{\sigma} \in \mathrm{R}^{6\times 1}$, one may eliminate

1. incremental strains $\Delta\boldsymbol{\varepsilon} \in \mathrm{R}^{6\times 1}$ from the material's constitutive laws, by substitution of strain displacement relations, and
2. incremental displacements $\Delta\mathbf{d}$ from the latter, by the use of *strain compatibility relations.*[10]

The boundary conditions on $\mathfrak{B}_\sigma$ can be used as they are; however, the boundary conditions on $\mathfrak{B}_d$ need to be expressed in terms of $\Delta\boldsymbol{\sigma}$.[11]

The strain compatibility relations in the Cartesian coordinates may be expressed as

$$\mathcal{M}\mathcal{L}^T\Delta\mathbf{d} = \mathbf{o} \tag{12.68}$$

where the second order differential matrix operator $\mathcal{M} \in \mathrm{R}^{6\times 6}$ is

$$\mathcal{M} = \begin{bmatrix} \frac{\partial^2}{\partial y^2} & \frac{\partial^2}{\partial x^2} & \cdot & -\frac{\partial^2}{\partial x\partial y} & \cdot & \cdot \\ \cdot & \frac{\partial^2}{\partial z^2} & \frac{\partial^2}{\partial y^2} & \cdot & \cdot & -\frac{\partial^2}{\partial z\partial y} \\ \frac{\partial^2}{\partial z^2} & \cdot & \frac{\partial^2}{\partial x^2} & \cdot & -\frac{\partial^2}{\partial x\partial z} & \cdot \\ 2\frac{\partial^2}{\partial z\partial y} & \cdot & \cdot & -\frac{\partial^2}{\partial x\partial z} & -\frac{\partial^2}{\partial x\partial y} & \frac{\partial^2}{\partial x^2} \\ \cdot & 2\frac{\partial^2}{\partial x\partial z} & \cdot & -\frac{\partial^2}{\partial z\partial y} & \frac{\partial^2}{\partial y^2} & -\frac{\partial^2}{\partial x\partial y} \\ \cdot & \cdot & 2\frac{\partial^2}{\partial x\partial y} & \frac{\partial^2}{\partial z^2} & -\frac{\partial^2}{\partial z\partial y} & -\frac{\partial^2}{\partial x\partial z} \end{bmatrix} \tag{12.69}$$

From Eq.(12.55), one may write for equilibrium equations

$$\mathcal{L}'\,\Delta\boldsymbol{\sigma} = \mathbf{o} \qquad \mathbf{x} \text{ in } \mathfrak{D} \tag{12.70}$$

for strain displacement relations

$$\Delta\boldsymbol{\varepsilon} = \mathcal{L}'^T\Delta\mathbf{d} + \Delta\boldsymbol{\varepsilon}_o \tag{12.71}$$

---

[10]The deformed geometry of structure, i.e., $\mathfrak{D}$ after deformations, can be uniquely identified either by the knowledge of incremental displacements $\Delta\mathbf{d} \in \mathrm{R}^{3\times 1}$ or by the knowledge of incremental strains $\Delta\boldsymbol{\varepsilon} \in \mathrm{R}^{6\times 1}$in $\mathfrak{D}$. Since $\Delta\boldsymbol{\varepsilon}$ has three more scalar information than $\Delta\mathbf{d}$, the incremental strain components must satisfy among themselves the *strain compatibility relations.* As observed by A. E. H. Love (in his *The Mathematical Theory of Elasticity*, *4th* edition, Dover, 1946, pages 49, 17, 16) these relations and their proof are given by Saint-Venant (in his edition of Navier's *Résumé des Leçons sur l'application de la Méchanique*, Paris, 1864, Appendice 3) and E. Beltrami [*Paris, C. R.*, t. 108 (1889), cf. Koenigs, *Leçons de Cinématique*, Paris, 1897, p. 411]. A more modern proof is given by I. S. Sokolnikoff (page 24 of his *Mathematical Theory of Elasticity*, 1946 which was based on his 1941 and 1942 lectures in the Program of Advanced Instruction and Research in Mechanics at the Graduate School of Brown University).

[11]A difficult problem which becomes trivial if the structure is supported in a statically determinate fashion.

and for constitutive laws of material

$$\Delta\boldsymbol{\varepsilon} = \mathbf{C}''\Delta\boldsymbol{\sigma} \tag{12.72}$$

where $\mathbf{D}^{-1} = \mathbf{C}''$ from Eq.(12.38) is used. Substituting $\Delta\boldsymbol{\varepsilon}$ from Eq.(12.71) into Eq.(12.72), the latter may be rewritten as

$$\mathcal{L}'^{T}\Delta\mathbf{d} + \Delta\boldsymbol{\varepsilon}_o = \mathbf{C}''\Delta\boldsymbol{\sigma} \tag{12.73}$$

Operating both sides of this equation with $\mathcal{M}\in\mathrm{R}^{6\times 6}$, and using Eq.(12.69), i.e.,

$$\mathcal{M}\mathcal{L}^{T}\Delta\mathbf{d} = \mathbf{o} \tag{12.74}$$

one may obtain

$$\mathcal{M}\mathbf{C}''\Delta\boldsymbol{\sigma} = \mathcal{M}\Delta\boldsymbol{\varepsilon}_o \qquad \text{for } \mathbf{x} \text{ in } \mathfrak{D} \tag{12.75}$$

which are basically six strain compatibility equations in terms of the incremental stress response $\Delta\boldsymbol{\sigma} \in \mathrm{R}^{6\times 1}$. Combining this with the three equilibrium equations in Eq.(12.70), the six partial differential equations to be satisfied by $\Delta\boldsymbol{\sigma} \in \mathrm{R}^{6\times 1}$ at all $\mathbf{x}$ in $\mathfrak{D}$ may be obtained. Since equilibrium equations are of first order and compatibility equations are of second order, this is not a case of an overly defined problem for $\Delta\boldsymbol{\sigma}$. It is true that at a domain point we have nine equations in six unknowns; however, the first order equilibrium equations are merely constraint equations for the second order compatibility equations. For isotropic materials, the combined equations may be obtained from the literature[12] as

$$(1+\nu)\nabla^2\Delta\boldsymbol{\sigma} + \mathit{diag}[1,1,1,\tfrac{1}{2},\tfrac{1}{2},\tfrac{1}{2}]\ \mathcal{L}'^{T}\nabla\ \Delta\Theta = \Delta\boldsymbol{\tau} \qquad \text{for } \mathbf{x} \text{ in } \mathfrak{D} \tag{12.76}$$

where vector $\Delta\boldsymbol{\tau} \in \mathrm{R}^{6\times 1}$ is a linear function of $\mathcal{M}\Delta\boldsymbol{\varepsilon}_o$, matrix differential operator $\nabla \in \mathrm{R}^{3\times 1}$ is as defined in Eq.(12.59), and $\Delta\Theta = \mathit{trace}(\Delta\mathbf{S})$, i.e.,

$$\Delta\Theta = \Delta\sigma_{xx} + \Delta\sigma_{yy} + \Delta\sigma_{zz} \tag{12.77}$$

Note that Eq.(12.76) constitutes six second order partial differential equations for the six components of incremental stress response $\Delta\boldsymbol{\sigma}$.

The boundary conditions associated with the partial differential equations in Eq.(12.76) are:

$$\Delta\boldsymbol{\sigma} = \mathbf{o} \quad \text{for } \mathbf{x} \text{ on } \mathfrak{B} \tag{12.78}$$

assuming that $\Delta\boldsymbol{\varepsilon}_o = \mathbf{o}$ on $\mathfrak{B}$, and the structure is supported in a statically determinate manner.

---

[12]See S. Timoshenko and J. N. Goodier, *Theory of Elasticity*, 2nd edition, 1951, McGraw-Hill Book Co., New York (Art. 77, pp. 229-232) for the case where $\Delta\boldsymbol{\varkappa} \neq \mathbf{o}$ and $\Delta\boldsymbol{\varepsilon}_o = \mathbf{o}$. For the present case $\Delta\boldsymbol{\varkappa} = \mathbf{o}$ and $\Delta\boldsymbol{\varepsilon}_o \neq \mathbf{o}$.

Alternately one may solve $\Delta\boldsymbol{\sigma}$ from compatibility equations in Eq.(12.75) with constraints in Eq.(12.70) and boundary conditions in Eq.(12.78), by invoking the *stress function* concept.[13] Let $\Delta\dot{\boldsymbol{\psi}} \in \mathrm{R}^{6\times 1}$ be such that

$$\Delta\boldsymbol{\sigma} = \mathcal{M}^T \Delta\dot{\boldsymbol{\psi}} \tag{12.79}$$

holds. Here $\Delta\dot{\boldsymbol{\psi}} \in \mathrm{R}^{6\times 1}$ is different than the conventional *extended Airy stress function.*[14] Vector $\Delta\boldsymbol{\sigma}$ computed from Eq.(12.79) satisfies the equilibrium equations in Eq.(12.70), since, according to Eq.(12.68), we have:

$$\mathcal{L}'\mathcal{M}^T = \mathbf{0} \tag{12.80}$$

By substituting $\Delta\boldsymbol{\sigma}$ from Eq.(12.79) into compatibility equations given by Eq.(12.75) and using boundary conditions in Eq.(12.78), one obtains the following boundary value problem:

$$\left.\begin{array}{ll} \mathcal{M}\mathrm{C}''\mathcal{M}^T\Delta\dot{\boldsymbol{\psi}} = \mathcal{M}\Delta\boldsymbol{\varepsilon}_o & \text{for } \mathbf{x} \text{ in } \mathfrak{D} \\ \Delta\dot{\boldsymbol{\psi}} = \text{arbitrary linear function of } \mathbf{x} & \text{for } \mathbf{x} \text{ on } \mathfrak{B} \end{array}\right\} \tag{12.81}$$

for $\Delta\dot{\boldsymbol{\psi}}$.

Once $\Delta\boldsymbol{\sigma}$ is obtained at the material points in $\mathfrak{D}$, by using the material's constitutive laws in Eq.(12.38), one may compute $\Delta\boldsymbol{\varepsilon}$, and then integrating Eq.(12.35) $\Delta\mathbf{d}$ may be determined.[15]

12.3.3 Compatibility of Induced Strains

One may observe that the boundary value problem defined by Eqs.(12.76 and 12.78) or the boundary value problem defined by Eqs.(12.81) would have $\Delta\boldsymbol{\sigma} = \mathbf{o}$ as the solution were the right-hand side of the differential equation in Eq.(12.76) or Eq.(12.81) zero, that is, if the controls $\Delta\boldsymbol{\varepsilon}_o$ satisfy the strain compatibility condition

$$\mathcal{M}\Delta\boldsymbol{\varepsilon}_o = \mathbf{o} \quad \text{for } \mathbf{x} \text{ in } \mathfrak{D} \tag{12.82}$$

where differential matrix operator $\mathcal{M} \in \mathrm{R}^{6\times 6}$ is as defined in Eq.(12.69).[16] When control strains $\Delta\boldsymbol{\varepsilon}_o$ satisfy the strain compatibility requirement,

---

[13]Proposed for two-dimensional problems by B. B. Airy in *Brit. Assoc. Rep.* 1862, and *Phil. Trans. Roy. Soc.*, vol. 153 (1863), p.49. Extension for three-dimensional problems using $\Delta\boldsymbol{\psi} \in \mathrm{R}^{3\times 1}$ by Maxwell in *Edinburgh Roy. Soc. Trans.*, vol. 26 (1870), or *Scientific Papers*, vol. 2, p. 161.

[14]This treatment is from the paper "Shape Control of Plate Structures with Geometrically Compatible Inserted Strains" by Ahmet Suner and Senol Utku, being prepared for the *Second World Conference on Structural Control, 28 June - 1 July 1998, Kyoto, Japan.*

[15]See S. Timoshenko and J. N. Goodier, *Theory of Elasticity*, 2nd edition, 1951, McGraw-Hill Book Co., New York (Art. 78, pp.232-233).

[16]For compatibility conditions in shells as expressed in shell coordinates (i.e., curvilinear coordinates which are coincident with principal curvature directions of middle surface), see *Theory of Thin Shells*, V. V. Novozhilov (English translation by Lowe and Radok), Art. §5., pp. 27-29, P. Noordhoff Ltd., Groningen, the Netherlands, 1959.

there will be only displacement response but not any stress or strain response; that is, no energy would be required to insert the control strains $\Delta\boldsymbol{\varepsilon}_o$, i.e., the strain energy ΔU defined by Eq.(12.67) would be zero.

As an example,[17] consider a rectangular plate of unit thickness deforming in the (x,y) plane. Suppose that the plate is of isotropic material with thermal expansion coefficient α, supported in a statically determinate manner, and its plane deformations in the (x,y) plane are controlled by controlling the temperature change field, $\Delta\grave{T}(x,y)$ in domain $\mathfrak{D}$ and boundary $\mathfrak{B}$ of the plate. Then the control strains are

$$\Delta\boldsymbol{\varepsilon}_o^T = [\Delta\varepsilon_{xx_o}, \Delta\varepsilon_{yy_o}, \Delta\gamma_{xy_o}] = a\ \Delta\grave{T}\ [1,1,0] \tag{12.83}$$

and Eq.(12.82) for this case becomes

$$[\frac{\partial^2}{\partial x^2}, \frac{\partial^2}{\partial y^2}, -2\frac{\partial^2}{\partial x \partial y}]\Delta\boldsymbol{\varepsilon}_o = 0 \tag{12.84}$$

Substitution of $\Delta\boldsymbol{\varepsilon}_o \in R^{3\times 1}$ from Eq.(12.83) into Eq.(12.84) yields

$$\nabla^2 \Delta\grave{T} = 0 \quad \text{in } \mathfrak{D} \tag{12.85}$$

and

$$\Delta\grave{T} \quad \text{prescribed on } \mathfrak{B} \tag{12.86}$$

which is a well-posed boundary value problem. The control temperatures $\Delta\grave{T}(x,y)$ obtained from this problem will yield compatible control strains which will not create stresses in the material. By changing boundary temperatures, from Eqs.(12.85) and 12.86), one may obtain a family of temperature fields $\Delta\grave{T}$ that satisfy strain compatibility and may contain the one required by the control law.

Note that the compatibility condition given by Eq.(12.82) for distributed parameter adaptive structures corresponds to the one given in Eq.(3.78) for discrete parameter adaptive structures. When possible, the control strains should satisfy the compatibility requirement for economies in the energy budget of the control system.[18]

12.3.4 Control

The static control of distributed parameter systems can be done by a feedback system similar to the one sketched in Fig. (4.1) for discrete parameter systems. The basic issues may be listed as follows:

---

[17]This example is taken from the paper "On Control of Stresses in Silicon Web Growth," Senol Utku, Sujit K. Ray, and Ben K. Wada, *Computers and Structures*, Vol. 23, No.5, pp.657-664, 1986.

[18]For an application in thin shells, see "Adaptive Inflatable Space Structures - Shape Control of Reflector Surface," S. Utku, C. P. Kuo, J. Garba, M. Salama, and B. K. Wada, *Proceedings of Fourth International Conference on Adaptive Structures*, Nov. 2-4, 1993, Cologne, Federal Republic of Germany, Breitback, Wada, Natori, eds., Technomic Publishing Co., Inc., Lancaster, PA, pp. 359-372, 1994.

1. Distribution of actuators (placement problem).
2. Distribution of observed response components (identification of observed quantities).
3. Control law that enables one to determine the magnitude and distribution of inserted control strains as a function of observed response components for a *robust control.*
4. Control time.
5. Optimality.

Because of difficulties in handling differential equations, current trends are in the direction of treating distributed parameter adaptive structures as suitably discretized discrete parameter structures. To have a good theoretical handle on the control of distributed parameter structures requires more time. Since the hardware for the feedback control of distributed parameter systems and the new horizons they open appear ahead of the theory, many engineers will probably design their control systems initially by trial and error. Such experiences may not only bring about practical systems but also help in the development of reliable control theories for the distributed parameter systems.

12.4 Dynamic Case

In the previous section, it is assumed that the excitations are slowly applied on the distributed parameter structure such that no appreciable inertial forces are developed at the material particles. In this section no such restriction exists. However, we assume that the excitation-response relations are linear and the structural system is time invariant.

12.4.1 Excitation-Response Relations

As in the static case, the relationships are the quantitative descriptions of the equilibrium of stresses at a material particle according to *d'Alembert's principle*, the geometric compatibility of the displacement field with strains according to the rules of *Euclidean geometry*, and the connection between stresses and strains according to *Hooke's Law.*

Equilibrium Equations

The differences here from the static case are

- presence of inertial forces, and the assumption that

- variables are no longer incremental in order to ensure linearity.

Using the static case equations from Eq.(12.18) and assuming zero damping, one may write the equilibrium equations for material points in the domain $\mathfrak{D}$ as

$$\mathcal{L}'\boldsymbol{\sigma}+\rho\ddot{\mathbf{d}}+\dot{\mathbf{D}}\ \mathbf{d}=-\varkappa \quad \text{for all } \mathbf{x} \text{ in } \mathfrak{D} \tag{12.87}$$

and on the boundary where the stresses are prescribed as

$$\mathbf{L}'\ \boldsymbol{\sigma}=\boldsymbol{\sigma}_{\sigma} \quad \text{for all } \mathbf{x} \text{ in } \mathfrak{B}_{\sigma} \tag{12.88}$$

where differential matrix operator $\mathcal{L}' \in \mathrm{R}^{3\times 6}$ and matrix operator $\mathbf{L}' \in \mathrm{R}^{3\times 6}$ are as defined in Eqs.(12.17 and 12.22) in Cartesian coordinates, $\dot{\mathbf{D}} \in \mathrm{R}^{3\times 3}$ is the stiffness of the infinitesimal element due its preexisting stresses, $\mathbf{d} \in \mathrm{R}^{3\times 1}$ is the description of the displacement vector in Cartesian coordinates

$$\mathbf{d}^T=[d_x,d_y,d_z]=[u,v,w] \tag{12.89}$$

$\boldsymbol{\sigma} \in \mathrm{R}^{6\times 1}$ is the stress vector with Cartesian stress components

$$\boldsymbol{\sigma}^T=[\sigma_{xx},\sigma_{yy},\sigma_{zz},\sigma_{xy},\sigma_{xz},\sigma_{yz}] \tag{12.90}$$

ρ is the unit mass, $\varkappa \in \mathrm{R}^{3\times 1}$ is the Cartesian description of the body force per unit volume, and $\boldsymbol{\sigma}_{\sigma} \in \mathrm{R}^{3\times 1}$ is the Cartesian description of the prescribed boundary traction per unit boundary area. Note that the equilibrium equations above are valid so long as the deformed configuration of structure is well represented by its undeformed configuration.

Geometric Relations

These are the same as in the previous section. Using Eq.(12.35) by dropping the increment symbol, these relations may be given in Cartesian coordinates as

$$\boldsymbol{\varepsilon}=\mathcal{L}'^T\mathbf{d}+\boldsymbol{\varepsilon}_o \tag{12.91}$$

where differential matrix operator $\mathcal{L}' \in \mathrm{R}^{3\times 6}$ is as in Eq.(12.17), displacement vector $\mathbf{d} \in \mathrm{R}^{3\times 1}$ is as in Eq.(12.89), strain vector $\boldsymbol{\varepsilon} \in \mathrm{R}^{6\times 1}$ is

$$\boldsymbol{\varepsilon}^T=[\varepsilon_{xx},\varepsilon_{yy},\varepsilon_{zz},\gamma_{xy},\gamma_{xz},\gamma_{yz}] \tag{12.92}$$

and prescribed strain vector $\boldsymbol{\varepsilon}_o \in \mathrm{R}^{6\times 1}$ is

$$\boldsymbol{\varepsilon}_o^T=[\varepsilon_{o_{xx}},\varepsilon_{o_{yy}},\varepsilon_{o_{zz}},\gamma_{o_{xy}},\gamma_{o_{xz}},\gamma_{o_{yz}}] \tag{12.93}$$

which may represent control strains or strains induced by temperature change. These are the linear strain displacement relations.

Material's Constitutive Law

For linearly elastic material, from Eq.(12.36) one may write

$$\boldsymbol{\sigma} = \mathbf{D}\,\boldsymbol{\varepsilon} \tag{12.94}$$

where material matrix $\mathbf{D} \in R^{6\times 6}$ is real, symmetric and positive definite. For the isotropic case, $\mathbf{D}$ and its inverse are as in Eqs.(12.37 and 12.40).

Excitation Response Relations

For the material points in domain $\mathfrak{D}$, by combining Eqs.(12.87, 12.91, and 12.94), one may write:

$$\begin{bmatrix} \rho\frac{d^2}{dt^2} + \grave{\mathbf{D}} & \mathcal{L}' & \cdot \\ \mathcal{L}'^T & \cdot & -\mathbf{I} \\ \cdot & -\mathbf{I} & \mathbf{D} \end{bmatrix} \begin{Bmatrix} \mathbf{d} \\ \boldsymbol{\sigma} \\ \boldsymbol{\varepsilon} \end{Bmatrix} = \begin{Bmatrix} -\boldsymbol{\varkappa} \\ -\boldsymbol{\varepsilon}_o \\ \cdot \end{Bmatrix} \quad \text{for } \mathbf{x} \text{ in } \mathfrak{D} \tag{12.95}$$

In the control of adaptive structures one is interested in response due to $\boldsymbol{\varepsilon}_o$; hence, one may rewrite these equations as

$$\begin{bmatrix} \rho\frac{d^2}{dt^2} & \mathcal{L}' & \cdot \\ \mathcal{L}'^T & \cdot & -\mathbf{I} \\ \cdot & -\mathbf{I} & \mathbf{D} \end{bmatrix} \begin{Bmatrix} \mathbf{d} \\ \boldsymbol{\sigma} \\ \boldsymbol{\varepsilon} \end{Bmatrix} = \begin{Bmatrix} \cdot \\ -\boldsymbol{\varepsilon}_o \\ \cdot \end{Bmatrix} \quad \text{for } \mathbf{x} \text{ in } \mathfrak{D} \tag{12.96}$$

where, as in the static case, the contributions from the preexisting stresses are assumed zero, i.e.,

$$\grave{\mathbf{D}}\mathbf{d} = \mathbf{o} \tag{12.97}$$

is assumed.

For points on boundary $\mathfrak{B} = \mathfrak{B}_\sigma \cup \mathfrak{B}_d$, one may write from Eq.(12.56) for a structure on statically determinate unyielding supports:

$$\begin{bmatrix} \cdot & \mathbf{L}' & \cdot \\ \mathbf{I} & \cdot & \cdot \end{bmatrix} \begin{Bmatrix} \mathbf{d} \\ \boldsymbol{\sigma} \\ \boldsymbol{\varepsilon} \end{Bmatrix} = \begin{Bmatrix} \mathbf{o} \\ \mathbf{o} \end{Bmatrix} \quad \begin{matrix} \text{for } \mathbf{x} \text{ in } \mathfrak{B}_\sigma \\ \text{for } \mathbf{x} \text{ in } \mathfrak{B}_d \end{matrix} \tag{12.98}$$

where no control excitation on boundary $\mathfrak{B}$ is assumed.

12.4.2 Inverse Relations

By eliminating $\boldsymbol{\sigma}$ and $\boldsymbol{\varepsilon}$, the problem defined by Eqs.(12.96 and 12.98) can be expressed in terms of $\mathbf{d}$ as

$$\begin{matrix} \rho\frac{d^2}{dt^2}\mathbf{d} + \mathcal{L}'\mathbf{D}\mathcal{L}'^T\mathbf{d} = -\mathcal{L}'\mathbf{D}\,\boldsymbol{\varepsilon}_o \quad \text{for } \mathbf{x} \text{ in } \mathfrak{D} \\ \begin{bmatrix} \mathbf{L}'\mathbf{D}\mathcal{L}'^T \\ \mathbf{I} \end{bmatrix}\mathbf{d} = \mathbf{o} \quad \text{for } \mathbf{x} \text{ on } \mathfrak{B} \end{matrix} \qquad 0 \le t \le T \tag{12.99}$$

The solution of $\mathbf{d}(\mathbf{x}, t)$ from these equations in terms of $\boldsymbol{\varepsilon}_o(\mathbf{x}, t)$ constitutes the inverse relations. Once $\mathbf{d}(\mathbf{x}, t)$ is obtained, one may use

$$\left\{ \begin{array}{c} \mathbf{d} \\ \boldsymbol{\sigma} \\ \boldsymbol{\varepsilon} \end{array} \right\} = \left[\begin{array}{c} \mathbf{I} \\ \mathbf{D}\mathcal{L}'^T \\ \mathcal{L}'^T \end{array} \right] \mathbf{d} + \left\{ \begin{array}{c} \cdot \\ \mathbf{D}\boldsymbol{\varepsilon}_o \\ \boldsymbol{\varepsilon}_o \end{array} \right\} \qquad (12.100)$$

which is basically the restatement of Eq.(12.66).

For the purpose of actuator placement, actuator selection, and control computation problems, the inverse relations are best expressed in terms of Green's function $\mathbf{G}(\mathbf{x}, t, \mathbf{x}_o, t_o)$ which is the replacement of $\mathbf{d}$ corresponding to the unit value of $\boldsymbol{\varepsilon}_o$, i.e., unit strain pulse $\boldsymbol{\delta}(\mathbf{x}, t, \mathbf{x}_o, t_o)$ inserted at time t_o at location $\mathbf{x}_o$ in $\mathfrak{D}$, namely, the solution of the problem

$$\left. \begin{array}{ll} \mathcal{L}'\mathbf{D}\mathcal{L}'^T\mathbf{G}(\mathbf{x}, t, \mathbf{x}_o, t_o) = -\mathcal{L}'\mathbf{D}\ \boldsymbol{\delta}(\mathbf{x}, t, \mathbf{x}_o, t_o) & \text{for } \mathbf{x} \text{ in } \mathfrak{D} \\ \left[\begin{array}{c} \mathbf{L}'\mathbf{D}\mathcal{L}'^T \\ \mathbf{I} \end{array} \right] \mathbf{G}(\mathbf{x}, t, \mathbf{x}_o, t_o) = \mathbf{o} \quad \text{for } \mathbf{x} \text{ on } \mathfrak{B} & \end{array} \right\} 0 \le t \le T \qquad (12.101)$$

which is a restatement of Eqs.(12.99) for

$$\boldsymbol{\varepsilon}_o = \boldsymbol{\delta}(\mathbf{x}, t, \mathbf{x}_o, t_o) \qquad (12.102)$$

In these expressions $\boldsymbol{\delta}(\mathbf{x}, t, \mathbf{x}_o, t_o)$ is the Dirac delta function in $(\mathbf{x}, t)$ space with singularity at $(\mathbf{x}_o, t_o)$.Green's function $\mathbf{G}(\mathbf{x}, t, \mathbf{x}_o, t_o)$ may be obtained analytically or numerically by solving Eq.(12.101), or experimentally by measuring in laboratory models. Once it is available, $\mathbf{d}$ for any $\boldsymbol{\varepsilon}_o$ may be obtained by superposition

$$\mathbf{d}(\mathbf{x}, t) = \int_0^t \int_{\mathfrak{D}} \mathbf{G}(\mathbf{x}, t, \mathbf{x}_o, \tau)\ \boldsymbol{\varepsilon}_o(\mathbf{x}_o, \tau)\ d\mathfrak{D}_o\ d\tau \qquad (12.103)$$

since the incremental problem is linear. Having obtained $\mathbf{d}(\mathbf{x}, t)$ by using Eq.(12.103), one may use Eq.(12.100) for the determination of other response quantities, i.e., for the determinations of $\boldsymbol{\sigma}(\mathbf{x}, t)$, and $\boldsymbol{\varepsilon}(\mathbf{x}, t)$.

12.4.3 Control

The dynamic control of distributed parameter systems can be done by a feedback system similar to the ones sketched in Figs. (8.3, 8.5, 8.7) for discrete parameter systems. The basic issues may be listed as follows:

1. Distribution of actuators (placement problem).

2. Distribution of observed response components (identification of observed quantities).

3. Control law that enables one to determine the magnitude and distribution of inserted control strains as a function of observed response components for a *robust control.*

4. Control time.

5. Optimality.

Because of difficulties in handling differential equations, current trends are in the direction of treating distributed parameter adaptive structures as suitably discretized discrete parameter structures. To have a good theoretical handle on control of distributed parameter structures requires more time. Since the hardware for the feedback control of distributed parameter systems and the new horizons they open appear ahead of the theory, many engineers will probably design their control systems initially by trial and error. Such experiences may not only bring about practical systems but also help the development of reliable control theories for the distributed parameter systems.

12.5 Recapitulation

In this introductory chapter of distributed parameter adaptive structures, the basic problems associated with actuator placement, actuator selection, and feedback control are discussed for static and dynamic cases. These basic problems are the identification of linear excitation-response relations and their inverses and compatibility of actuator inserted strains. They are given in Cartesian coordinates, in incremental form for static control, and in non-incremental form for dynamic control. The inverse relations are expressed in terms of Green's functions. Conditions for stress-free control are also discussed. However, no discussions are given for the determination of control laws.

References

[1] AIRY, B. B., 1862, *Brit. Assoc. Rep.*; also, *Phil. Trans. Roy. Soc.*, vol. 153 (1863), p.49.

[2] ANDERSON, W. and NEWSOM, J. R., 1991, "Control-Structure Interaction Research at NASA Langley Research Center," pp. 43-55, *Fourth International Conference on Adaptive Structures, Nov. 2-4, 1990, Cologne, Federal Republic of Germany*, Breithbach, Wada, Natori (editors), Technomic Publishing Co., Inc., Lancaster, PA.

[3] ATHANS, M. and FALB, P. L., 1966, *Optimal Control, An Introduction to the Theory and Its Applications*, McGraw-Hill Book Co., New York.

[4] BAYCAN, C. M. and UTKU, S., 1995, "Energy Control in Active Vibration Control of Statically Indeterminate Structures," Paper AIAA-95-1134, *Proceedings of 36th Structures, Structural Dynamics and Materials Conference,* New Orleans, LA.

[5] BAYCAN, C. M., 1996, *Effect of Compatibility of Inserted Strains on Energy Consumption in Active Control of Adaptive Structures*, doctoral dissertation, Civil and Environmental Engineering Department, Duke University, Durham, NC.

[6] BELTRAMI, E., 1889, *Paris, C. R.*, t. 108, cf. Koenigs, *Leçons de Cinématique*, Paris, 1897 p. 411.

[7] BREITBACH, E. J., 1992, "Adaptive Structures in Europe," pp. 32-48, *Second Joint Japan/US Conference on Adaptive Structures, Nov. 12-14, 1991, Nagoya, Japan*, Matsuzaki, Wada (editors), Technomic Publishing Co., Lancaster, PA.

[8] CAUGHEY, T. K., 1960, *Jour. Appl. Mech.*, vol. 27.

[9] CHENEY, E. W.,1966, *Introduction to Approximation Theory*, In'l. Series in Pure and Appld. Math., McGraw-Hill Book Co., New York.

[10] CONNOR, J. J. and KLINK, B. S. A., 1966, *Introduction to Motion Based Design*, Computational Mechanics Publications, Ashurst, Southampton, U.K.

[11] CRANDALL, S. H., 1956, *Engineering Analysis, A Survey of Numerical Procedures*, McGraw-Hill Book Co., New York.

[12] FUJITA, T. (editor), 1991, *Seismic Isolation and Response Control for Nuclear and Non-Nuclear Structures*, for SMIRT 11, Institute of Industrial Science, University of Tokyo, Japan.

[13] GAVIN, H. P., 1997, "ER Material Models and Vibration Control," *Proceedings of 11th Symposium on Structural Dynamics and Control*, VPI&SU, Blacksburg, VA.

[14] GOLUB, G. H. and VAN LOAN, C. F., 1989, Matrix Computations, 2nd. ed., The Johns Hopkins Press, Baltimore.

[15] HENNEBERG, L., 1886, *Static der starren Systeme*, Darmstadt.

[16] HOUSNER, G. W., SOONG, T. T. and MASRI, S. F., 1994, panel 3-18, vol. 1, *Proceedings of First World Conference on Structural Control*, Los Angeles, CA.

[17] HURTY, W. C. and RUBINSTEIN, M. F., 1964, *Dynamics of Structures*, Prentice-Hall, Inc., Englewood Cliffs, NJ.

[18] KOBORI, T., 1994, "Future Direction on Research and Development on Seismic-Response-Controlled Structure," panel 19-31, vol. 1, *Proceedings of First World Conference on Structural Control*, Los Angeles, CA.

[19] JALIHAL, P., 1994, *Active Vibration Control in Buildings with Passive Base Isolation*, doctoral dissertation, Civil and Environmental Engineering Department, Duke University, Durham, NC.

[20] KALMAN, R. E., 1963, "The Theory of Optimal Control and the Calculus of Variations," *Mathematical Optimization Techniques*, R. Bellman (editor), University of California Press, Berkeley, CA.

[21] LA BARBERA, M., 1983, "Why the Wheels Won't Go," *The American Naturalist*, vol. 121, no. 3, pp. 395-408.

[22] LOVE, A. E. H., 1946, *The Mathematical Theory of Elasticity*, 4th edition, Dover Publications, New York.

[23] LIN, T. Y., 1955, Design of Prestressed Concrete Structures, John Wiley & Sons, New York.

[24] LU, L.-Y., 1991, *On the Placement of Active Members in Adaptive Truss Structures*, doctoral dissertation, Civil and Environmental Engineering Department, Duke University, Durham, NC.

[25] LU, L.-Y., UTKU. S., and WADA, B. K., 1992, "On the Placement of Active Members in Adaptive Truss Structures for Vibration Control," *Journal of Smart Materials and Structures*, no. 1, pp.8-23.

[26] MAXWELL, J. C., 1870, *Edinburgh Roy. Soc.*, vol. 26 or *Scientific Papers*, vol. 2, p. 161.

[27] MIURA, K., 1992, "Adaptive Structures Research at ISAS, 1984-1990," *Journal of Intelligent Material Systems and Structures*, vol. 3, pp. 54-74.

[28] NOVOZHILOV, V. V., 1951, *Theory of Thin Shells*, English translation by Lowe and Radok, 1959, P. Noordhoff Ltd., Groningen, The Netherlands.

[29] RAMESH, A. V., DAS, S. K., UTKU, S., LU, L.-Y. and SALAMA, M., 1988, "Real-Time Computation of Control Torques for Mechanical Manipulators Using Concurrent Processors," *Computational Mechanics '88: Proceedings of International Conference of Computational Engineering Science, Georgia Tech., Atlanta, GA*, pp. 43.vii.1 - 43.vii.6., Springer-Verlag, New York.

[30] RAMESH, A. V., UTKU, S. and GARBA, J., 1989, "Computational Complexities and Storage Requirements of Some Riccati Equation Solvers," *AIAA J. of Guidance, Control and Dynamics*, vol. 12, no. 4, pp. 469-479.

[31] RAMESH, A. V., UTKU, S., and LU, L.-Y., 1989, "DETRANS: A Fast and Storage Efficient Algorithm for Static Analysis of Determinate Trusses," *JPL D-6194*, also *Journal of Aerospace Engineering, ASCE*, vol. 4, no. 3, pp. 274-285, 1991.

[32] RAMESH, A. V., UTKU, S. and WADA, B. K., 1991, "Real-Time Control of Geometry and Stiffness in Adaptive Structures," *Proceedings of Second World Congress on Computational Mechanics, August*

1990, Stuttgart, FRG, pp. I/L4 59-62, also *Computer Methods in Applied Mechanics and Engineering 90*, pp. 761-779, North-Holland.

[33] RAMESH, A. V., UTKU, S. and WADA, B. K., 1991, "Parallel Computation of Geometry Control in Adaptive Truss Structures," *Proceedings of First US National Congress on Computational Mechanics*, Chicago.

[34] RAMESH, A. V., 1991, *Geometry Control in Adaptive Truss Structures*, doctoral dissertation, Civil and Environmental Engineering Department, Duke University, Durham, NC.

[35] SAINT-VENANT, B. DE, 1864, *Résumé de Leçons sur l'application de la Méchanique (of Navier)* Paris, Appendice 3.

[36] SENER, M., UTKU, S. and WADA, B. K., 1994, "Geometry Control in Prestressed Adaptive Space Trusses", *Journal of Smart Materials and Structures*, vol. 3, no. 2, pp. 219-225.

[37] SENER, M. and UTKU, S., 1995 "Active-Passive Base Isolation System for Seismic Response Controlled Structures," paper AIAA-95-1088, *Proceedings of 36th Structures, Structural Dynamics and Materials Conference*, New Orleans, LA.

[38] SENER, M., 1996, *Adaptive Base Isolation: A System to Control Seismic Energy Flow into Buildings*, doctoral dissertation, Civil and Environmental Engineering Department, Duke University, Durham, NC.

[39] SOKOLNIKOFF, I. S., 1956, *Mathematical Theory of Elasticity*, 2nd edition, McGraw-Hill Book Co., Inc., New York.

[40] SUNER, A., NAGCHAUDHURI, A. and UTKU, S., 1995, "Effect of Time Lag and Use of Compensators in the Active Control of Buildings Subjected to Earthquake Excitations," *Journal of Structural Control*, vol. 2, no. 2, pp. 79-91.

[41] SUNER, A. and UTKU, S., 1997, "Shape Control of Plate Structures with Geometrically Compatible Inserted Strains," draft of paper for the Second World Conference on Structural Control, July 1998, Kyoto, Japan.

[42] TIMOSHENKO, S. and GOODIER, J. N., 1951, *Theory of Elasticity*, 2nd edition, McGraw-Hill Book Co., New York.

[43] UTKU, S., 1968, ELAS- A General Purpose Computer Program for the Equilibrium Problems of Linear Structures, User's Manual (with F. A. Akyuz): *JPL TR-32-1240, vol. 1*, vol. 2 Documentation: *JPL TR 32-1240, vol. 1 addendum*, formulation: *Concrete Thin Shells*, ACI Publication SP-28, pp. 383-417, American Concrete Institute, Detroit, MI, 1971, code: *COSMIC*, University of Georgia, Athens, GA, 1968.

[44] UTKU, S., 1973, "Sturm Sequences or Law of Inertia of Quadratic Forms?" *Computers and Structures*, vol. 3, pp. 419-420.

[45] UTKU, S., LESTINGI, J. and SALAMA, M., 1980, "On Impact of Distributive Computing in Education," *Proceedings of ASCE National Convention, 1980, Portland, OR.*

[46] UTKU, S., 1981, Numerical Solutions of Partial Differential Equations, *CPS/Math 222 Lecture Notes*, Civil Engineering Department, Duke University, Durham, NC.

[47] UTKU, S., KUO, C. P. and SALAMA, M., 1985, "Direct Computation of Optimal Control of Forced Linear Systems," paper 85-0753, *Proceedings of 26th Structures, Structural Dynamics and Materials Conference*, 4/15-17/85, Orlando FL.

[48] UTKU, S., RAY, S. K. and WADA, B. K., 1986, "On Control of Stresses in Silicon Web Growth," *Computers and Structures*, vol. 23, no. 5, pp. 657-664.

[49] UTKU, S., RAMESH, A. V., DAS, S. K., WADA, B. K. and CHEN, G. S., 1989, "Control of a Slow-Moving Space Crane as an Adaptive Structure," paper 89-1286, *Proceedings of 30th Structures, Structural Dynamics and Materials Conference*, 4/89, Mobile, AL; also *AIAA Journal*, vol. 29, no. 6, pp. 961-967, June 1991.

[50] UTKU, S., NORRIS, C. H. and WILBUR, J. B., 1991, *Elementary Structural Analysis*, 4th edition, McGraw-Hill Inc., New York.

[51] UTKU, S., 1991, *Instructor's Manual of Elementary Structural Analysis, 4th edition*, McGraw-Hill Inc., New York.

[52] UTKU, S., 1991, *LADS: Linear Analysis of Discrete Structures, A Study Manual for Elementary Structural Analysis*, 4th edition, Dr. Utku & Associates, Durham, NC 27705-5754.

[53] UTKU, S. and WADA, B. K., 1991, "Adaptive Structures in Japan," *Journal of Intelligent Material Systems and Structures*, vol. 4, pp. 437-451.

[54] UTKU, S., FREELAND, R. E., HELMS, H., KUO, C. P., McELROY, P. M. and PORTER, C., 1991, "Design and Material Representation of Lightweight Composite Panels for Large Deployable Reflectors," *JPL-D8912*, Jet Propulsion Laboratory, Pasadena, CA.

[55] UTKU, S., KUO, C. P., GARBA, J., SALAMA, M., WADA, B. K., 1993, "Adaptive Inflatable Space Structures - Shape Control of Reflector Surface," *Fourth International Conference on Adaptive Structures, Nov. 2-4, 1993, Cologne, Federal Republic of Germany*, Breithbach,

Wada, Natori (editors), pp. 359-372, Technomic Publishing Co., Inc., Lancaster, PA, 1994.

[56] WIBERG, D. W., 1971, *Theory and Problems of State Space and Linear Systems*, Schaum's Outline Series, McGraw-Hill Book Co., New York.

[57] WILKINSON, J. H., 1965, *The Algebraic Eigenvalue Problem*, Oxford University Press, New York.

[58] ZILL, S. N. and SEYFARTH, E.-A., 1996, "Exoskeletal Sensors for Walking," *Scientific American*, July 96.

Index

Milton Keynes UK
Ingram Content Group UK Ltd.
UKHW031129141024
449569UK00006B/333